Progress in Gene Expression

Series Editor

Michael Karin
Department of Pharmacology
School of Medicine
University of California, San Diego
La Jolla, CA 92093-0636

Books in the Series:

Gene Expression: General and Cell-Type-Specific
M. Karin, editor
ISBN 0-8176-3605-6

Inducible Gene Expression, Volume I: Environmental Stresses and Nutrients
P. A. Baeuerle, editor
ISBN 0-8176-3728-1

Inducible Gene Expression, Volume II: Hormonal Signals
P. A. Baeuerle, editor
ISBN 0-8176-3734-6

Oncogenes as Transcriptional Regulators, Volume 1: Retroviral Oncogenes
M. Yaniv and J. Ghysdael, editors
ISBN 3-7643-5486-0

Oncogenes as Transcriptional Regulators, Volume II: Cell Cycle Regulators
and Chromosomal Translocation Products
M. Yaniv and J. Ghysdael, editors
ISBN 3-7643-5709-6

Molecular Biology of Steroid and Nuclear Hormone Receptors
L. Freedman, editor
ISBN 0-8176-3952-7

Molecular Biology of Steroid and Nuclear Hormone Receptors

Leonard P. Freedman
Editor

Birkhäuser
Boston • Basel • Berlin

Leonard P. Freedman
Cell Biology and Genetics
Memorial Sloan-Kettering Cancer Center
New York, NY 10021

Library of Congress Cataloging In-Publication Data

Molecular biology of steroid and nuclear hormone receptors / Leonard
  P. Freedman, editor.
     p.   cm. -- (Progress in gene expression)
    Includes bibliographical references and index.
    ISBN 0-8176-3952-7. -- ISBN 3-7643-3952-7
    1. Hormone receptors.   2. Steroid hormones--Receptors.
    3. Transcription factors.   I. Freedman, Leonard P., 1958- .
    II. Series.
    QP571.7.M65   1997
    573.4'48845--dc21   97-21837
                                   CIP

QP
571
.7
.M65
1998

Printed on acid-free paper
© Birkhäuser Boston 1998

Birkhäuser

ISBN 0-8176-3952-7
ISBN 3-7643-3952-7
Typeset by Alden Bookset, Oxford, England
Printed and bound by Hamilton Printing Co., Rensselaer, New York
Printed in the United States of America
9 8 7 6 5 4 3 2 1

Contents

Foreword
Intracellular Receptors:
New Instruments for a Symphony of Signals

In the late eighteenth century, it was proposed on theoretical grounds that each of the body's organs, beginning with the brain, must be "a factory and laboratory of a specific humor which it returns to the blood", and that these circulating signals "are indispensable for the life of the whole" (Bordeu 1775). During the nineteenth century, some remarkable physiological experiments revealed the actions of humoral factors that affected the for and function of multiple tissues, organs and organ systems within the body (Berthold 1849); much later, the chemical and molecular nature of some of those factors was determined.

Against this deep historical backdrop of the founding studies of intercellular signaling, molecular biology sprang into existence a mere forty years ago, rooted in the revelation of regulable gene expression in bacteria. But contemporaneous with those classical analyses of transcriptional regulation of the lactose operon, the modern era of signal transduction was inaugurated by the identification of cAMP as a second messenger --- an intracellular mediator of hormonal activation of glycogen catabolism (Sutherland and Rall 1960). Later in that same decade, it emerged that cAMP is a critical signal not only in metazoans, but even in bacteria, where it serves an analogous function as a critical switch that activates expression of genes required for catabolism of complex carbon sources, including those of the lactose operon. Indeed, Tomkins proposed the existence of a "metabolic code" in which certain small molecules acquire "symbolic value", representing and thereby evoking global physiological states, and that these symbolic values are stabilized in evolution by selective forces similar to those that maintain the near-universality of the genetic code (Tomkins 1975).

Of course, the linkage between signal transduction and transcriptional regulation is far more profound than their shared histories. Transcriptional regulatory factors are the endpoints of many, probably most, signal transduction pathways --- that is, signaling typically triggers specific changes in gene expression. And nowhere is that linkage more direct than with the molecules whose signals are mediated by proteins encoded by a gene superfamily commonly termed the nuclear receptors, or, in a broader nomenclature that I favor, the intracellular receptors (IRs).

These receptors were first described in the 1960s in studies showing that radiolabeled estradiol is bound selectively in target cells --- specifically, saturably and noncovalently --- to a protein recovered initially in the cytosol and subsequently in the nuclear fraction (Jensen and Jacobson 1962; Gorski et al. 1968). Reports of analogous presumptive receptors for other steroid hormones were quick to follow. Thus, steroid receptors were inferred using a standard endocrinological and pharmacological strategy: begin with a known physiological signaling molecule, and identify in responsive tissues a protein, a putative receptor, that binds the signal with high affinity and specificity.

The cloning of the glucocorticoid receptor (Miesfeld et al. 1984), the first mammalian transcriptional regulatory factor so isolated, led eventually to elucidation of the largest superfamily of transcriptional regulatory factors (Mangelsdorf et al. 1995). At present, the bona fide receptors in the IR superfamily-- those with known signaling ligands --- are outnumbered by the so-called orphan receptors, for which ligands remain unidentified. In principle, a given orphan might be constitutive, or even nonfunctional.However, an intriguing pair of evolution-based arguments persuade me that cognate ligands exist for most or all of the orphans:

First, efficient intercellular communication is essential for all multicellular organisms. Individual highly differentiated cells must be informed of the status of the whole organism, and multiple cell types must in turn collaborate to produce appropriate and coordinate physiological effects. Lipophilic small molecules seem ideal as signals: they are simple to synthesize or acquire from the environment, relatively stable in circulation, and may readily enter target cells. Structural simplicity, however, invokes an "information capacity paradox": how can molecules that lack chemical complexity symbolize complex physiological states? Receptors resolve the paradox: their association with a ligand can be viewed as a large-scale "chemical modification" of the ligand, providing it with sufficient complexity to specify within target cells complex programs of gene expression. This general logic, binding of macromolecular adapters to small molecules as a way to impart complexity, is not unique in biology. For example, the interaction of a transfer RNA with its cognate amino acid could be considered as a chemical modification of the amino acid that confers on that simple molecule sufficient chemical complexity to read the genetic code (Yamamoto 1985). The point here is that intercellular signals, and therefore receptors, are essential for all multicellular organisms --- and that complex multicellular organisms likely require more signals and more receptors than do simple organisms .

The second evolution-based observation is that IRs, remarkably, are a new gene family, found only in metazoans. IR-related genes are not found in fungi or in higher plants --- the eukaryotes from which the metazoans most recently diverged --- or more distantly related eukaryote, in eubacteria, or in archae. Moreover, the steroid receptor subfamily of IRs is found only among the most complex metazoans, the vertebrates. Finally, the DNA binding and ligand binding domains appear to have evolved as a unit (albeit at different rates, indicating different evolutionary pressures on the two domains) (Amero et al. 1992); notably, all orphans include both domains. Thus, the IRs do not represent an ancient family of regulators in which a subset was recently co-opted to mediate signaling by lipophilic ligands. Rather, it appears that IRs evolved in metazoans specifically to exploit simple lipophilic molecules for intercellular signaling. Interestingly, recent studies hint that lipophilic ligands may signal in higher plants through another new gene family, unrelated to IRs, that is unique to plants (Ulmasov et al. 1997). According to this line of reasoning, then, each IR will bind to a small metabolite, a nutrient, an environmental compound, that has acquired a signaling role relatively recently --- after the evolutionary divergence of metazoans, fungi and higher plants.

These ideas imply that the IRs, and the orphans in particular, could be used as biological probes to identify cognate ligands and to uncover the "symbolic values" of those ligands, i.e., the physiological consequences of their signaling. As described in this volume, this "reverse endocrinology" strategy, in which a cloned orphan receptor is used to probe for its ligand and biological action, is indeed effective. Perhaps most notably, the investigations described in this monograph, taken collectively, validate this strategy more broadly, for we learn that it is generally illuminating to identify, isolate and characterize a broad range of factors that interact with IRs, and that in doing so, we learn both about receptor action per se and about cellular function. Examples of this "receptor as probe" approach are abundant and well represented in this volume:

• a multicomponent eukaryotic molecular chaperone complex was shown to potentiate the ligand responsiveness of IRs and other signaling molecules;
• the first genomic response elements were isolated, revealing a general role for transcriptional enhancers;
• composite response elements were described, demonstrating a strategy for producing combinatorial regulation at "nodes" at which different signaling pathways intersect and communicate;
• IR phosphorylation revealed a separate type of crosstalk node, and showed that IR activities are determined by integrating multiple signaling inputs;
• specification of receptor oligomerization state --- monomer, homodimer or heterodimer --- demonstrated another means of combinatorial regulation;
• studies of intracellular trafficking of IRs revealed a class of proteins that could be targeted to different cellular compartments or subcompartments, depending on interactions with chaperones, signals and structural components;
• IR interactions with chromatin remodeling machineries and histone modification enzymes suggest how regulators contend with chromosome packaging , and how they may exploit chromatin in regulatory mechanisms;
• the chemical and regulatory subtleties of ligand-receptor interactions are providing in atomic detail a view of allosteric transitions, and a primary role for hydrophobic core rearrangements in transmitting those changes;
• the role of IRs in physiology, development and disease has opened a window on molecular interactions that drive complex phenomena.

From the vantage point of the receptors, many of its molecular interactions can be considered to be signaling interactions (Yamamoto 1997) that "inform" the receptor both of the state of the cell and of the organism in order to specify the precise receptor activity appropriate for a given context. Thus, the IR superfamily functions in metazoans to coordinate a symphony of signals; indeed, the entire superfamily appears to have co-evolved in metazoans in parallel with a cognate family of new signals.

Beginning with a bold but unsubstantiated pronouncement over two centuries ago, it is apparent that signal transduction now holds center stage in studies of fundamental cellular processes, such as cell division and gene expression, in sweeping analyses of development and the nervous system, and in discovery of the molecular the molecular bases of disease. Fittingly, *The Molecular Biology of Steroid and Nuclear Hormone Receptors* maps out the present-day position on that stage for a large and important family of signal integrators and gene regulators, and it does so through the voices of some of the next generation of leading investigators in that field.

Keith R. Yamamoto
Department of Cellular and Molecular Pharmacology
University of California
San Francisco, California 94143-0450

Literature Cited

Amero, S.A., Kretsinger, R.H., Moncrief, N.D., Yamamoto, K.R., and Pearson, W.R. (1992): *Mol. Endocrinol.* **6**, 3-7.

Berthold, A.A. (1849): *Arch. Anat. Physiol. Wiss. Med.* **16**, 42-46.

Bordeu, T. (1775), Analyse medicinale du Sang: Recherches sur les malade chronique, Ruault, Paris.

Gorski, J., Toft, D.O., Shyamala, G., Smith, D., and Notides, A. (1968): *Rec. Progr. Horm. Res.* **24**, 45-80.

Jensen, E. and Jacobson (1962): *Rec. Progr. Horm. Res.* **18**, 387.

Mangelsdorf, D.J., Thummel, C., Beato, M., Herrlich, P., Schutz, G., Umesono, K., Blumberg, B., Kastner, P., Mark, M., Chambon, P. and Evans, R.M. (1995): *Cell* **83**, 835-839.

Miesfeld, R., Okret, S., Wikstrom, A.-C., Wrange, Y., Gustafsson, J.-P., and Yamamoto, K.R. (1984): *Nature* **312**, 779-781.

Sutherland, E.W., and Rall, T.W. (1960): *Pharmacol. Rev.* **12**, 265.

Tomkins, G.M. (1975): *Science* **189**, 960-763.

Ulmasov T., Hagen G., and Guilfoyle T.J. (1997): *Science* **276**, 1865-1868.

Yamamoto, K.R. (1985): *Ann. Rev. Genetics* **19**, 209-252.

Yamamoto, K.R. (1997): *The Harvey Lectures* **91**, 1-19.

Series Preface

The control of gene expression is a central-most topic in molecular biology as it deals with the utilization and regulation of gene information. As we see huge efforts mounting all over the developed world to understand the structure and organization of several complex eukaryotic genomes in the form of Gene Projects and Genome Centers, we have to remember that without understanding the basic mechanisms that govern the use of genetic information, much of this effort will not be very productive. Fortunately, however, research during the past decade on the mechanisms that control gene expression in eukaryotes has been extremely successful in generating a wealth of information on the basic strategies of transcriptional control.

The progress in understanding the control of eukaryotic transcription can only be appreciated by realizing that twelve years ago we did not know the primary structure of a single sequence specific transcriptional activator, and those whose primary structures were available (e.g. homeo-domain proteins) were not yet recognized to function in this capacity. Also, ten years ago transcription was thought to be carried out by an abstract assembly of transcription factors and RNA polymerases referred to as the "transcriptional machinery," while nowadays many of these basic components have been purified to homogeneity and are available as molecular clones. While the progress in this field has been incredible, it is far from reaching a plateau and it is likely that the next ten years will result in an even greater and faster increase in our understanding of gene regulation. However, we have reached a point at which some generalizations can be made, recurrent themes can be identified, and unifying hypotheses formulated. The purpose of this series is to summarize this overwhelming amount of information in a small number of volumes, each containing chapters written by well-recognized experts dealing with highly related topics. By studying the progress made in a select number of model systems, it is hoped that the reader will be able to apply this knowledge to his or her own favorite experimental system.

It is our goal that the *Progress in Gene Expression* series serves as an important resource for graduate students and experienced researchers alike, in the fields of molecular biology, cell biology, biochemistry, biotechnology, cell physiology, endrocrinology, and related fields. More exciting volumes are in the planning stages, and suggestions for future volumes are appreciated, and should be directed to the Series Editor.

Michael Karin
March, 1997

Dedicated to the memory of my parents,
David and Tilla Freedman

Preface

Among the panoply of signal transduction systems, the members of the steroid/ nuclear hormone receptor superfamily stand out as examples of a deceptively simple, direct signaling pathway. These receptors are soluble, intracellular proteins that can both bind hormonal ligands with high affinity and interact with specific DNA target elements. In the process, they regulate the transcription levels of a vast array of genes, several of which remain to be identified, playing central roles in normal physiological processes as well as in many important diseases. In fact, the steroid and nuclear hormone receptors constitute the largest single family of transcription factors in eukaryotes. But as we have learned more and more about these hormone-regulated transcription factors, the overlapping complexities of their multiple functions have become increasingly clear. It is the purpose of this book to treat the various functions encoded by the steroid/nuclear receptors in a systematic way, whereby a particular function, such as DNA-binding, is dealt with in detail in a single chapter, enabling the reader to compare and contrast such functions among the many members of this superfamily.

In this volume of *Progress in Gene Expression*, experts in the various areas covered focus on specific functions of both steroid and nuclear hormone receptors, rather than present individual reviews of each receptor, as has been typically done in previous published books on this topic. In the opening chapter, Picard discusses the role of heat-shock proteins in the regulation of steroid receptor functions, with a particular emphasis on how yeast genetic approaches can provide valuable information that is as yet difficult to glean from mammalian systems. DeFranco then reviews the regulation of nuclear localization and delocalization of steroid and nuclear receptors. Simons presents a comprehensive overview of the structure and function of the receptor ligand binding domain, comparing and contrasting many such domains among the receptor family members. Incorporated into this review are important insights from recent crystal structures and the way in which these structures relate to the additional functions encoded within this domain, such as dimerization and transactivation. Similarly, Rastinejad describes the known three-dimensional structures of the steroid and nuclear receptor DNA-binding domains, and explains how that information provides the functional basis for the way these proteins interact specifically with DNA. Cheskis and Freedman discuss how cognate ligands for steroid and nuclear receptors can have major effects on the dimerization status of the receptors, which, in the case of heterodimerizing nuclear receptors, provides the basis for cross-talk between distinct endocrine systems.

Three chapters focus specifically on how steroid and nuclear receptors regulate transcription. Bagchi reviews the rapidly expanding literature on transcriptional activation and basal repression by the receptors, with a particular emphasis on a newly discovered class of ligand-dependent interacting factors that appear to function as coactivators of steroid/nuclear receptor transcriptional control. Herrlich and Göttlicher discuss the variety of ways receptors can actually repress activated transcription, an important aspect of receptor function that appears to be mechanistically distinct from their transactivation properties. Watson and Archer describe the increasingly important and central role of chromatin in the way we view receptor transcriptional regulation. Garabedian and colleagues review the functional significance of phosphorylation for steroid receptor function.

The final two chapters focus on the largest subclass of nuclear receptors – the orphans. Lazar and Harding describe a particular class of orphan receptors defined by the fact that they bind to DNA as monomers, rather than the typical dimeric complexes exhibited by most of the known members of the receptor superfamily. Forman completes the volume with a review of the approaches taken to identify cognate ligands for the many orphan receptors that have been isolated by molecular cloning approaches. The topic of this final chapter will undoubtedly define, to a large extent, the future direction of nuclear receptor research as we approach the twenty-first century.

It is my hope that the concept and organization of this book will make it a key information resource for readers, in their own work, as well as a thought-provoking tool for the task of discovering how this large family of hormone-modulated gene regulators, clearly all derived from some primordial precursor, have evolved both similar and distinct strategies to carry out their multiple functions.

Leonard Freedman
New York, June 1997

List of Contributors

Trevor K. Archer, Departments of Obstetrics & Gynaecology, Biochemistry and Oncology, The University of Western Ontario, London, Ontario, Canada N6A 4L6

Milan K. Bagchi, The Population Council & The Rockefeller University, New York, NY 10021

Boris Cheskis, Department of Nuclear Receptors, Wyeth-Ayerst Research, Radnor, PA 19087

Donald B. DeFranco, PhD, Department of Biological Sciences, University of Pittsburgh, Pittsburgh, PA 15260

Barry Marc Forman, The City of Hope National Medical Center, Beckman Research Institute, Duarte, CA 91010-0269

Leonard Freedman, Cell Biology and Genetics Program, Memorial Sloan-Kettering Cancer Center, New York, NY 10021

Michael J. Garabedian, Department of Microbiology, The Kaplan Cancer Center New York University Medical Center, New York, NY 10016

Martin Göttlicher, Forschungszentrum Karlsruhe, Institute of Genetics, 76021 Karlsruhe, Germany

Heather P. Harding, Skirball Institute, New York University School of Medicine New York, NY 10016

Peter Herrlich, Forschungszentrum Karlsruhe, Institute of Genetics, 76021 Karlsruhe, Germany

Adam Hittelman, Department of Microbiology, The Kaplan Cancer Center, New York University Medical Center, New York, NY 10016

Roland Knoblauch, Department of Microbiology, The Kaplan Cancer Center, New York University Medical Center, New York, NY 10016

Marija Krstic, Division of Biochemistry and Molecular Biology, Institute of Biochemical and Life Sciences, Davidson Building, University of Glasglow, Glasglow G12 8QQ, Scotland

Mitchell A. Lazar, Departments of Medicine and Genetics, University of Pennsylvania School of Medicine, 611 CRB, Philadelphia, PA 19104-6149

Didier Picard, Département de Biologie Cellulaire, Université de Genève, Sciences III, CH-1211 Genève 4, Switzerland

Fraydoon Rastinejad, Department of Pharmacology, University of Virginia, Charlottesville, Virginia 22908

Inez Rogatsky, Department of Microbiology, The Kaplan Cancer Center, New York University Medical Center, New York, NY 10016

S. Stoney Simons, Jr., Chief, Steroid Hormones Section, Bldg. 8, NIDDK, NIH, Bethesda, MD 20892-0805

Janet Trowbridge, Department of Microbiology, The Kaplan Cancer Center, New York University Medical Center, New York, NY 10016

Catherine E. Watson, Department of Obstetrics & Gynaecology, The University of Western Ontario, London, Ontario, Canada

1

The Role of Heat-Shock Proteins in the Regulation of Steroid Receptor Function

Didier Picard

Introductory Remark

It has been known for almost three decades that steroid aporeceptors exist in a large complex sedimenting at 9S which becomes reduced to about 4S upon ligand binding. Ever-improving tools have helped to provide overwhelming evidence that steroid receptors form hormone-reversible complexes with the heat-shock protein 90 (Hsp90) plus several Hsp90 accessory proteins (Pratt 1993, Smith and Toft 1993, Pratt and Toft 1997, Toft 1997). The protein components of the aporeceptor complex have been characterized and cloned. Assembly can now be recapitulated *in vitro* with purified components. The vast majority of the evidence that this heterocomplex exists and is relevant to the regulation of steroid receptor activity in cells is based on biochemical experiments in cell-free systems. In the last few years, a variety of experiments including genetic experiments, both in mammalian tissue culture cells and in budding yeast, have begun to fill in this gap. In this review, I will focus on the progress that has been achieved with *in vivo* experiments.

Lessons from Biochemical Studies

Of the enormous number of members of the steroid and nuclear receptor superfamily, only the five vertebrate steroid receptors (estrogen, progesterone, androgen, glucocorticoid and mineralocorticoid receptors, hereafter

Molecular Biology of Steroid and Nuclear Hormone Receptors
Leonard P. Freedman, Editor
© 1998 Birkhäuser Boston

abbreviated ER, PR, AR, GR and MR, respectively) are known, and they are thought to form complexes with the Hsp90 molecular chaperone complex in the absence of ligand (Pratt and Toft 1997, Toft 1997) (see Figure 1). Ligand binding *per se* does not displace the Hsp90 complex. Rather it induces a conformational change in the hormone binding domain (HBD), the "binding site" of the complex, which in turn results in the release of the Hsp90 complex. This release correlates with the conversion of the receptor to a transcription factor, and is a key step in this signal transduction pathway. The Hsp90 complex appears to play a dual role by ensuring that steroid receptors are high affinity receptors for the relatively hydrophobic ligand (Bresnick et al 1989, Rafestin-Oblin et al 1989, Smith 1993), and that they are kept inactive as transcription factors in the absence of ligand (for discussion, see Picard et al 1988, Yamamoto et al 1988, Picard 1993, Pratt 1993).

In vitro reconstitution of steroid receptor-Hsp90 complexes has proven much more challenging than for many other protein complexes. This is essentially due to the following characteristics (see Figure 1): (1) Assembly

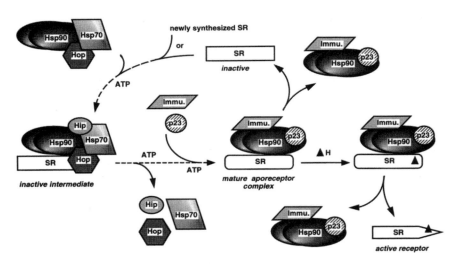

Figure 1.—Model for assembly of steroid receptors with the Hsp90 molecular chaperone complex.

This simplified model was inspired by previously published versions (Smith et al 1995, Pratt and Toft 1997) that were based on *in vitro* reconstitution experiments. Dashed arrows represent more complex and/or incompletely understood steps. The following interactions are known to be direct: Hsp90-Hop, Hop-Hsp70, Hip-Hsp70, p23-Hsp90, Immu.-Hsp90. It is not clear at which point the Hsp70 binding protein Hip (=p48) enters the complex (for discussion, see Prapapanich et al 1996). SR indicates steroid receptor; Hsp90, heat-shock protein 90; Hsp70, heat-shock protein 70; H, steroid hormone; Immu., immunophilin (either FKBP51 or FKBP52 or cyclophilin-40).

is energy-dependent and proceeds via several steps including relatively short-lived intermediates (see Dittmar et al 1996, Johnson et al 1996, Nair et al 1996 and references therein); this gives assembly a directionality. (2) The mature complex is inherently unstable. Smith has recently introduced a dynamic view of assembly and disassembly (Smith 1993, Smith et al 1995, see also Pratt and Toft 1997). According to his "disactivation loop model", receptor-Hsp90 complexes continuously fall apart, but are rapidly and efficiently reassembled in the presence of ATP. Therefore, in a steady-state situation complexes would appear to be stable. Upon ligand binding, the conformational change would preclude the receptor from entering the assembly pathway again. Some implications of this model are worth noting: Defined roles are attributed to the components of the Hsp90 complex; therefore, altering/depleting specific proteins biochemically or genetically in cells should give defined phenotypes (see below) resembling those seen *in vitro*. Furthermore, the steroid receptor molecule as well as components of the Hsp90 complex could be targets for other regulatory signals at different points in the assembly-disassembly pathway. This notion could, for example, help to elucidate the molecular mechanism by which growth factors can activate certain steroid receptors in the absence of their cognate ligand (O'Malley et al 1995, Picard et al 1997). In the case of the activation of the ER by epidermal growth factor, direct phosphorylation of the unliganded ER by MAP kinase appears to be required (Bunone et al 1996). However, it remains to be determined at which step in the assembly-disassembly pathway the ER is the substrate for this kinase, and whether and how this phosphorylation results in the release of the Hsp90 complex.

CELL BIOLOGY OF STEROID RECEPTOR-HSP90 COMPLEXES

The location where assembly and disassembly occurs in the cell is completely unknown. However, it must be emphasized that the subcellular distribution of the unliganded steroid receptor appears to be irrelevant for its assembly with the Hsp90 complex. There is ample evidence that even those steroid receptors (ER and PR) that are nuclear in the absence of hormone form complexes with Hsp90, even though the bulk of Hsp90 is cytoplasmic (Pratt and Toft 1997, see also below). Since these nuclear aporeceptors continuously shuttle between the nucleus and the cytoplasm (Guiochon-Mantel et al 1991, Chandran and DeFranco 1992, Dauvois et al 1993), it is possible that assembly could be confined to the cytoplasm. However, the possibility cannot be excluded that assembly proceeds in both compartments.

IN VIVO STUDIES OF STEROID RECEPTOR-HSP90 COMPLEXES

The steroid receptor community has not always been a "unanimous believer" in the physiological significance of receptor-Hsp90 complexes. For years the unrelenting efforts of biochemists were viewed with suspicion by those with a more genetic or molecular inclination. Even today, critical reports have not subsided. Mainly on the basis of immunohistochemistry, it has been claimed that the nuclear progesterone receptor (PR) cannot be bound to Hsp90 because there is not enough Hsp90 in the nucleus (Tuohimaa et al 1993, Pekki et al 1994, Pekki et al 1995). However, they seem to have disregarded their own results showing that a fraction of the total Hsp90 *does* reside in the nucleus (Tuohimaa et al 1993), and their interpretation is difficult to reconcile with all the other evidence (see also below). Nevertheless, it is important to appreciate the difficulty of extending *in vitro* results to the *in vivo* situation. It is not trivial to rule out that molecular chaperones only bind to steroid receptors upon cell fractionation. In the following, I will review what has been done to address this issue and to discover the role of the chaperone complex in cells (Table 1).

The current evidence is essentially based on the following approaches: (1) *in vivo* cross-linking of complexes, (2) subcellular relocalization of Hsp90 or steroid receptors upon overexpression in mammalian COS-7 cells, (3) pharmacological experiments in mammalian cells, (4) genetic experiments in the budding yeast *Saccharomyces cerevisiae* (*S. cerevisiae*). I would like to mention at this point that indirect evidence also derives from the observation that heterologous proteins can be subjected to hormonal control by fusion to an HBD (Picard 1993, Mattioni et al 1994, Picard 1994). We have argued that this is most simply explained by hypothesizing a hormone-reversible steric hindrance by Hsp90.

Steroid Receptor-Hsp90 Complexes Can Be Cross-linked in Intact Cells

Cell-permeable chemical cross-linkers have been used to show that both the cytoplasmic GR and the nuclear ER can be stabilized as Hsp90 chaperone complexes in intact cells resembling those described for cell-free systems (Rexin et al 1992, Segnitz and Gehring 1995). Although this constitutes some of the best evidence for the existence of such complexes *in vivo*, a word of caution should be added. It cannot be completely ruled out that some cell lysis, which occurred during the treatment, contributed to this result, or that the association with chaperones occurred during the early phase of the treatment with cross-linkers, as a result of it.

Subcellular Relocalization upon Overexpression of Hsp90 or Steroid Receptors

Reasoning that overexpression of either Hsp90 or a steroid receptor should affect the subcellular localization of the steroid receptor or Hsp90, respectively, Catelli and collaborators performed a series of transient transfection experiments in COS-7 cells. They could easily distinguish exogenous chicken Hsp90 from endogenous monkey Hsp90 with a species-specific monoclonal antibody. They found that overexpression of a chicken Hsp90 derivative with a nuclear localization signal leads to the relocalization of coexpressed cytoplasmic GR and PR mutants to the nucleus (Kang et al 1994). Conversely, overexpression of nuclear ER and PR can increase the fraction of coexpressed wild-type Hsp90 in the nucleus (Meng et al 1996). They have already begun to use this system to characterize mutants of Hsp90 with respect to their ability to interact with steroid receptors (Meng et al 1996). Mutants that have previously been found to be defective for *in vitro* association (Sullivan and Toft 1993) also fail to be shifted to the nucleus in this assay. What is most surprising about these experiments is that it was possible to outcompete the enormous levels of endogenous monkey Hsp90 with the exogenously expressed chicken Hsp90. Although these results show that steroid receptors and Hsp90 *can* interact *in vivo*, one must wonder to what extent data obtained with vastly overexpressed molecules reflect the physiological situation. It should also be emphasized that these findings cannot reveal where assembly occurs in the cell. The steady-state distribution of a protein can be altered either because it is "dragged" to another compartment as part of a complex, or because it is trapped there after moving there by itself.

Pharmacological Experiments

As pharmacological agents become available that specifically inhibit or alter components of the Hsp90 complex or interfere with its assembly, it will be possible to study the regulation of endogenous steroid receptor-Hsp90 complexes in wild-type cells. One of the most promising compounds that has very recently been introduced into steroid receptor research is geldanamycin (GA). This benzoquinone ansamycin appears to bind Hsp90 specifically (Whitesell et al 1994) and blocks binding of Hsp90 to p23 (Johnson and Toft 1995). As a result, GA blocks the step requiring p23 in reconstitution experiments with PR such that PR-Hsp70-Hop-Hsp90 complexes (denoted "inactive intermediate" in Figure 1) accumulate that are defective for hormone binding (Johnson and Toft 1995, Smith et al 1995). Additional effects of GA on Hsp90 function should not be ruled out. There is evidence from the work at the Pratt laboratory, on the reconstitution of GR, that GA can also block the conversion of the GR-Hsp90-Hop-Hsp70 complex from an

Table 1.—*In vivo* Studies of Steroid Receptor-Hsp90 Heterocomplexes

Experimental System	Observations	Reference
In vivo cross-linking	Cytoplasmic GR and nuclear ER heterocomplexes same as *in vitro*.	Rexin et al 1992; Segnitz and Gehring 1995
Colocalization in COS cells	Cytoplasmic steroid receptors accumulate in the nucleus upon overexpression of a nuclear version of Hsp90.	Kang et al 1994
	More Hsp90 accumulates in the nucleus upon overexpression of nuclear steroid receptors.	Meng et al 1996
Hsp90 inhibitory drug (GA)	Nonfunctional intermediates of receptor-Hsp90 complex accumulate; hormone binding and transactivation defective.	Smith et al 1995; Whitesell and Cook 1996
Immunosuppressive drugs (FK506, CsA) [1]	Subtle increase in hormone binding affinity of PR.	Renoir et al 1994
	Potentiation of hormonal response.	Ning and Sanchez 1993; Tai et al 1994; Milad et al 1995; Renoir et al 1995; Ratajczak et al 1996
Calmodulin antagonist [2]	Reduces number of hormone-binding-competent GR molecules.	Ning and Sánchez 1996
Overexpression of TPR domain of PP5	Partial inhibition of GR-mediated transactivation.	Chen et al 1996a

Yeast (_S. cerevisiae_)

Hsp90 mutants	Hormonal response less efficient; hormone binding defective.	Picard et al 1990a; Bohen and Yamamoto 1993; Bohen 1995; Nathan and Lindquist 1995; Fang et al 1996
	N-terminal charged domain and C-terminal pentapeptide MEEVD of Hsp90 dispensable as _in vitro_.	Louvion et al 1996
Hop (Sti1)	Required for efficient hormonal response [3].	Chang et al 1997
cyclophilin-40	Cpr7 required for hormonal response [3].	Duina et al 1996a
	Cyp40 (= Cpr6) dispensable.	Duina et al 1996a; Warth et al 1997
p23	Increased sensitivity to GA analog Macbecin I in $\Delta p23$ strain.	S. Bohen, personal communication
DnaJ (Ydj1)	High basal activity without hormone in _ydj1_ mutant strains [3].	Kimura et al 1995
	Reduced transactivation in _ydj1_ mutant strains.	Caplan et al 1995

[1] _In vivo_ effects may be indirect since no significant effects have been observed _in vitro_.
[2] _In vivo_ effects may be indirect since no role has been found for calmodulin _in vitro_.
[3] No significant effects in our assays (J.-F. Louvion, J. W. Liu, R. Warth and D. Picard, unpublished results).

inactive form to a form that is unstable (not stabilized by p23) but able to bind ligand (Dittmar and Pratt 1997). In the model of Figure 1, this functionally more mature form between the "inactive intermediate" and the mature complex has been omitted for simplification. Whatever the precise molecular mechanism is, the effects of GA on the composition and ligand binding activity of PR and GR heterocomplexes can also be observed upon treatment of intact cells with GA (Smith et al 1995, Whitesell and Cook 1996). Functionally, GA has been shown to abolish the transcriptional response of GR (Whitesell and Cook 1996). Thus, this finding provides pharmacological evidence that a particular form of the steroid receptor-Hsp90 complex *is* important for ligand binding and response in intact cells. Although there is a perfect correlation so far between the *in vitro* and *in vivo* effects of GA, results obtained with this drug should be interpreted with caution. Interfering with Hsp90 functions is likely to have pleiotropic effects. Indeed, it has been shown that GA interferes with the *in vivo* function, folding, and/or stability of several other proteins that interact with Hsp90, notably $pp60^{v\text{-}src}$ (Whitesell et al 1994), Raf (Schulte et al 1995, Schulte et al 1996), mutant forms of p53 (Blagosklonny et al 1995, Blagosklonny et al 1996), and reverse transcriptase of hepatitis B virus (Hu and Seeger 1996, Hu et al 1997).

Since the large immunophilins FKBP51, FKBP52, and cyclophilin-40 are part of the mature aporeceptor complex, immunosuppressive drugs have been examined for an effect on steroid receptor regulation. However, studies with FK506 and cyclosporin A (CsA), which bind FKBPs and cyclophilins, respectively, have yielded conflicting results. FK506 may very slightly increase the hormone binding affinity of the progesterone receptor (PR) (Renoir et al 1994), but assembly of steroid receptor complexes *in vitro* is not affected by FK506 or CsA (Hutchison et al 1993, Owens-Grillo et al 1995) and does not even require any immunophilin (Dittmar et al 1996). In contrast, both drugs potentiate the steroid receptor response to hormone *in vivo*, including in yeast (Ning and Sanchez 1993, Tai et al 1994, Milad et al 1995, Renoir et al 1995, Ratajczak et al 1996). While there is no evidence that these effects are direct, there is evidence that FK506 can modify the hormonal response by inhibition of an outward-directed transporter that decreases the effective hormone concentration by pumping hormones out of the cell (Kralli and Yamamoto 1996).

The calmodulin antagonist phenoxybenzamine has been found to inhibit transactivation by GR (Ning and Sánchez 1996). It appears to reduce the number of hormone-binding-competent GR molecules without affecting GR levels or hormone binding affinity. Since calmodulin has not been found in steroid receptor-Hsp90 complexes, it is difficult to draw any conclusions from these observations. The issue of specificity also has to be considered when the inhibitor is a protein domain expressed at "pharmacological" doses. Protein phosphatase 5 (PP5) has recently been discovered as a novel component of the GR-Hsp90 heterocomplex, and it has been shown to bind Hsp90 through its tetratricopeptide repeat (TPR) domain. A dominant-negative mutant of PP5 consisting of only the TPR domain could partially inhibit GR-mediated transactivation (Chen et al 1996a). While it may have competed with wild-type PP5, it could also interfere with binding of other proteins containing TPR domains, such as Hop, FKBP51, FKBP52, and cyclophilin-40, whose binding to Hsp90 depends on their TPR domains and is mutually exclusive (Owens-Grillo et al 1995, Owens-Grillo et al 1996). In this case, again, pleiotropic effects could not be excluded.

Yeast as a Genetic Test Tube For Steroid Receptor Studies

Steroid receptors from vertebrates can function in a hormone-dependent fashion in yeast (Garabedian 1993, see also Picard et al 1990b). Since yeast seems to provide all the other components of this signaling system, yeast genetics can be applied to dissect the heterocomplex. Several homologs of the proteins of the vertebrate Hsp90 complex have been recognized in yeast. They include: (1) the Hsp90 homologs Hsp82 and Hsc82, whose function is essential for viability, although expression of one of the two virtually identical homologs is sufficient (Borkovich et al 1989), (2) the yeast Hop homolog Sti1, which is only essential for viability at low and high temperatures (Nicolet and Craig 1989), (3) a family of Hsp70 proteins with considerable apparent redundancy (see Craig et al 1993), (4) the cyclophilin-40 homologs Cyp40 (= Cpr6) and Cpr7, which are not essential for viability (Duina et al 1996b; Warth et al 1997), and (5) a putative p23 homolog (28% sequence identity and 54% similarity compared to human p23), which appears to be dispensable for viability (S. Bohen, personal communication). It is remarkable that homologs of mammalian Hip (Prapapanich et al 1996) and the Hsp90-associated FKBPs have not been identified, despite the availability of the complete yeast genome sequence. That yeast may be a useful test tube is attested by the finding that several vertebrate steroid receptors can also be coprecipitated with Hsp90 from yeast extracts (Picard et al 1990a, Bohen and Yamamoto 1993, Bohen 1995, Kimura et al 1995, Fang et al 1996). However, the biochemical characterization of the steroid receptor-Hsp90 heterocomplex in yeast remains rather sketchy (see below).

Hsp90

The use of yeast to dissect this signaling complex began when it was shown that reducing the levels of Hsp90 20-fold relative to wild-type severely compromises the hormonal response of GR, ER, and MR in yeast (Picard et al 1990a). Transactivation remains hormone-dependent but requires much higher doses of ligand. Biochemically, this correlates with an increase in the proportion of steroid receptor molecules that are not complexed with Hsp90.

This was the first indication that Hsp90 may be required for the efficiency of the hormonal response. For the GR, the defect has recently been attributed to a decrease in ligand binding affinity (Bohen 1995). It remains to be established that this is also true for other receptors, such as the ER, that can regain wild-type affinity even after disruption of the Hsp90 complex and denaturation (see for example Sakai and Gorski 1984). Interestingly, low levels of Hsp90 in yeast also compromise signaling by retinoid receptors (Holley and Yamamoto 1995), which are not known to form "stable" complexes with Hsp90 suggesting that they may require it transiently.

Bohen and Yamamoto (1993) subsequently screened for viable Hsp82 mutants that are defective for steroid receptor signaling at intermediate hormone concentrations. For some, induction can still be obtained at very high concentrations. These mutations turned out to be single- or double-point mutations that differentially affect the hormonal responses of GR, MR, PR, and ER. For example, the mutation G313N severely impairs induction of all four, whereas E431K specifically affects GR. These findings not only confirmed that Hsp90 functions for viability and that steroid receptor regulation can be separated genetically, but also revealed different steroid receptors may have subtly different Hsp90 requirements. More recently, Bohen has provided evidence that hormone binding of GR is defective in strains with *HSP82* point mutations (Bohen 1995). While it had previously been claimed that the binding of these Hsp82 mutants to GR is unperturbed (Bohen and Yamamoto 1993), he has now demonstrated that complexes with mutant Hsp82 proteins are less stable. Thus, it could be speculated that the hormone binding defect is due to a failure to form or maintain "stable" complexes.

Using a screen for temperature-sensitive (ts) mutations in *HSP82*, Nathan and Lindquist (1995) recovered another set of interesting mutations. Seven of the eight mutations, which are ts for viability, also display a defect in GR signaling at permissive temperature. The eighth mutation, G170D, behaves like a classical ts mutation, affecting both viability and GR signaling only upon a shift to the restrictive temperature. With temperature-shift experiments, they could show that Hsp90 function is required for establishment *and* maintenance of the hormone-responsive state of GR. Caplan and collaborators examined the response of the AR in the same mutant strain. They found that the AR loses its high-affinity hormone binding at the restrictive temperature (Fang et al 1996).

To delineate functional domains of Hsp90, we have performed a deletion analysis of the isoform Hsp82 (Louvion et al 1996). All viable truncations were wild-type for steroid receptor regulation. They allowed us to demonstrate that two eukaryote-specific regions of Hsp82, the N-terminal charged domain and the extremely conserved C-terminal pentapeptide MEEVD, are dispensable for both functions. Since this finding agrees with results obtained *in vitro* (Sullivan and Toft 1993) and in the colocalization assay in mammalian cells (Meng et al 1996), it further validates yeast as a model system. We have previously argued (Louvion et al 1996) that the power of the yeast system could be considerably increased if nonviable Hsp90 mutants could be tested for specific Hsp90 functions that are not required for viability. Viability could be ensured by one of the Hsp82 mutants that is defective for steroid receptor signaling (Picard et al 1990a, Bohen and Yamamoto 1993, Nathan and Lindquist 1995). One could then attempt to complement this particular defect with a series of derivatives of yeast or vertebrate Hsp90.

Hsp90 Accessory Proteins

Since Hop (= Sti1), which is essential for *in vitro* reconstitution of receptor-Hsp90 heterocomplexes (Chen et al 1996b, Dittmar et al 1996), is not essential for viability of yeast at intermediate temperatures (Nicolet and Craig 1989), steroid signaling could be analyzed in yeast knock-out strains. Lindquist and her collaborators have reported that the activation of GR in a Δsti1 strain requires much higher hormone concentrations and never reaches wild-type levels (Chang et al 1997). The effects on hormone binding affinity have not been assessed. Using the same strain as well as a different one, we have found only a very subtle shift in the dose-response curve with several steroid receptors (J.-F. Louvion, J. W. Liu and D. Picard, unpublished results). Our assays differ from those of Lindquist et al by an important technical detail: the Lindquist laboratory does the hormonal induction of steroid receptors for only one hour, whereas we determine transactivation during an overnight induction. A caveat with the former approach is that mutants that have pleiotropic defects affecting growth and protein folding will fare poorly in a short-term assay, simply because everything takes more time. Our protocol does not allow us to rule out the possibility that bypass mechanisms are activated, but it does yield a very clear answer to a stringent genetic question: Is a particular gene absolutely required or not? In this particular case, it remains to be elucidated why Hop is clearly less important in yeast than *in vitro*, and why Hop cannot be detected in GR complexes from yeast (Chang et al 1997).

In a very similar set of experiments, GR and ER activation have been found to be defective in a strain lacking Cpr7 (Duina et al 1996a), one of the two cyclophilin-40 homologs. Using our long-term assay we found that the other yeast homolog, Cyp40 (= Cpr6), is dispensable for normal regulation of PR, ER, and GR (Warth et al 1997). As far as Cyp40 goes, this finding agrees with the results of Duina et al. (1996a), but we could not confirm that Cpr7 is required (R. Warth and D. Picard, unpublished results). Apart from reminding the reader here that *in vitro* experiments have failed to suggest any role for cyclophilins (and FKBPs), we should also note that Cyp40 (= Cpr6) could not be detected in GR complexes (Chang et al 1997). Cpr7 has yet to be examined. Thus, it could be speculated that immunophilins are not genuine components of the Hsp90 heterocomplex *in vivo*, and that the effects seen with the deletion of the *CPR7* gene in certain assays are indirect.

Like Hop, p23 could have been expected to be absolutely essential for proper regulation of steroid receptors, or even for viability. However, the only phenotype of the deletion appears to be an increased sensitivity of the hormonal response of steroid receptors to the GA analog macbecin I (S. Bohen, personal communication). It will be interesting to see whether the sensitive step in the absence of the p23 homolog is related to the one discovered by Pratt and collaborators in their *in vitro* reconstitution system involving the conversion of the "inactive intermediate" to a hormone-binding-competent intermediate (see above). This phenomenology might be explained with two *ad hoc* assumptions: (1) Yeast p23 *does* stabilize steroid receptor-Hsp90 complexes, but much less efficiently than its vertebrate homolog. (2) Macbecin I only partially inhibits either one of the two sensitive steps (conversion of the receptor-Hsp90-Hop-Hsp70 complex to the ligand-binding form and stabilization of the complex by binding of p23 to Hsp90). This hypothesis could explain why there is no effect upon deleting the p23 gene and yet an increase in the sensitivity to macbecin I. It is also compatible with the observation that only a small fraction of all steroid receptor molecules in a yeast extract are able to bind hormone *in vitro* (see, for example, Metzger et al 1988, Kralli et al 1995).

In contrast to the Hsp70-binding protein Hip (Prapapanich et al 1996), homologs of the bacterial Hsp70 cochaperone DnaJ have never been detected in steroid receptor-Hsp90 heterocomplexes in vertebrate systems. However, mutations in *YDJ1*, which encodes a cytosolic DnaJ homolog in yeast, have been reported to display synthetic lethality with *HSP90* mutations (Kimura et al 1995) and to affect the regulation of steroid receptor activity. Caplan and his collaborators found that the maximal transcriptional responses of AR and GR are reduced several-fold in *ydj1* mutant strains and that the hormone binding capacity of AR remains unchanged (Caplan et al 1995). In contrast, Lindquist and her collaborators observed a dramatic increase (up to 200-fold in the case of GR) in the basal transcriptional activity of GR and ER in the absence of hormone (Kimura et al 1995). The induced levels in the presence of hormone were not significantly different from those in a wild-type strain. They also showed that wild-type Ydj1 coprecipitates with GR from yeast extracts along with Hsp90 and Hsp70. Using a different assay system, we have not seen any evidence of increased activity of ER and PR in the absence of hormone in *ydj1* mutant strains (J.-F. Louvion, J. W. Liu and D. Picard, unpublished results), which agrees with the findings from the Caplan laboratory. The differences in the reported basal levels are particularly surprising, ranging from a mere 2-fold increase (Caplan et al 1995) to a 200-fold increase (Kimura et al 1995). At present, it is difficult to reconcile all these contradictory findings, since there is only a partial overlap in the receptors and *ydj1* alleles that were used. One would have expected that complete knock-out strains

($\Delta ydj1$), used by all three groups, would have yielded comparable results. It has been speculated that Ydj1 could be involved both in preventing spontaneous activation of steroid receptors without ligand and in chaperoning the final steps of folding upon ligand activation (Bohen et al 1995). Further biochemical and genetic characterization is necessary since *ydj1* mutant strains are very sick, and thus, indirect effects cannot easily be excluded, despite the fact that Ydj1 coprecipitates with GR. It would also be useful to search for the presence and potential role of DnaJ proteins in steroid receptor-Hsp90 complexes in the vertebrate system.

CONCLUSIONS

Each of the above-mentioned approaches to studying the steroid receptor-Hsp90 heterocomplex *in vivo* has its shortcomings. But if they are taken together, they offer very strong evidence for a dual role of Hsp90 in regulating steroid receptor activity both in cells and in the test tube. Hsp90 helps to keep steroid receptors inactive as transcription factors in the absence of ligand, and to ensure efficient hormone responsiveness. The roles of Hsp90 accessory proteins such as Hop, p23, and immunophilins remain to be further characterized. The yeast experiments suggest that their functions may be more complex and/or more redundant than anticipated from the *in vitro* systems. Some may eventually even lose their claim to being genuine components of the heterocomplex. In the future, we need a more solid and comprehensive analysis of the biochemical properties of the heterocomplex in yeast, and novel approaches to studying this signaling complex in mammalian cells.

ACKNOWLEDGMENTS

I am very grateful to all those colleagues who kindly communicated unpublished data, and to Olivier Donzé for critical comments on the manuscript. Work in my laboratory has been supported by the Swiss National Science Foundation and the Canton de Genève.

REFERENCES

Blagosklonny MV, Toretsky J, Bohen S, and Neckers L (1996): Mutant conformation of p53 translated *in vitro* or *in vivo* requires functional HSP90. *Proc. Natl. Acad. Sci. USA* 93: 8379–8383.
Blagosklonny MV, Toretsky J, and Neckers L (1995): Geldanamycin selectively destabilizes and conformationally alters mutated p53. *Oncogene* 11: 933–939.

Bohen SP (1995): Hsp90 mutants disrupt glucocorticoid receptor ligand binding and destabilize aporeceptor complexes. *J. Biol. Chem.* 270: 29433–29438.

Bohen SP, Kralli A, and Yamamoto KR (1995): Hold'em and fold'em: chaperones and signal transduction. *Science* 268: 1303–1304.

Bohen SP, and Yamamoto KR (1993): Isolation of Hsp90 mutants by screening for decreased steroid receptor function. *Proc. Natl. Acad. Sci. USA* 90: 11424–11428.

Borkovich KA, Farrelly FW, Finkelstein DB, Taulien J, and Lindquist S (1989): hsp82 is an essential protein that is required in higher concentrations for growth of cells at higher temperatures. *Mol. Cell. Biol.* 9: 3919–3930.

Bresnick EH, Dalman FC, Sanchez ER, and Pratt WB (1989): Evidence that the 90-kDa heat shock protein is necessary for the steroid binding conformation of the L cell glucocorticoid receptor. *J. Biol. Chem.* 264: 4992–4997.

Bunone G, Briand P-A, Miksicek RJ, and Picard D (1996): Activation of the unliganded estrogen receptor by EGF involves the MAP kinase pathway and direct phosphorylation. *EMBO J.* 15: 2174–2183.

Caplan AJ, Langley E, Wilson EM, and Vidal J (1995): Hormone-dependent transactivation by the human androgen receptor is regulated by a dnaJ protein. *J. Biol. Chem.* 270: 5251–5257.

Chandran UR, and DeFranco DB (1992): Internuclear migration of chicken progesterone receptor, but not simian virus-40 large tumor antigen, in transient heterokaryons. *Mol. Endocrinol.* 6: 837–844.

Chang H-CJ, Nathan DF, and Lindquist S (1997): In vivo analysis of the Hsp90 cochaperone Sti1 (p60). *Mol. Cell Biol.* 17: 318–325.

Chen M-S, Silverstein AM, Pratt WB, and Chinkers M (1996a): The tetratricopeptide repeat domain of protein phosphatase 5 mediates binding to glucocorticoid receptor heterocomplexes and acts as a dominant negative mutant. *J. Biol. Chem.* 271: 32315–32320.

Chen S, Prapapanich V, Rimerman RA, Honoré B, and Smith DF (1996b): Interactions of p60, a mediator of progesterone receptor assembly, with heat shock proteins Hsp90 and Hsp70. *Mol. Endocrinol.* 10: 682–693.

Craig EA, Gambill BD, and Nelson RJ (1993): Heat shock proteins: molecular chaperones of protein biogenesis. *Microbiol. Rev.* 57: 402–414.

Dauvois S, White R, and Parker MG (1993): The antiestrogen ICI182780 disrupts estrogen receptor nucleocytoplasmic shuttling. *J. Cell Sci.* 106: 1377–1388.

Dittmar KD, Hutchison KA, Owens-Grillo JK, and Pratt W B (1996): Reconstitution of the steroid receptor · hsp90 heterocomplex assembly system of rabbit reticulocyte lysate. *J. Biol. Chem.* 271: 12833–12839.

Dittmar KD and Pratt WB (1997): Folding of the glucocorticoid receptor by the reconstituted hsp90–based chaperone machinery. *J. Biol. Chem.* 272: 13047–13054.

Duina AA, Chang H-CJ, Marsh JA, Lindquist S, and Gaber R F (1996a): A cyclophilin function in Hsp90-dependent signal transduction. *Science* 274: 1713–1715.

Duina AA, Marsh JA, and Gaber RF (1996b): Identification of two CyP-40-like cyclophilins in *Saccharomyces cerevisiae*, one of which is required for normal growth. *Yeast* 12: 943–952.

Fang Y, Fliss AE, Robins DM, and Caplan AJ (1996): Hsp90 regulates androgen receptor hormone binding affinity *in vivo*. *J. Biol. Chem.* 271: 28697–28702.

Garabedian M J (1993): Genetic approaches to mammalian nuclear receptor function in yeast. *Companion Methods Enzymol.* 5: 138–146.

Guiochon-Mantel A, Lescop P, Christin-Maitre S, Loosfelt H, Perrot-Applanat M and Milgrom E (1991): Nucleocytoplasmic shuttling of the progesterone receptor. *EMBO J.* 10: 3851–3859.

Holley SJ, and Yamamoto KR (1995): A role for Hsp90 in retinoid receptor signal transduction. *Mol. Biol. Cell* 6: 1833–1842.

Hu J, and Seeger C (1996): Hsp90 is required for the activity of a hepatitis B virus reverse transcriptase. *Proc. Natl. Acad. Sci. USA* 93: 1060–1064.

Hu J, Toft DO, and Seeger C (1997): Hepadnavirus assembly and reverse transcription require a multi-component chaperone complex which is incorporated into nucleocapsids. *EMBO J.* 16: 59–68.

Hutchison KA, Scherrer LC, Czar MJ, Ning Y, Sanchez ER, Leach KL, Deibel MRJ, and Pratt WB (1993): FK506 binding to the 56-kilodalton immunophilin (Hsp56) in the glucocorticoid receptor heterocomplex has no effect on receptor folding or function. *Biochem.* 32: 3953–3957.

Johnson J, Corbisier R, Stensgard B, and Toft DO (1996): The involvement of p23, hsp90, and immunophilins in the assembly of progesterone receptor complexes. *J. Steroid Biochem. Molec. Biol.* 56: 31–37.

Johnson J L and Toft D O (1995): Binding of p23 and hsp90 during assembly with the progesterone receptor. *Mol. Endocrinol.* 9: 670–678.

Kang KI, Devin J, Cadepond F, Jibard N, Guiochon-Mantel A, Baulieu EE, and Catelli MG (1994): In vivo functional protein-protein interaction: nuclear targeted hsp90 shifts cytoplasmic steroid receptor mutants into the nucleus. *Proc. Natl. Acad. Sci. USA* 91: 340–344.

Kimura Y, Yahara I, and Lindquist S (1995): Role of the protein chaperone YDJ1 in establishing Hsp90-mediated signal transduction pathways. *Science* 268: 1362–1365.

Kralli A, Bohen SP, and Yamamoto KR (1995): LEM1, an ATP-binding-cassette transporter, selectively modulates the biological potency of steroid hormones. *Proc. Natl. Acad. Sci. USA* 92: 4701–4705.

Kralli A, and Yamamoto KR (1996): An FK506-sensitive transporter selectively decreases intracellular levels and potency of steroid hormones. *J. Biol. Chem.* 271: 17152–17156.

Louvion J-F, Warth R, and Picard D (1996): Two eukaryote-specific regions of Hsp82 are dispensable for its viability and signal transduction functions in yeast. *Proc. Natl. Acad. Sci. USA* 93: 13937–13942.

Mattioni T, Louvion J-F, and Picard D (1994): Regulation of protein activities by fusion to steroid binding domains. In: *Protein expression in animal cells*, Methods in Cell Biology, Roth M, ed., Vol. 43. Academic Press, pp. 335–352.

Meng X, Devin J, Sullivan WP, Toft D, Baulieu E-E, and Catelli M-G (1996): Mutational analysis of Hsp90a dimerization and subcellular localization: dimer disruption does not impede "in vivo" interaction with estrogen receptor. *J. Cell Sci.* 109: 1677–1687.

Metzger D, White JH, and Chambon P (1988): The human estrogen receptor functions in yeast. *Nature* 334: 31–36.

Milad M, Sullivan W, Diehl E, Altmann M, Nordeen S, Edwards DP, and Toft DO

(1995): Interaction of the progesterone receptor with binding proteins for FK506 and cyclosporin A. *Mol. Endocrinol.* 9: 838–847.

Nair SC, Toran EJ, Rimerman RA, Hjermstad S, Smithgall TE, and Smith DF (1996): A pathway of multi-chaperone interactions common to diverse regulatory proteins: Estrogen receptor, Fes tyrosine kinase, heat shock transcription factor HSF1, and the arylhydrocarbon receptor. *Cell Stress Chap.* 1: 237–250.

Nathan DF, and Lindquist S (1995): Mutational analysis of Hsp90 function: interactions with a steroid receptor and a protein kinase. *Mol. Cell. Biol.* 15: 3917–3925.

Nicolet CM, and Craig EA (1989): Isolation and characterization of STI1, a stress-inducible gene from *Saccharomyces cerevisiae*. *Mol. Cell Biol.* 9: 3638–3646.

Ning Y-M, and Sánchez ER (1996): *In vivo* evidence for the generation of a glucocorticoid receptor-heat shock protein-90 complex incapable of binding hormone by the calmodulin antagonist phenoxybenzamine. *Mol. Endocrinol.* 10: 14–23.

Ning YM, and Sanchez ER (1993): Potentiation of glucocorticoid receptor-mediated gene expression by the immunophilin ligands FK506 and rapamycin. *J. Biol. Chem.* 268: 6073–6076.

O'Malley BW, Schrader WT, Mani S, Smith C, Weigel NL, Conneely OM, and Clark JH (1995): An alternative ligand-independent pathway for activation of steroid receptors. *Recent Prog. Horm. Res.* 50: 333–347.

Owens-Grillo JK, Czar MJ, Hutchison KA, Hoffman K, Perdew GH, and Pratt WB (1996): A model of protein targeting mediated by immunophilins and other proteins that bind to hsp90 via tetratricopeptide repeat domains. *J. Biol. Chem.* 271: 13468–13475.

Owens-Grillo JK, Hoffmann K, Hutchison KA, Yem AW, Deibel MR, Handschumacher RE, and Pratt WB (1995): The cyclosporin A-binding immunophilin CyP-40 and the FK506-binding immunophilin Hsp56 bind to a common site on Hsp90 and exist in independent cytosolic heterocomplexes with the untransformed glucocorticoid receptor. *J. Biol. Chem.* 270: 20479–20484.

Pekki A, Ylikomi T, Syvala H, and Tuohimaa P (1994): Progesterone receptor and hsp90 are not complexed in intact nuclei. *J. Steroid Biochem. Mol. Biol.* 48: 475–479.

Pekki A, Ylikomi T, Syvala H, and Tuohimaa P (1995): Progesterone receptor does not form oligomeric (8S), non-DNA-binding complex in intact cell nuclei. *J. Cell Biochem.* 58: 95–104.

Picard D (1993): Steroid-binding domains for regulating the functions of heterologous proteins *in cis*. *Trends Cell Biol.* 3: 278–280.

Picard D (1994): Regulation of protein function through expression of chimaeric proteins. *Curr. Opin. Biotech.* 5: 511–515.

Picard D, Bunone G, Liu JW, and Donzé O (1997): Steroid-independent activation of steroid receptors in mammalian and yeast cells and in breast cancer. *Biochem. Soc. Trans.* 25: 597–602.

Picard D, Khursheed B, Garabedian MJ, Fortin MG, Lindquist S, and Yamamoto KR (1990a): Reduced levels of hsp90 compromise steroid receptor action *in vivo*. *Nature* 348: 166–168.

Picard D, Salser SJ, and Yamamoto KR (1988): A movable and regulable inactivation function within the steroid binding domain of the glucocorticoid receptor. *Cell* 54: 1073–1080.

Picard D, Schena M, and Yamamoto KR (1990b): An inducible expression vector for both fission and budding yeasts. *Gene* 86: 257–261.

Prapapanich V, Chen S, Nair SC, Rimerman RA, and Smith DF (1996): Molecular cloning of human p48, a transient component of progesterone receptor complexes and an hsp70-binding protein. *Mol. Endocrinol.* 10: 420–431.

Pratt WB (1993): The role of heat shock proteins in regulating the function, folding, and trafficking of the glucocorticoid receptor. *J. Biol. Chem.* 268: 21455–21458.

Pratt WB, and Toft DO (1997): Steroid receptor interactions with heat shock protein and immunophilin chaperones. *Endocr. Rev.* 18: 306–360.

Rafestin-Oblin M-E, Couette B, Radanyi C, Lombes M, and Baulieu E-E (1989): Mineralocorticosteroid receptor of the chick intestine. *J. Biol. Chem.* 264: 9304–9309.

Ratajczak T, Mark PJ, Martin RL, and Minchin RF (1996): Cyclosporin A potentiates estradiol-induced expression of the cathepsin D gene in MCF7 breast cancer cells. *Biochem. Biophys. Res. Commun.* 220: 208–212.

Renoir J-M, Bihan SL, Mercier-Bodard C, Gold A, Arjomandi M, Radanyi C, and Baulieu E-E (1994): Effects of immunosuppressants FK506 and rapamycin on the heterooligomeric form of the progesterone receptor. *J. Steroid Biochem. Molec. Biol.* 48: 101–110.

Renoir JM, Mercier Bodard C, Hoffmann K, Le Bihan S, Ning YM, Sanchez ER, Handschumacher RE, and Baulieu EE (1995): Cyclosporin A potentiates the dexamethasone-induced mouse mammary tumor virus-chloramphenicol acetyl-transferase activity in LMCAT cells: a possible role for different heat shock protein-binding immunophilins in glucocorticosteroid receptor-mediated gene expression. *Proc. Natl. Acad. Sci. USA* 92: 4977–4981.

Rexin M, Busch W, Segnitz B, and Gehring U (1992): Structure of the glucocorticoid receptor in intact cells in the absence of hormone. *J. Biol. Chem.* 267: 9619–9621.

Sakai D, and Gorski J (1984): Reversible denaturation of the estrogen receptor and estimation of polypeptide chain molecular weight. *Endocrinology* 115: 2379–2383.

Schulte TW, Blagosklonny MV, Ingui C, and Neckers L (1995): Disruption of the Raf-1-Hsp90 molecular complex results in destabilization of Raf-1 and loss of Raf-1-Ras association. *J. Biol. Chem.* 270: 24585–24588.

Schulte TW, Blagosklonny MV, Romanova L, Mushinski JF, Monia BP, Johnston JF, Nguyen P, Trepel J, and Neckers LM (1996): Destabilization of Raf-1 by geldanamycin leads to disruption of the Raf-1-MEK-mitogen-activated protein kinase signalling pathway. *Mol. Cell. Biol.* 16: 5839–5845.

Segnitz B, and Gehring U (1995): Subunit structure of the nonactivated human estrogen receptor. *Proc. Natl. Acad. Sci. USA* 92: 2179–2183.

Smith DF (1993): Dynamics of heat shock protein 90-progesterone receptor binding and the disactivation loop model for steroid receptor complexes. *Mol. Endocrinol.* 7: 1418–1429.

Smith DF, and Toft DO (1993): Steroid receptors and their associated proteins. *Mol. Endocrinol.* 7: 4–11.

Smith DF, Whitesell L, Nair SC, Chen S, Prapapanich V, and Rimerman RA (1995): Progesterone receptor structure and function altered by geldanamycin, an hsp90-binding agent. *Mol. Cell. Biol.* 15: 6804–6812.

Sullivan WP, and Toft DO (1993): Mutational analysis of hsp90 binding to the progesterone receptor. *J. Biol. Chem.* 268: 20373–20379.

Tai PK, Albers MW, McDonnell DP, Chang H, Schreiber SL, and Faber LE (1994): Potentiation of progesterone receptor-mediated transcription by the immunosuppressant FK506. *Biochem.* 33: 10666–10671.

Toft DO (1997): Protein activity control: hormone receptor function. In: *Molecular biology of chaperones and folding catalysts*, Bukau B, ed. Harwood Academic Publications.

Tuohimaa P, Pekki A, Blauer M, Joensuu T, Vilja P, and Ylikomi T (1993): Nuclear progesterone receptor is mainly heat shock protein 90-free in vivo. *Proc. Natl Acad. Sci. USA* 90: 5848–5852.

Warth R, Briand P-A, and Picard D (1997): Functional analysis of the yeast 40 kDa cyclophilin Cyp40 and its role for viability and steroid receptor regulation. *Biol. Chem.* 378: 381–391.

Whitesell L, and Cook P (1996): Stable and specific binding of heat shock protein 90 by geldanamycin disrupts glucocorticoid receptor function in intact cells. *Mol. Endocrinol.* 10: 705–712.

Whitesell L, Mimnaugh E G, De Costa B, Myers C E and Neckers L M (1994): Inhibition of heat shock protein HSP90-pp60v-src heteroprotein complex formation by benzoquinone ansamycins: essential role for stress proteins in oncogenic transformation. *Proc. Natl. Acad. Sci. USA* 91: 8324–8328.

Yamamoto KR, Godowski PJ, and Picard D (1988): Ligand regulated nonspecific inactivation of receptor function: A versatile mechanism for signal transduction. *Cold Spring Harbor Symp. Quant. Biol.* 53: 803–811.

2

Subcellular and Subnuclear Trafficking of Steroid Receptors

Donald B. DeFranco

Many external signals affect cell physiology through alterations in gene expression. Cell surface receptors, once they have been activated by the binding of appropriate ligands, mobilize signal transduction cascades that may ultimately affect the activity of defined sets of transcription factors. Receptors for steroid hormones, in contrast, are soluble, intracellular proteins that function as transcription factors to directly regulate the transcriptional activity of target genes. Thus, steroid hormones utilize a streamlined signal transduction system in which a single protein, the steroid hormone receptor, has the capacity both to recognize an external signal and to transduce that signal to alterations in specific gene expression. The delivery of activated steroid receptors to genomic target sites must be efficient to account for the rapidity and selectivity of many transcriptional responses to steroid hormones. This review will focus on recent advances in subcellular trafficking of steroid receptors, and will include discussions both of receptor trafficking between different subcellular compartments (i.e., the cytoplasm and nucleus) and of the trafficking of receptors within a specific compartment (i.e., the nucleus).

I. Localization of Unliganded Steroid Receptors

The tight association of ligand-bound steroid receptors with nuclei was recognized in the late 1960s (Gorski et al 1968, Jensen et al 1968). However,

Molecular Biology of Steroid and Nuclear Hormone Receptors
Leonard P. Freedman, Editor
© 1998 Birkhäuser Boston

a controversy has persisted for over 20 years concerning the subcellular localization of unliganded steroid receptors. As a result, many fundamental questions regarding steroid receptor trafficking remained unanswered. In which subcellular compartment do these signal transduction proteins first encounter ligand? Is ligand binding required for the accumulation of steroid receptors within the nucleus? Once specific anti-receptor antibodies became available, cell biological approaches were used to directly visualize steroid receptor compartmentalization. However, initial results from cell biological studies did not provide definitive conclusions regarding steroid receptor subcellular localization; instead, they implied the possible existence of separate and distinct subcellular trafficking pathways for different receptors. Unoccupied progesterone receptors (PRs) and estrogen receptors (ERs) appeared to be localized predominantly within nuclei (Perrot-Applanat et al 1985, Welshons et al 1984), while in most studies unliganded glucocorticoid receptors (GRs) (Antakly and Eisen 1984, Picard and Yamamoto 1987, Wikström et al 1987), mineralocorticoid receptors (MRs) (Lombes et al 1990), and androgen receptors (ARs) (Jenster et al 1993, Zhou et al 1994) were localized within the cytoplasm. This view, however, was not uniformly accepted. Legitimate concerns were raised about the effects of different fixation conditions on receptor compartmentalization and the specificity of the available antibodies (e.g., see Welshons and Judy 1995).

The resolution of this controversy did not result from general agreement regarding the most appropriate fixation conditions, or from the development of definitive immunological probes, but from the recognition that steroid receptor nuclear transport is bidirectional. Thus, as first demonstrated by Milgrom and co-workers for rabbit PR and human ER (Guiochon-Mantel et al 1991), and later by other investigators for chicken PR (Chandran and DeFranco 1992), rat GR (Madan and DeFranco 1993), and mouse ER (Dauvois et al 1993), steroid receptors have the capacity to shuttle between the nuclear and cytoplasmic compartments. Thus, steroid receptors are not statically confined to either the cytoplasmic or nuclear compartment; they establish an equilibrium distribution between these compartments based upon their relative rates of nuclear import and nuclear export. Receptors will accumulate within the cytoplasm if nuclear import is rate limiting, while a limitation in the rate of nuclear export will lead to the preferential accumulation of receptors within nuclei (DeFranco et al 1995). These rates are likely to be limited not by differences in inherent rate of passage through the nuclear pore complex (NPC), but rather by the rate of receptor release from compartment-specific anchoring complexes (see below).

II. Signal Sequences for Nuclear Import of Steroid Receptors

One of the first questions that arose from the recognition that nuclear transport of steroid receptors is bidirectional concerned the nature of signals that direct receptors to nuclear import, versus export, pathways. Picard and Yamamoto first identified a constitutive nuclear localization signal sequence (NLS) located within the DNA-binding domain (Picard and Yamamoto 1987) of the rat GR. This signal has been mapped in more detail in the human (Cadepond et al 1992) GR, rat (Tang et al 1997) GR, and other steroid receptors (Ylikomi et al 1992, Zhou et al 1994), and has been shown to be comprised of multiple protosignals. These protosignals are characterized by a prevalence of basic amino acids, and they function when present in unique pairs (Ylikomi et al 1992, Tang et al 1997). In this way, steroid receptor NLSs resemble the prototype bipartite NLS first identified within the nucleoplasmin protein (Dingwall and Laskey 1991). *In vitro* studies have established that the GR most likely utilizes a common GTP- and ATP-dependent pathway for nuclear import (Yang and DeFranco 1994, DeFranco et al 1995), although interactions between steroid receptor NLSs and the bona fide NLS receptor, karyopherin/importin (Görlich et al 1994), have yet to be definitively demonstrated (LaCasse et al 1993).

A separate hormone-dependent NLS was identified within the ligand binding domain (LBD) of rat GR (Picard and Yamamoto 1987), but this signal does not appear to resemble the bipartite, basic NLS found in the receptor's DBD. Fine structure mapping of this putative NLS has not been performed. There has been some speculation that the NLS activity in steroid receptor LBDs may only serve an accessory role in nuclear transport (Cadepond, Gasc et al 1992), but even if this is true, the precise mechanism involved in hormone-dependent facilitation of steroid receptor nuclear transport remains unknown. Since a number of proteins that lack NLSs can be transported into nuclei in association with an NLS-containing protein (Sommer et al 1991, Zhao and Padmanabhan 1988), it is conceivable that the apparent hormone-dependent NLS activity associated with some steroid receptor LBDs is imparted by LBD-associated proteins that possess bona fide NLSs. If this is indeed the case, the LBD co-transporting partner might remain associated with ligand-bound, and presumably activated, receptors during their passage through the NPC. Although there are some attractive candidates for such LBD-co-transporting substrates (e.g., hsp70 and hsp90; see Okuno et al 1993, Kang et al 1994, respectively), definitive proof that this co-transport operates in physiologically relevant contexts has not been provided (see below).

III. CYTOPLASMIC RETENTION OF STEROID RECEPTORS

Unliganded, cytoplasmic GRs exist as heteromeric complexes that minimally contains hsp90 and particular immunophilin proteins (Bohen and Yamamoto 1994). While the constitutive nuclear localization of LBD-deleted GRs suggested that nuclear import of the receptor is restricted by its association within heteromeric complexes (Picard and Yamamoto 1987, Pratt 1993), a number of studies raised questions about the simplicity of this model. Other steroid receptors that are assembled into heteromeric complexes when unliganded appear to be localized predominantly within the nucleus (Smith and Toft 1993). Furthermore, in some cells, unliganded GRs also appeared to localize predominantly within nuclei (Sanchez et al 1990). Are these results in conflict with a presumed role for steroid receptor heteromeric complexes in limiting nuclear import? An important point to consider is that the assembly of steroid receptor heteromeric complexes is a dynamic process. This has been shown most clearly for PR *in vitro* (Smith 1993), whose association with chaperones such as hsp90 is transient even in the absence of hormone binding. Thus, there is likely to be a constant turnover of individual components of steroid receptor heteromeric complexes. If steroid receptor NLSs are transiently exposed to NLS receptors (which are predominantly cytoplasmic; Görlich et al 1994), a productive interaction may ensue that would commit steroid receptor-NLS receptor complexes to subsequent interactions with the NPC. It follows that the stability of steroid receptor-heteromeric complexes, which probably varies for individual receptors, and perhaps within different cell types, could have a direct impact on the cytoplasmic retention of unliganded receptors.

We have confirmed that the hormone dependence of GR import reflects a requirement for receptor activation, or more precisely the dissociation of hsp90 from a GR-heteromeric complex (Yang and DeFranco 1996). This was shown using a novel delivery system to enable sufficient accumulation of sodium molybdate in live cells to stabilize GR-hsp90 complexes. Stabilization of GR-hsp90 complexes led to a dramatic reduction in hormone-dependent nuclear import of the receptor *in vivo* (Yang and DeFranco 1996). In this study the composition of GR heteromeric complexes in sodium molybdate-treated cells was not examined, so it is unclear whether this treatment led to the generation of novel GR-hsp90-containing complexes or captured a particular intermediate complex (Yang and DeFranco 1996). Nonetheless, these data imply that the efficiency of nuclear import is governed by the relative stability of receptor heteromeric complexes. This hypothesis was also supported by the analysis of PR nuclear import in sodium molybdate-treated cells. In this case, both hormone-independent and -dependent nuclear import of PR was inhibited by sodium molybdate treatment (Yang and DeFranco 1996). One interpretation of these results is that stabilization of PR heteromeric complexes likewise has a detrimental effect on the ability of the receptor to import into nuclei. Thus, the differential localization of unliganded GR (i.e., cytoplasmic) versus PR (i.e., nuclear) observed in most cell types may simply result from differences in the inherent stability of receptor heteromeric complexes.

In contrast to this view, it has been suggested that PR and GR may remain associated with hsp90 during the nuclear import process (Kang et al 1994). This conclusion was based upon studies with hsp90 chimeras that possessed a linked, heterologous NLS. PR and GR derivatives that lacked their own NLS were found to import into nuclei when co-expressed with NLS-hsp90 chimeras (Kang et al 1994). While these results provide a convincing demonstration of the association between hsp90 and GR or PR *in vivo*, they do not address the question of whether the co-transport of hsp90 with steroid receptors occurs when native signal sequences within receptors are used to direct their import into nuclei. Hsp90 engineered with a potent NLS at its amino terminus may efficiently engage the nuclear import machinery and stabilize steroid receptor complexes in a manner that is not utilized by native receptors.

IV. SIGNAL SEQUENCES FOR NUCLEAR EXPORT OF STEROID RECEPTORS

Is there an analogous signal that directs the nuclear export of steroid receptors? This has been a more difficult question to address, because it is only in recent years that some mechanistic insight been gained into nuclear protein export (Schmidt-Zachmann et al 1993, Gerace 1995). At present, two distinct nuclear export signal sequences (NESs) have been best characterized. A leucine-rich NES has been identified that is responsible for the rapid, energy-dependent export of the protein kinase A inhibitor peptide (PKI) (Wen et al 1995) and of the HIV-1 Rev protein (Fischer et al 1995). This type of NES has been found to interact with a novel NPC protein which may participate in some aspect of the nuclear export process (Stutz et al 1995). In contrast to this leucine-rich NES, the hnRNP A1 protein possesses an unrelated NES that is interdigitated with its NLS (Michael et al 1995). This unique bifunctional NLS/NES is recognized by a 90 kDa protein termed transportin, which has been shown by *in vitro* assays to function in nuclear protein import (Pollard et al 1996). Steroid receptors do not appear to utilize either of these established NESs for nuclear export, since their NLS is not homologous to the hnRNP A1 NLS/NES (Michael et al 1995). Likewise, truncated receptors, which lack any apparent matches to leucine-rich NES, export from nuclei in transient heterokaryon assays (Madan and DeFranco 1993).

It has recently been hypothesized that the basic NLS of steroid receptors may serve a dual role in nuclear transport, mediating both nuclear import and export (Guiochon-Mantel et al 1994). This hypothesis was supported by *in vitro* experiments that provided evidence for ATP-independent, GTP-dependent nuclear export of NLS-conjugates (Moroianu and Blobel 1995). However, this hypothesis has been challenged by more recent experiments that demonstrated a signal-independent nuclear export of injected cytoplasmic proteins and a lack of temperature-dependence for nuclear export of NLS-conjugates (Michael et al 1995). The nuclear export of bona fide NES conjugates is typically rapid (t1/2 ~ 10 minutes; see Wen et al 1995) and characterized by its temperature- and energy-dependence (Michael et al 1995, Wen et al 1995). Thus, the relatively slow nuclear export of steroid receptors may proceed through a default pathway, such as one that has been hypothesized to lack any signal sequence requirement (Schmidt-Zachmann et al 1993). In this case, the nuclear retention capacity of steroid receptors may be the critical variable that regulates the efficiency of receptor nuclear export.

V. NUCLEAR RETENTION OF STEROID RECEPTORS

In the absence of any definitive evidence for the presence of bona fide NESs within steroid receptors, it is reasonable to assume—at least presently—that steroid receptor nuclear export may proceed via a default pathway that is signal sequence- independent (Schmidt-Zachmann, Dargemont et al 1993). In this case, nuclear export of these proteins may be regulated strictly by the efficiency of receptor retention within nuclei. What factors might regulate steroid receptor nuclear retention? It has recently been proposed that nuclear retention of GR is regulated primarily by its DNA-binding capacity (Sackey et al 1996). However, this hypothesis was not supported by earlier experiments which did not reveal any obvious differences between nuclear export capacity of wild-type GR and that of a DNA-binding mutant as measured by transient heterokaryon assays (Madan and DeFranco 1993). Given the differences in methodology utilized in these two studies, and the inherent limitations of both assays, the role of steroid receptor DNA-binding in nuclear retention remains unresolved. In addition, mutations within steroid receptor DBDs may not only affect receptor-DNA binding, but also alter receptor interactions with proteins that influence nuclear retention. For example, the DBD of AR and that of GR appear to possess important determinants for nuclear matrix binding (van Steensel et al 1995b, Tang and DeFranco 1996). If distinct segments of steroid receptor DBDs constitute binding sites for chromatin- or nuclear matrix-associated proteins, such mutations may have an effect on nuclear retention of receptors irrespective of their influence on receptor DNA binding.

We have utilized an *in vitro* approach to examine the relationship between GR nuclear retention and export. Preliminary results indicate that the dissociation of ligand from occupied nuclear GRs is accompanied by their rapid release from high-affinity chromatin binding sites. Interestingly, GR release from chromatin was not associated with a correspondingly rapid export from nuclei. Thus, nuclear retention of GRs does not appear to be strictly governed by their tight association with chromatin. Rather, receptors may be retained within a unique nuclear export staging area following their release from chromatin. The transfer of receptors from this staging area to the NPC appears to be the rate-limiting step in GR nuclear export. GRs that release from the chromatin do not appear to collect at the nucleoplasmic face of the NPC, implying that export staging areas may not be physically linked to the NPC. It remains to be established whether this putative nuclear export staging area represents a novel subnuclear compartment or is a component of chromatin that is distinguished simply by its relatively low affinity for GR. Irrespective of the identity of this nuclear export staging area, it would also be of interest to discover whether or not nuclear export of other proteins is limited by analogous interactions with this putative, novel subnuclear compartment.

Additional mechanistic details of steroid receptor nuclear export are currently very limited. There have been claims that nuclear export of steroid receptors is energy-independent, based primarily on an observed efflux of nuclear receptors from cells subjected to prolonged ATP depletion (Guiochon-Mantel et al 1991). However, pioneering studies from Munck's group first showed that ATP depletion in fact led to a dramatic increase in GR nuclear retention (Mendel et al 1986, Hu et al 1994). These "null" GRs, as they were termed by Munck and co-workers (Mendel et al 1986), accumulated within nuclei of ATP-depleted cells and were relatively resistant to high salt and detergent extraction. More recently we have extended Munck's studies and shown that ATP depletion leads to a dramatic increase in the binding of GR (and PR) to the nuclear matrix (Tang and DeFranco 1996). The increased association of GRs with the nuclear matrix of ATP-depleted cells is reversed upon restoration of cellular ATP levels (Tang and DeFranco 1996). As the binding of steroid receptors to the nuclear matrix clearly limits their nuclear export capacity, at least one step in the overall nuclear export of steroid receptors, i.e., release from the nuclear matrix, appears to be ATP-dependent.

VI. SUBNUCLEAR TRAFFICKING OF STEROID RECEPTORS

What is the fate of steroid receptors that emerge in the nucleus following passage through the NPC? *In vivo* footprinting analyses have established that steroid receptors can rapidly locate specific target sites within the genome

(Zaret and Yamamoto 1984, Reik et al 1991). How does this occur, and what is the mechanism by which steroid receptors traffic within the nucleus? When sophisticated cell imaging techniques were applied to visualize steroid receptors within the nucleus, receptors were found to be localized within discrete regions (i.e., speckles or foci), and not randomly distributed. For example, Van Steensel and co-workers found that endogenous GRs in a variety of cultured cell types collected into 1,000–2,000 small nuclear speckles in hormone-treated cells (van Steensel et al 1995a). Importantly, this staining pattern was observed irrespective of fixation conditions. In cells treated with hormone antagonist, an indistinguishable speckling pattern was noted, which suggests that these nuclear speckles may not represent receptors actively engaged in RNA-polymerase-II-directed transcription (van Steensel et al 1995a). In support of this conclusion is the fact that newly synthesized RNA or RNA polymerase II did not colocalize with GR speckles (van Steensel et al 1995a). In direct contrast to these results, different nuclear staining patterns of agonist-bound GR, as compared to antagonist-bound GR, were observed in living cells by Htun and co-workers, using a green fluorescent protein-GR fusion protein (Htun et al 1996). Since Htun and co-workers did not visualize sites of active transcription, it is not known what proportion of agonist-bound GRs visualized in their studies, if any, are actively involved in transcription.

If discrete steroid receptor nuclear foci visualized at the light microscope level do not represent receptors actively engaged in transcription, what is their physiological relevance? RNA splicing factors have been found to localize within discrete subnuclear compartments that are visualized at the light microscope level as speckles (Spector 1993). These speckles may correspond to nuclear structures designated as interchromatin granule clusters (ICGCs) (Spector 1993) and represent storage sites for splicing factors that are not actively engaged in RNA processing (Huang and Spector 1996). RNA splicing factors are not confined to the ICGCs and can be recruited to sites of active transcription (Huang and Spector 1996), which appear to be associated with distinct nuclear structures termed perichromatin fibrils (PCFs) (Spector 1993). Given the close proximity of PCFs and ICGCs, it may be difficult to make precise assignments of subnuclear compartmentalization based strictly by the appearance of speckles at the light microscope level. Thus, functionally distinct speckles may exist that could represent various intermediate stages of subnuclear trafficking. Steroid receptor nuclear speckles are distinct from speckles that possess splicing factors such as SC-35 (van Steensel et al 1995a), which suggests that there may be discrete subsets of storage sites within the nucleus that differ in their composition and function.

Nuclear steroid receptors have also been examined at high resolution using electron microscopy (Perrot-Applanat et al 1986, Vasquez-Nin et al 1991). However, even in this case, conflicting results were generated concerning the effects of hormone on precise subnuclear localization of receptors. It is difficult to make comparisons between these different studies, since the technology to assess ultrastructure of the nucleus has advanced dramatically in recent years (He et al 1990, Spector 1993). Advances in both sample preparation and imaging have permitted a much more sophisticated view of the compartmentalization of transcription domains within the nucleus (Spector 1993). Once these sophisticated techniques are applied to the analysis of steroid receptor localization within the nucleus, controversies surrounding the identity and functional significance of receptor foci may finally be resolved.

VII. STEROID RECEPTOR INTERACTIONS WITH THE NUCLEAR MATRIX

The interior of the nucleus is highly organized into discrete structural and functional domains that serve to compartmentalize the processes of DNA replication, transcription, and RNA splicing (Spector 1993). The nuclear matrix supplies a molecular scaffold for such subnuclear organization and utilizes both protein and RNA components for this purpose (Berezney 1991). The composition of the nuclear matrix varies among different cell and tissue types, and it can change as cells respond to various external signals (Getzenberg and Coffey 1990, Getzenberg 1994). The differential partitioning of some transcription factors between the nuclear matrix and soluble compartments of the nucleus (Bidwell et al 1993, Van Wijnen et al 1993, Sun et al 1994) illustrates the dynamic nature of protein association with the nuclear matrix. Thus, there is some plasticity in nuclear matrix composition which may be relevant either to the establishment of cell-type-specific gene expression programs during development, or to the regulation of such established programs by various environmental factors.

The tumor suppresser Rb protein provides an excellent example of conditional association with the nuclear matrix. Rb is associated with the nuclear matrix predominantly during the G_1 phase of the cell cycle, when Rb expresses its growth-suppressing activities (Mancini et al 1994). Nuclear matrix-bound Rb is relatively hypophosphorylated, but upon its hyperphosphorylation as cells progress through G_1, Rb association with the nuclear matrix is diminished (Mittnacht and Weinberg 1991, Mittnacht et al 1994). Phosphorylation of Rb by cyclin-dependent protein kinases appears to be a trigger for its release from the nuclear matrix (Mittnacht et al 1994). Thus, the differential phosphorylation of Rb plays a direct role in regulating its affinity for the nuclear matrix. Phosphorylation has also been found to influence the subnuclear compartmentalization of other nuclear proteins (Gui et al 1994, Ktistaki et al 1995, Colwill et al 1996). For example, nuclear compartmentalization of serine-arginine rich RNA splicing factors has been found to be regulated by members of the Clk/Sty and SRPK1 family of protein kinases (Gui et al 1994, Colwill et al 1996).

Using high-specific-activity radiolabeled steroids, a number of investigators found a fraction of steroid receptors associated with the nuclear matrix (Barrack and Coffey 1980). In fact, these studies provided the first demonstration of transcription factor binding to the nuclear matrix. The interaction between steroid receptors and the nuclear matrix is hormone-dependent, and involves saturable, high-affinity interactions (Barrack and Coffey 1980). Recently, discrete domains of steroid receptors required for nuclear matrix binding have been identified. For AR and GR, the DBD and LBD contribute to nuclear matrix binding, although the relative contributions of these domains differ between these two highly-related proteins (van Steensel et al 1995b). The relative proportion of steroid receptors associated with the nuclear matrix varies in different target tissues, particularly with the sex steroid receptors (Barrack 1987). It has been proposed that this cell-type- or tissue-specific binding of steroid receptors to the nuclear matrix may be mediated by specific acceptor proteins. A candidate steroid receptor nuclear matrix acceptor protein has been isolated from chick oviduct (Schuchard et al 1991).

Despite the recognition nearly 20 years ago that steroid receptors can associate with the nuclear matrix, the mechanisms that serve to regulate receptor binding to the matrix remain enigmatic. However, we have recently found that GR interactions with the nuclear matrix are dynamic in intact cells and potentially regulated by an ATP-driven process. This was revealed in an analysis of GR subcellular localization in cells depleted of cellular ATP. Under our energy deprivation conditions, GR binding to the nuclear matrix was found to be dramatically increased (Tang and DeFranco 1996). Increased nuclear matrix binding of another protein, the SV40 large tumor antigen, did not result from ATP depletion, which suggests that a nonspecific collapse of nuclear proteins onto the nuclear matrix was not occurring under these conditions (Tang and DeFranco 1996). In addition, receptors that collected on the nuclear matrix of ATP-depleted cells were not permanently trapped there and could be released upon restoration of cellular ATP pools (Tang and DeFranco 1996). Given these results we hypothesized that although GR *binding* to the nuclear matrix is *ATP-independent*, receptor *release* from the matrix is *ATP-dependent* (Tang and DeFranco 1996).

What is the physiological significance of dynamic interactions between steroid receptors and the nuclear matrix? It is tempting to speculate that this subnuclear trafficking pathway may play a role in receptor localization of specific high-affinity binding sites. Thus, the rapid binding and ATP-dependent release of steroid receptors from the nuclear matrix may allow them to more effectively scan the genome in search of specific target sites. Since either cell-type-specific transcription factors (Dworetzky et al 1992) or the genes themselves (Gasser and Laemmli 1986, Getzenberg 1994) can be associated with specific regions of the matrix, both receptor-DNA and receptor-protein interactions could be involved in this matrix scanning. As suggested previously (Barrack 1987), receptor targeting to the specific nuclear matrix sites may be influenced by specific matrix-associated acceptor proteins (Schuchard et al 1991). This being the case, our model is an elaboration of the previously proposed nuclear matrix acceptor hypothesis, but it adds the notion that the search for such sites is a dynamic one that utilizes the energy of ATP hydrolysis.

How is subnuclear trafficking of steroid receptors regulated? A number of molecular chaperones use ATP binding and/or hydrolysis to regulate their ability to deliver proteins to and from distinct compartments within the cytoplasm (Georgopoulos and Welch 1993, Hendrick and Hartl 1993). Perhaps these proteins serve analogous roles in the nucleus to direct proteins, such as steroid receptors, to distinct subnuclear compartments. There are numerous proteins within the nucleus that participate in the assembly of multisubunit complexes, and thus may be considered as molecular chaperones (Csermely et al 1995). In addition, some molecular chaperones such as the 70 kDa heat shock protein, hsp70, have been found to have some resident time within the nucleus (Gasc et al 1990, Mandell and Feldherr 1990). In recent studies from our laboratory (Tang et al 1997), mistargeting of a mutant GR was found to be corrected upon overexpression of a mammalian homologue of the DnaJ protein, an hsp70 partner. These results provide the first demonstration of a molecular chaperone influencing trafficking within the nucleus and may have broad implications for participation of chaperones in various nuclear processes (Tang et al 1997). These studies highlight the utility of steroid receptors in providing a useful model system for future studies devised to probe the mechanisms of protein trafficking within the nucleus.

REFERENCES

Antakly T, and Eisen HJ (1984): Immunocytochemical localization of glucocorticoid receptors in target cells. *Endocrinology* 115: 1984–1989.

Barrack ER (1987): Localization of steroid receptors in the nuclear matrix. In: *Steroid Hormone Receptors. Their Intracellular Localisation*, Clark CR, ed. Chichester (England): Ellis Horwood Ltd.

Barrack ER, and Coffey DS (1980): The specific binding of estrogens and androgens to the nuclear matrix of sex responsive tissues. *J. Biol. Chem.* 255: 7265–7275.

Berezney R (1991): The nuclear matrix: A heuristic model for investigating genomic organization and function in the cell nucleus. *J. Cell. Biochem.* 47: 109–123.

Bidwell JP, van Wijnen AJ, Fey AG, Dworetzky S, Penman S, Stein JL, Lian JB, and Stein GS (1993): Osteocalcin gene promoter-binding factors are tissue-specific nuclear matrix components. *Proc. Natl. Acad. Sci. USA* 90: 3162–3166.

Bohen SP, and Yamamoto KR (1994): Modulation of steroid receptor signal transduction by heat shock proteins. In: *The Biology of Heat Shock Proteins and*

Molecular Chaperones, Morimoto, RI, Tissieres, A, Georgopoulos, C, eds. Cold Spring Harbor, NY: Cold Spring Harbor Laboratory Press.

Cadepond F, Gasc JM, Delahaye F, Jibard N, Schweizer GG, Segard MI, Evans R, and Baulieu EE (1992): Hormonal regulation of the nuclear localization signals of the human glucocorticosteroid receptor. *Exp. Cell Res.* 201: 99–108.

Chandran UR, and DeFranco DB (1992): Internuclear migration of chicken progesterone receptor, but not simian virus-40 large tumor antigen, in transient heterokaryons. *Mol. Endocrinol.* 6: 837–844.

Colwill K, Pawson T, Andrews B, Prasad J, Manley JL, Bell JC, and Duncan PI (1996): The Clk/Sty protein kinase phosphorylates SR splicing factors and regulates their intranuclear distribution. *EMBO J.* 15: 265–275.

Csermely P, Schnaider T, and Szanto I (1995): Signaling and transport through the nuclear membrane. *Biochim. Biophys. Acta* 1241: 425–452.

Dauvois S, White R, and Parker MG (1993): The antiestrogen ICI 182780 disrupts estrogen receptor nucleocytoplasmic shuttling. *J. Cell Sci.* 106: 1377–1388.

DeFranco DB, Madan AP, Tang Y, Chandran UR, Xiao N, and Yang J (1995) Nucleocytoplasmic Shuttling of Steroid Receptors. In: *Vitamins and Hormones*, Litwack, G, ed. New York: Academic Press.

Dingwall C, and Laskey RA (1991): Nuclear targeting sequences—a consensus? *TIBS* 16: 478–481.

Dworetzky SI, Wright KL, Fey EG, Penman S, Lian JB, Stein JL, and Stein GS (1992): Sequence-specific DNA binding proteins are components of a nuclear matrix attachment site. *Proc. Natl. Acad. Sci. USA* 89: 4178–4182.

Fischer U, Huber J, Boelens WC, Mattaj IW, and Luhrmann R (1995): The HIV-1 Rev activation domain is a nuclear export signal that accesses an export pathway used by specific cellular RNAs. *Cell* 82: 475–483.

Gasc J-M, Renoir J-M, Delahaye F, and Baulieu E-E (1990): Nuclear localization of two steroid receptor-associated proteins, hsp90 and p59. *Exp. Cell Res.* 186: 362–367.

Gasser SM, and Laemmli UK (1986): Cohabitation of scaffold binding regions with upstream/enhancer elements of three developmentally regulated *Drosophila melanogaster* genes. *Cell* 46: 521–530.

Georgopoulos C, and Welch WJ (1993): Role of the major heat shock proteins as molecular chaperones. *Ann. Rev. Cell Biol.* 9: 601–634.

Gerace L (1995): Nuclear export signals and the fast track to the cytoplasm. *Cell* 82: 341–344.

Getzenberg RH (1994): Nuclear matrix and the regulation of gene expression: tissue specificity. *J. Cell. Biochem.* 55: 22–31.

Getzenberg RH, and Coffey DS (1990): Tissue specificity of the hormonal response in sex accessory tissues is associated with nuclear matrix protein patterns. *Mol. Endocrinol.* 4: 1336–1342.

Görlich D, Prehn S, Laskey RA, and Hartmann E (1994): Isolation of a protein that is essential for the first step of nuclear protein import. *Cell* 79: 767–778.

Gorski J, Toft D, Shyamala G, Smith D, and Notides A (1968): Hormone receptors: studies on the interaction of estrogen with the uterus. *Recent Prog. Hormone Res.* 24: 45–80.

Gui J-F, Lane WS, and Fu X-D (1994): A serine kinase regulates intracellular localization of splicing factors in the cell cycle. *Nature* 369: 678–682.

Guiochon-Mantel A, Delabre K, Lescop P, and Milgrom E (1994): Nuclear localization signals also mediate the outward movement of proteins from the nucleus. *Proc. Natl. Acad. Sci. USA* 91: 7179–7183.

Guiochon-Mantel A, Lescop P, Christin-Maitre S, Loosefelt H, Perrot-Applanat M, and Milgrom E (1991): Nucleocytoplasmic shuttling of the progesterone receptor. *EMBO J.* 10: 3851–3859.

He D, Nickerson JA, and Penman S (1990): Core filaments of the nuclear matrix. *J. Cell Biol.* 110: 569–580.

Hendrick JP, and Hartl F-U (1993): Molecular chaperone functions of heat-shock proteins. *Ann. Rev. Biochem.* 62: 349–384.

Htun H, Barsony J, Renyi I, Gould DL, and Hager GL (1996): Visualization of glucocorticoid receptor translocation and intranuclear organization in living cells with a green fluorescent protein chimera. *Proc. Natl. Acad. Sci. USA* 93: 4845–4850.

Hu L-M, Bodwell J, Hu J-M, Orti E, and Munck A (1994): Glucocorticoid receptors in ATP-depleted cells. Dephosphorylation, loss of hormone binding, Hsp90 dissociation and ATP-dependent cycling. *J. Biol. Chem.* 269: 6571–6577.

Huang S, and Spector DL (1996): Intron-dependent recruitment of pre-mRNA splicing factors to sites of transcription. *J. Cell Biol.* 133: 719–732.

Jensen EV, Suzuki T, Kawashima T, Stumpf WE, Jungblut PW, and DeSombre ER (1968): A two step mechanism for the interaction of estradiol with rat uterus. *Proc. Natl. Acad. Sci. USA* 59: 632–638.

Jenster G, Trapman J, and Brinkmann AO (1993): Nuclear import of the human androgen receptor. *Biochem. J.* 293: 761–768.

Kang KI, Devin J, Cadepond F, Jibard N, Guiochon-Mantel A, Baulieu E-E, and Catelli M-G (1994): *In vivo* functional protein-protein interaction: Nuclear targeted hsp90 shifts cytoplasmic steroid receptor mutants into the nucleus. *Proc. Natl. Acad. Sci. USA* 91: 340–344.

Ktistaki E, Ktistakis NT, Papadogerogaki E, and Talianidis I (1995): Recruitment of hepatocyte nuclear factor 4 into specific intranuclear compartments depends on tyrosine phosphorylation that affects its DNA-binding and transactivation potential. *Proc. Natl. Acad. Sci. USA* 92: 9876–9880.

LaCasse EC, Lochnan HA, Walker P, and Lefebvre YA (1993): Identification of binding proteins for nuclear localization signals of the glucocorticoid and thyroid receptors. *Endocrinology* 132: 1017–1025.

Lombes M, Farman N, Oblin ME, Baulieu EE, Bonvalet JP, Erlanger BF, and Gasc JM (1990): Immunohistochemical localization of rat renal mineralocorticoid receptor by using an anti-idiotypic antibody that is an internal image of aldosterone. *Proc. Natl. Acad. Sci. USA* 87: 1086–1088.

Madan AP, and DeFranco DB (1993): Bidirectional transport of glucocorticoid receptors across the nuclear envelope. *Proc. Natl. Acad. Sci. USA* 90: 3588–3592.

Mancini MA, Shan B, Nickerson JA, Penman S, and Lee W-H (1994): The retinoblastoma gene product is a cell cycle-dependent, nuclear matrix-associated protein. *Proc. Natl. Acad. Sci. USA* 91: 418–422.

Mandell RB, and Feldherr CM (1990): Identification of two HSP70-related *Xenopus* oocyte proteins that are capable of recycling across the nuclear envelope. *J. Cell Biol.* 111: 1775–1783.

Mendel DB, Bodwell JE, and Munck A (1986): Glucocorticoid receptors lacking hormone-binding activity are bound in nuclei of ATP-depleted cells. *Nature* 324: 478–480.

Michael WM, Choi M, and Dreyfuss G (1995): A nuclear export signal in hnRNP A1: a signal-mediated, temperature-dependent nuclear protein export pathway. *Cell* 83: 415–422.

Mittnacht S, Lees JA, Desai D, Harlow E, Morgan DO, and Weinberg RA (1994): Distinct sub-populations of the retinoblastoma protein show a distinct pattern of phosphorylation. *EMBO J.* 13: 118–127.

Mittnacht S, and Weinberg RA (1991): G1/S phosphorylation of the retinoblastoma protein is associated with an altered affinity for the nuclear compartment. *Cell* 65: 381–393.

Moroianu J, and Blobel G (1995): Protein export from the nucleus requires the GTPase Ran and GTP hydrolysis. *Proc. Natl. Acad. Sci. USA* 92: 4318–4322.

Okuno Y, Imamoto N, and Yoneda Y (1993): 70-kDa heat-shock cognate protein colocalizes with karyophilic proteins into the nucleus during their transport *in vitro*. *Exp. Cell Res.* 206: 134–142.

Perrot-Applanat M, Groyer-Picard M-T, Logeat F, and Milgrom E (1986): Ultrastructural localization of the progesterone receptor by an immunogold method: effect of hormone administration. *J. Cell Biol.* 102: 1191–1199.

Perrot-Applanat M, Logeat F, Groyer-Picard MT, and Milgrom E (1985): Immunocytochemical study of mammalian progesterone receptor using monoclonal antibodies. *Endocrinology* 116: 1473–1484.

Picard D, and Yamamoto KR (1987): Two signals mediate hormone-dependent nuclear localization of the glucocorticoid receptor. *EMBO J.* 6: 3333–3340.

Pollard VW, Michael WM, Nakielny S, Siomi MC, Wang F, and Dreyfuss G (1996): A novel receptor-mediated nuclear protein import pathway. *Cell* 86: 985–994.

Pratt WB (1993): The role of heat shock proteins in regulating the function, folding, and trafficking of the glucocorticoid receptor. *J. Biol. Chem.* 268: 21455–21458.

Reik A, Schütz G, and Stewart AF (1991): Glucocorticoids are required for establishment and maintenance of an alteration in chromatin structure: induction leads to a reversible disruption of nucleosomes over an enhancer. *EMBO J.* 10: 2569–2576.

Sackey FNA, Hache RJG, Reich T, Kwast-Welfeld J, and Lefebvre YA (1996): Determinants of subcellular distribution of the glucocorticoid receptor. *Mol. Endocrinol.* 10: 1191–1205.

Sanchez ER, Hirst M, Scherrer LC, Tang HY, Welsh MJ, Harmon JM, Simons SSJ, Ringold GM, and Pratt WB (1990): Hormone-free mouse glucocorticoid receptors overexpressed in Chinese hamster ovary cells are localized to the nucleus and are associated with both hsp70 and hsp90. *J. Biol. Chem.* 265: 20123–20130.

Schmidt-Zachmann MS, Dargemont C, Kuhn LC, and Nigg EA (1993): Nuclear export of proteins: the role of nuclear retention. *Cell* 74: 493–504.

Schuchard M, Subramaniam M, Ruesink T, and Spelsberg TC (1991): Nuclear matrix localization and specific matrix DNA binding by Receptor Binding Factor-1 of the avian progesterone receptor. *Biochemistry* 30: 9516–9522.

Smith DF (1993): Dynamics of heat shock protein 90-progesterone receptor binding and the disactivation loop model for steroid receptor complexes. *Mol. Endocrinol.* 7: 1418–1429.

Smith DF, and Toft DO (1993): Steroid receptors and their associated proteins. *Mol. Endocrinol.* 7: 4–11.

Sommer L, Hagenbüchle O, Wellauer PK, and Strubin M (1991): Nuclear targeting of the transcription factor PTF1 is mediated by a protein subunit that does not bind to the PTF1 cognate sequence. *Cell* 67: 987–994.

Spector DL (1993): Macromolecular domains within the cell nucleus. *Ann. Rev. Cell Biol.* 9: 265–315.

Stutz F, Neville M, and Rosbash M (1995): Identification of a novel nuclear pore-associated protein as a functional target of the HIV-1 Rev protein in yeast. *Cell* 82: 495–506.

Sun J-M, Chen HY, and Davie JR (1994): Nuclear factor 1 is a component of the nuclear matrix. *J. Cell. Biochem.* 55: 252–263.

Tang Y, and DeFranco DB (1996): ATP-dependent release of glucocorticoid receptors from the nuclear matrix. *Mol. Cell. Biol.* 16: 1989–2001.

Tang Y, Ramakrishnan C, Thomas J, and DeFranco DB (1997): A role for HDJ-2/ HSDJ in correcting subnuclear trafficking, transactivation and transrepression defects of a glucocorticoid receptor zinc finger mutant. *Mol. Biol. Cell* 8, in press.

van Steensel B, Brink M, van der Meulen K, van Binnendijl EP, Wansink DG, de Jong L, de Kloet ER, and van Driel R (1995a): Localization of the glucocorticoid receptor in discrete clusters in the cell nucleus. *J. Cell Sci.* 108: 3003–3011.

van Steensel B, Jenster G, Damm K, Brinkmann AO, and van Driel R (1995b): Domains of the human androgen receptor and glucocorticoid receptor involved in binding to the nuclear matrix. *J. Cell. Biochem.* 57: 465–478.

Van Wijnen AJ, Bidwell JP, Fey EG, Penman S, Lian JB, Stein JL, and Stein GS (1993): Nuclear matrix association of multiple sequence-specific DNA binding activities related to SP1, ATF, CCAAT, C/EBP, OCT-1, and AP-1. *Biochemistry* 32: 8397–8402.

Vasquez-Nin GH, Echeverria OM, Fakan S, Traish AM, Wotiz HH, and Martin TE (1991): Immunoelectron microscopic localization of estrogen receptor on pre-mRNA containing constituents of rat uterine cell nuclei. *Exp. Cell Res.* 192: 396–404.

Welshons W, Lieberman ME, and Gorski J (1984): Nuclear localization of unoccupied estrogen receptors: cytochalasin enucleation of GH3 cells. *Nature* 307: 747–749.

Welshons WV, and Judy BM (1995): Nuclear vs translocating receptor models and the excluded middle. *Endocrine* 3: 1–4.

Wen W, Meinkoth JL, Tsien RY, and Taylor SS (1995): Identification of a signal for rapid export of proteins from the nucleus. *Cell* 82: 463–473.

Wikström A-C, Bakke O, Okret S, Bronnegard M, and Gustafsson J-Å (1987): Intracellular localization of the glucocorticoid receptor: evidence for cytoplasmic and nuclear localization. *Endocrinology* 120: 1232–1242.

Yang J, and DeFranco DB (1994): Differential roles of heat shock protein 70 in the *in vitro* nuclear import of glucocorticoid receptor and simian virus 40 large tumor antigen. *Mol. Cell. Biol.* 14: 5088–5098.

Yang J, and DeFranco DB (1996): Assessment of glucocorticoid receptor-heat shock protein 90 interactions *in vivo* during nucleocytoplasmic trafficking. *Mol. Endocrinol.* 10: 3–13.

Ylikomi T, Bocquel MT, Berry M, Gronemeyer H, and Chambon P (1992): Cooperation of proto-signals for nuclear accumulation of estrogen and progesterone receptors. *EMBO J.* 11: 3681–3694.

Zaret KS, and Yamamoto KR (1984): Hormonally induced alterations of chromatin structure in the polyadenylation and transcription termination regions of the chicken ovalbumin gene. *Cell* 38: 29–38.

Zhao L-J and Padmanabhan R (1988): Nuclear transport of Adenovirus DNA polymerase is facilitated by interaction with preterminal protein. *Cell* 55: 1005–1015.

Zhou Z-X, Sar M, Simental JA, Lane MV, and Wilson EM (1994): A ligand-dependent bipartite nuclear targeting signal in the human androgen receptor. *J. Biol. Chem.* 269: 13115–13123.

3

Structure and Function of the Steroid and Nuclear Receptor Ligand Binding Domain

S. STONEY SIMONS, JR.

I. RELEVANCE OF THE LIGAND BINDING DOMAIN

The superfamily of steroid and nuclear receptors is composed of over 100 members and continues to grow. This collection of related proteins has attracted widespread interest over the last 3 decades because of their properties as soluble, intracellular transcription factors. Furthermore, the transcriptional activity of at least the initially studied receptors is regulated by the binding of a ligand, which is usually produced by the cells of a different organ and travels to the receptor-containing target cell. This property has set these proteins apart from most other transcription factors and has made the steroid and nuclear receptors even more fascinating.

The original, and still primary, criterion for membership in the superfamily of steroid receptors is the presence of a highly homologous, approximately 65-amino-acid long domain, containing two "zinc fingers," that is required for DNA binding. Each finger binds one molecule of zinc by means of complexation/coordination with four cysteines (Freedman et al 1988) to produce a species with no net charge (Fabris et al 1996). Recently, molecules such as DAX-1 (Zanaria et al 1994) and SHP (Seol et al 1996a), which do not possess a zinc finger DNA binding domain, have also been included in the superfamily because of their homology with the ligand binding domain of steroid receptors.

It was soon evident that most receptors contained five to six evolutionarily conserved domains (A–F in Figure 1) (Laudet et al 1992), with various activities being associated with individual domains. Thus, domain C, with the

Molecular Biology of Steroid and Nuclear Hormone Receptors
Leonard P. Freedman, Editor
© 1998 Birkhäuser Boston

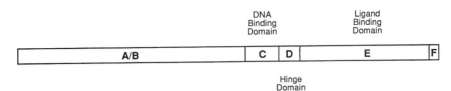

Figure 1.—The different domains of a sample receptor (absolute size of domains A, B, and F can vary significantly between receptors).

two zinc fingers, possesses intrinsic DNA binding activity and is called the DNA binding domain (DBD). Domain E, with moderate to low homology, is required for the binding of ligand and is called the ligand, or hormone, binding domain (LBD). However, most (> 90) members of the superfamily do not bind any known ligands, and thus are called "orphan receptors" (Mangelsdorf and Evans 1995, Enmark and Gustafsson 1996). Further studies have documented the presence of activities other than ligand binding in domain E. The purpose of this review is to summarize what is known about each of these LBD activities, with particular emphasis being devoted to the question of which of those properties are general for the superfamily members as a whole, and which are not. Given the explosion of results in several new directions under a multitude of conditions (e.g., whole cell, cell-free, over-expressed, and *in vitro* translated receptors of different isoforms from a variety of species), the identity of the receptor used in each study cited has usually been specified, to help determine whether differences in past or future experiments could be due to the precise receptor used. Subheadings of sections have been used liberally, as many readers may want to concentrate on specific topics. Finally, the vast majority of the receptors are of mammalian origin, but, given the power of yeast genetics, many studies are now being performed in yeast. Therefore, differences between the properties of receptor LBDs in yeast versus mammalian cells will be noted, when relevant.

Before embarking, though, I must apologize for the innumerable omissions of important studies, which inevitably characterize a review of this sort. With over 100 proteins to consider, it is impossible to do justice to each family member.

II. DEFINITION OF THE LIGAND BINDING DOMAIN (LBD)

A definition of the LBD has become infinitely more complicated with the identification of a huge number of proteins that are thought to contain an LBD but have not yet been found to bind any ligand, i.e., the orphan receptors. On the basis of our knowledge of those receptors that do bind ligands, it is assumed that the ~250 amino acids located about 40 amino acids carboxyl terminal to the DBD (domain C) constitute domain E, or an LBD. However, until a ligand is found, this aspect cannot be confirmed for the vast number of members of the superfamily.

For the purposes of this review, an LBD is defined as that region of a given receptor that has been demonstrated to bind several different ligands, including agonists and antagonists. For this reason, I will cover only those ligand binding receptors for which there is appreciable information. Thus, little or no attention will be devoted to some of the orphan receptors for which ligand binding has recently been observed, such as FXR, which is reported to bind farnesol (Forman et al 1995a), the relatively well known peroxisome proliferator-activated receptor (PPAR) (Issemann and Green 1990), or the newly described 54 kDa form of the estrogen receptor (ERβ) that is expressed in rat (Kuiper et al 1996) and human (Mosselman et al 1996) and binds numerous estrogenic steroids with affinities very similar to those of the original ER (ERα), despite a $< 55\%$ homology in the carboxyl terminal half of the receptor (Kuiper et al 1997). Orphan receptors and the search for their ligands are discussed in detail in Chapter 1.

What remains, and will be the focus of this review, is two general subfamilies of receptors and their isoforms. The first subfamily contains most of the originally described steroid receptors: androgen receptor (AR), estrogen receptor (ER), glucocorticoid receptor (GR), mineralocorticoid receptor (MR), and progesterone receptor (PR). The second subfamily has been called the nuclear receptors because the ligand-free receptors of this group have a much higher affinity for nuclei than do the steroid receptors. The nuclear receptors are ER, thyroid receptor (TR), vitamin D_3 receptor (VDR), and the two retinoid receptors RAR and RXR. It will be noted that ER has been included in both subfamilies. This is because the ER displays character-istics of both subfamilies and thus defies simple classification. For this reason, the ER will sometimes be treated as a member of either group. Most nuclear receptors exist as isoforms encoded by different genes, including the ER (Kuiper et al 1996, Mosselman et al 1996, Tremblay et al 1997) but this usually does not introduce much heterogeneity into the LBD of a given receptor (Zelent et al 1989, Lazar 1993, Leid et al 1995, Kuiper et al 1997). Therefore, the isoforms will not be treated individually.

A. Properties of Ligand Binding

Ligand binding to the steroid receptors generally occurs to a monomeric species of the soluble receptor that is associated with heat shock protein 90 (hsp90). For the nuclear receptors, ligand commonly binds to the dimeric form that is bound to DNA. The ER exhibits traits that are intermediate between those of the steroid and nuclear receptors. Thus, cytosolic solutions of ER display cooperative ligand binding, indicative of binding to an ER dimer (Notides et al 1981). On the other hand, steroid-free ER in intact cells was found to bind to DNA, as determined by a promoter interference assay (Pakdel et al 1993). This may simply reflect different aspects of cytoplasmic-nuclear shuttling of the ligand-free receptors (DeFranco et al 1991, Guiochon-Mantel et al 1991).

The rate-limiting step in the action of the steroid receptors such as GR is thought to be the binding of ligand to the receptor (Munck 1984), which appears to be a two-step process (Pratt et al 1975). Concerning the nuclear receptors other than ER, which can form both homo- and heterodimers (see Section VIII.C.), investigators disagree as to whether dimer formation affects ligand binding of either the free or DNA-bound receptors (Kersten et al 1996 versus Kurokawa et al 1994, Forman et al 1995b).

It has long been considered axiomatic that only one molecule of ligand is bound per monomeric receptor molecule, but the proof has been relatively recent. One of the first demonstrations was for the large form (B form) of purified hPR (see footnote 1), which was found to contain about 0.85 moles of steroid per receptor (Estes et al 1987). More recently, hER LBD (residues 282–595 with a 3-amino-acid leader sequence) that was overexpressed in $E.$ $coli$, but was composed of $> 95\%$ C-terminal truncated fragments (at positions 575, 571, and 569), bound steroid with nearly wild-type affinity and a ligand to receptor ratio of 0.84–0.95:1 in three different assays (Seielstad et al 1995). Another overexpressed hER (300–551 with an S305E mutation) generated from a fusion protein bound 0.98 moles of estradiol/mole of LBD (Brandt and Vickery 1997). A fluorometric titration method was used to determine that one molecule of ligand had bound to the D/E fragment of overexpressed hRARγ (Lupisella et al 1995) and hRXRα (Bourguet et al 1995a). These results have been supported by the observation of one ligand molecule in the x-ray structures of rTRα and hRARγ LBDs (Wagner et al 1995, Renaud et al 1995). However, several older reports suggest that more than one ligand binding site may exist per receptor monomer (Ruh and Baudendistel 1977, Martin et al 1988, Svec et al 1989). Interest in these observations (Simons 1996) has been rekindled by the recent, controversial report of a synergistic activation of ER by mixtures of environmental estrogens (Arnold et al 1996 versus Ramamoorthy et al 1997, Ashby et al 1997).

---

[1] The following lower-case abbreviations preceding the individual receptors are used to identify the species from which a specific steroid receptor derived: c (chicken), h (human), m (mouse), r (rat). PR-B and PR-A refer to the large and small isoforms respectively of PR, which result from alternative translational start sites (Conneely et al 1989, Kastner et al 1990).

1. Ligand Binding in Mammalian Cells Versus Yeast

In general, ligand binding to receptors is independent of whether the receptors were expressed in mammalian cells, yeast, or *E. coli* (Eul et al 1989, Wooge et al 1992, Caplan et al 1995, Lee et al 1994, Lupisella et al 1995). A worrisome exception is the GR. The low activity of the potent glucocorticoid dexamethasone (Dex) in yeast cells has usually been ascribed to permeability problems, or active transport of Dex out of the cell. However, there is an even more basic problem as demonstrated by the fact that the cell-free affinity of Dex for GR expressed in yeast is ~1/1000 that of GR from mammalian cells (Garabedian and Yamamoto 1992).

B. The Role of hsp90 in Ligand Binding

The involvement of hsp90 in ligand binding can be divided reasonably well along the lines of steroid versus nuclear receptors (see also Chapter 1). It has been well established that associated hsp90 is required for steroid binding to GR both *in vitro* (Bresnick et al 1989, Nemoto et al 1990, Pratt 1993) and *in vivo* (Picard et al 1990a). The binding of hsp90 to mGR caused a conformational change in the GR LBD such that there was more covalent labeling of cysteines by the thiol-specific derivatizing agent N-iodoacetyl-3-[^{125}I]iodotyrosine) (Stancato et al 1996). Steroid binding to MR *in vitro* also required bound hsp90 (Caamano et al 1993, Jalaguier et al 1996), although the *in vivo* requirement for biological activity did not appear to be as great as for GR (Picard et al 1990a). *In vitro* steroid binding to PR required bound hsp90 at 37°C, but not at 0°C (Smith 1993). Similarly, hsp90 was necessary for steroid binding to AR at 37°C in yeast (Fang et al 1996), but not at 0°C in a cell-free system (Nemoto et al 1992). In most cases, it appears that a 2:1 stoichiometry of hsp90:receptor exists (see Pratt and Toft 1997 for comprehensive review).

Hsp90 does not bind to TR, VDR, RAR, or RXR, and is not required for their binding of ligand (Dalman et al 1990, Dalman et al 1991b, Pratt 1990). This lack of required hsp90 may explain why the LBD-containing fragments of TR, VDR, RAR, and RXR expressed in *E. coli* retain ligand binding activity (Wooge et al 1992, Lupisella et al 1995), as does PR at 0°C (Eul et al 1989), while GR and MR do not (Nemoto et al 1990, Alnemri and Litwack 1993).

Hsp90 binding with ER appears to represent an intermediate situation that is complicated by the use of both the wild-type ER and the initially cloned temperature sensitive mutant (G400V) (Tora et al 1989). The LBD of the Val-400 mutant ER in a β-gal/hER (amino acids 282–595) construct was sufficient for a weak cell-free binding of hsp90 (Scherrer et al 1993). Nevertheless, the requirement of hsp90 for whole cell, steroid-induced activity of hER in some temperature-sensitive yeast cells was much less than for GR or MR (Picard et al 1990a). Other experiments in COS-7 cells indicated that the steroid-inducible chimera VP-16/GAL/hER (amino acids 300–595) did not associate with hsp90 (although the full-length hER did), while the corresponding Val-400 chimera did bind hsp90 (Lee et al 1996). Thus, further studies are needed to determine the role of hsp90 in the steroid binding and biological activity of ER.

1. Location of hsp90 Binding Sites

The most detailed studies have been performed with GR. Two probable sites of hsp90 association with mGR (residues 587–606 and 632–659, = 599–618 and 644–671 in rat) (see footnote 2) were identified from the ability of selected mGR peptides to block the cell-free reassociation of receptor and hsp90 (Dalman et al 1991a, see Pratt and Toft 1997 for comprehensive review). A comparable region of the rAR (positions 704–758), which is homologous to 597–651 of rGR, was found to be important for hsp90 interactions (Marivoet et al 1992). However, a different assay (formation of 8S complexes in sucrose gradients) indicated that no one region of the hGR LBD was crucial for hsp90 binding (Cadepond et al 1991).

As discussed above, the binding of hsp90 to ER in cell-free extracts depends both on whether the full-length or truncated receptor was used, and on whether the LBD contained wild-type sequences or the Val-400 mutation (Scherrer et al 1993, Lee et al 1996). Multiple sites within the LBD of ER appear to be sufficient for weak hsp90 binding, but additional sites in the DBD may be involved in the formation of complexes with the wild-type receptor *in vivo* (Chambraud et al 1990). Similarly, no one region of the cPR LBD is necessary for hsp90 binding, which instead may be a function of the general tertiary structure (Schowalter et al 1991).

2. Other Proteins Associated with LBD-Bound hsp90

Numerous GR- and PR-associated proteins are complexed with hsp90, as opposed to the LBD, in both the reconstitution of steroid binding activity (e.g., hsp90, hsp70, p60, p48, and p23) and the ordered maturation of steroid-free receptors (e.g, hsp56, FKBP54, and CyP-40) (Owens-Grillo et al 1995, Smith et al 1995, Dittmar et al 1996, see Pratt and Toft 1997 for comprehensive review). The greatly reduced transcriptional activity of full-length GR, but not of constitutively active GR lacking the LBD, in yeast cells deficient in the yeast homolog of CyP-40 (Cpr7) (Duina et al 1996) supported the conclusions from cell-free studies that an ordered assembly of multiple proteins is required for the steroid binding form of the steroid receptors. Protein-serine phosphatase, or PP5, interacts with GR through a tetratrico-peptide repeat (TPR) domain-mediated binding to hsp90. While the function

[2] Studies of the GR LBD have been conducted with receptors from human, mouse, and rat. The numbering in the LBD is as follows: mouse = human +12, rat = human +18.

of PP5 is not yet clear, overexpression of the TPR domain inhibited GR transactivation (Chen et al 1996b).

III. LOCATION AND SIZE OF RECEPTOR LBDs

Surprisingly, there is rarely a consensus on the boundaries of the LBD for any receptor. This lack of agreement has many causes, not the least of which is that the precise limits may not correspond to the border of a stable secondary structure, but may rather be influenced by the rest of the protein. Furthermore, while the homology between LBDs of related receptors, such as AR, GR, MR, and PR, can be 50%, it is usually low (30% for GR versus ER, and 15% for GR versus RAR) (Evans 1988), which impedes generalizations of LBD sequences. These situations also complicate this review because of the uncertainty as to whether a particular activity does or does not involve only sequences of the LBD. Generally the properties of the hinge region (domain D in Figure 1), which lies immediately amino terminal to the LBD, will not be covered. Consequently, several interesting reports will not be covered, such as the possible association of calmodulin with the rER (Bouhoute and Leclercq 1995) and the binding of RAP46 to the DNA binding form of numerous receptors (Zeiner and Gehring 1995).

A. Modular Activity of LBD

Historically, the LBD was considered to occupy the carboxyl terminal ∼250 amino acids of the receptor. This view arose from one of the most significant developments in the field of receptor research, namely the discovery that the receptors appeared to be constructed of modular functional units (Green and Chambon 1987, Giguere et al 1987). Thus, numerous chimeric proteins have been constructed containing the LBD of various steroid receptors. In some cases, the activity of the fusion protein is not regulated by added steroid (Picard et al 1990b), which permits a facile purification of the steroid-free molecules upon fusion to proteins such as glutathione S-transferase (Dallery et al 1993, Nemoto et al 1994, Schulman et al 1995, Jalaguier et al 1996) and the maltose binding protein (Nishikawa et al 1995, Jalaguier et al 1996), or to a leader sequence containing tandem arrays of histidine (Bourguet et al 1995b, Renaud et al 1995). Fusion of LBDs to DNA binding proteins such as the GAL4 DNA binding domain has permitted the development of immensely popular techniques like the yeast two-hybrid assay (Fields and Song; O.-k. 1989), although it is usually not clear whether the ligand influences the ability of the fusion protein to bind to DNA. In many other cases, the activity of the chimeric protein is seen only in the presence of added ligand (reviewed for GR in Simons 1994; see Drewes et al 1994, Fankhauser et al 1994, and Logie and Stewart 1995 for other examples). Recently this property has been extended to include either mutant LBDs (Wang et al 1994, Littlewood et al 1995) or LBDs of receptors from other species (Forman et al 1995a), thereby creating chimeric proteins that can be selectively regulated in the presence of endogenous receptors. In most cases, the chimeric proteins are positively regulated by added steroid. One apparent exception is the negative regulation of a chimeric dihydrofolate reductase/glucocorticoid receptor protein by added Dex (Israel and Kaufman 1993). However, this repressive activity of steroid is most likely due to a steroid-induced, cellular redistribution of the fusion protein to the nucleus, where its activity is ineffective.

B. The F Domain

Additional studies have revealed that the LBD of several receptors is composed of $\sim$250 amino acids but that they are not the last 250 residues of the receptor due to the presence of a variable-length F domain of generally unknown function (Figure 1). The hallmarks of the F domain are that it has even less sequence homology with the other receptors and that it is not required for ligand binding. Accordingly, the last 42 amino acids of hER constitute an F domain because they are not required for steroid binding but rather seem to modulate the transcriptional activity of estrogens and antiestrogens (Montano et al 1995). Interestingly, a novel estrogen receptor expressed in rat prostate and ovary (ERβ) has a shorter, nonhomologous F domain but responds just like the original ER to estrogens and antiestrogens (Kuiper et al 1996, Mosselman et al 1996). hRARα, β, and γ have F domains of 42, 35, and 34 amino acids respectively (Brand et al 1988, Krust et al 1989). A point mutation (M406A) amino terminal to the F domain of hRARα had a more detrimental effect on the transcriptional activity of the full-length receptor than of the truncated receptor lacking the F domain (Tate et al 1996). RXRs do not have F domains, and the ecdysone receptor has a mammoth F domain of 227 amino acids (Koelle et al 1991). cTRα has been proposed to have a small F domain (positions 396–408) (Forman et al 1989), but the loss of binding activity upon deletion of the carboxyl terminal 8 residues of hTRβ1 (Lin et al 1991) argues against an F domain in all isoforms. The terminal 11–14 amino acids of PR could represent a vestigial F domain (Montano et al 1995).

C. Precise Localization of LBDs

The presence of the F domain in steroid receptors has muddled the identification of the carboxyl terminus of receptor LBDs. While a retention of ligand binding activity is the operational definition for the minimal LBD sequence, even this is not straightforward when one is forced to consider reduced affinities and protein folding (see Section IV.A.). A further complication is illustrated by the hPR, where the C-terminal 42 residues are required for the binding of progesterone, but not for the binding of RU 486, Org 31,806, or Org 31,376 (Vegeto et al 1992). Conventionally, the LBD is considered to encompass those amino acids that are required for the high-affinity binding of any ligand. Perhaps the best studied is the GR, where the removal of the C-terminal 5 amino acids caused a 20-fold reduction of steroid binding affinity (Rusconi and Yamamoto 1987), while deletion of the C-terminal 14 (Zhang et al 1996a), 28 (Lanz and Rusconi 1994), 29 (Xu et al 1996b), or 42 (Zhang et al 1996a) amino acids eliminated the binding of agonists and the biological activity of agonists and antagonists. Similarly, agonist binding was lost upon removal of the C-terminal 40 amino acids of an hMR fusion protein (Jalaguier et al 1996) and the last 8 residues of the truncated hTRβ1 (201–456 [= 206–461]) (see footnote 3) (Lin et al 1991). In other cases, the situation is less clear. Thus, distinct binding determinants for 9-cis retinoic acid were reported to be located within residues 404–419 of hRARα (Tate et al 1994), but amino terminal deletions of this sequence in all three hRARs gave mutants that could still bind all-trans-retinoic acid (Damm et al 1993). Also, RXR is not thought to contain an F domain. However, removal of the C-terminal 29, but not the C-terminal 18 or 21 residues, caused loss of ligand binding (Leng et al 1995).

In addition to those receptor isoforms resulting from different genomic sequences (Zelent et al 1989, Lazar 1993, Leid et al 1995, Kuiper et al 1996), several naturally occurring receptor isoforms result from alternative splicing in the LBD. These isoforms usually, but not always (Gaitan et al 1995), do not bind ligand and are dominant negative receptors (Koenig et al 1989, Fuqua et al 1992, Damm et al 1993, Bamberger et al 1995, Matsui and Sashihara 1995, Oakley et al 1996). Moreover, the nature of the hormone responsive element can be a determining factor. Thus, most natural hTRβ mutants associated with generalized resistance to thyroid hormone (GRTH) were dominant negative receptors of wild-type TR on an inverted palindromic TRE, which binds homodimers, but not on TREs binding heterodimers (Piedrafita et al 1995). Also, the C-terminal truncation of 65 amino acids from mRARγ afforded a dominant negative repressor of RARα or γ in a cell- and promoter-specific manner, while less truncation (53 or 46 amino acids) gave dominant negative repressors under all circumstances (Matsui and Sashihara 1995).

---

[3] The length of hTRβ1 is alternatively listed as 456 (Wurtz et al 1996) and 461 amino acids (Lazar 1993) because of the presence of a second Met that is 5 amino acids upstream of the first (Koenig et al 1988, Murray et al 1988). Similarly, the size of hVDR in the literature varies by 3 (427 [Whitfield et al 1996] as opposed to 424 [Hilliard et al 1994]). Therefore, when these receptors are under discussion, both numberings will be given.

The amino terminal end of the LBD is the area of greatest imprecision. For example, the boundary for the rGR has been variously given as anywhere from 546 to 574 (Simons 1994). More recently, position 550 has been shown to be the amino terminal end of the rGR LBD in a fusion construct, since further deletions reduced steroid binding to below detection (Xu et al 1996b). Whether the homologous amino acid in the LBD of other receptors will also constitute the N-terminal boundary remains to be determined. Interestingly, steroid binding was obtained for a chimera of the hMR LBD (amino acids 729–984, = 541–795 of rGR), but not after an N-terminal deletion of 38 amino acids to yield residues 766–984 (Jalaguier et al 1996).

D. Antibodies to Receptor LBDs

The most common antigenic sequence of the steroid receptors is the amino terminal half of the protein. However, antibodies have been reported for the LBD of most receptors. A partial listing is as follows: AR (Chang et al 1989, Doesburg et al 1997), ER (Greene et al 1984, Giambiagi and Pasqualini 1990), GR (reviewed in Simons 1994), PR (Weigel et al 1992), RARγ (Driscoll et al 1996), and TRα (Forman et al 1989). A more complete online source of this information has recently been established at *http://nrr.georgetown.edu/NRR/ NRR.html* (Martinez et al 1997).

E. X-ray Structures of LBDs

The long-awaited x-ray structure of a receptor LBD appeared in 1995 with the publication of three structures: ligand-free hRXRα (Bourguet et al 1995b) and ligand-bound hRARγ (Renaud et al 1995, Rochel et al 1997) and rTRα1 (Wagner et al 1995) (Figure 2). Despite being for different receptors, all three structures displayed the same overall tertiary structure of an "antiparallel alpha-helical sandwich" consisting of 12 helices (Bourguet et al 1995b). In both ligand-bound structures, the ligand was completely buried in the interior of the LBD, leading to the conclusion that ligand binding caused a conformational change in the LBD (Renaud et al 1995, Wagner et al 1995). The inability to crystallize the ligand-free hRARγ LBD has impeded direct comparisons (Rochel et al 1997).

The core of the LBD is composed of five helices (3, 4, 5, 8, and 9). As a result of a comparison of the holo-RAR with the apo-RXR, it was proposed that ligand binding had the effect of setting off a mousetrap, with the W loop between helices 2 and 3 flipping over to cover the ligand binding cavity (Renaud et al 1995). At the same time, minor changes throughout the LBD caused it to assume a more compact structure. On the strength of the similar $3°$ structures for dissimilar receptors, it was postulated that the LBD of all of the members of the steroid receptor superfamily would have a similar structure and could be aligned (Figure 3) to predict what sequence would constitute which helix (Wurtz et al 1996). Because all of the x-ray structures are from the nuclear receptor subclass, it will be very interesting to see if the predicted structure applies to the LBDs of the steroid receptors.

A putative helix-turn-zipper motif, corresponding to 590–630 of rGR, had been noted in all receptors (Maksymowych et al 1992). It overlaps the 20 amino acid "signature" sequence (595–614 of rGR) that arose from a two-step alignment of receptor LBDs (Wurtz et al 1996). Probing the SwissProt 31.0 data base with the "signature" sequence of [(F,W,Y)(A,S,I)(K,R,E,G)-xxxx(F,L)xx(L,V,I)xxx(D,S)(Q,K)xx(L,V)(L,I,F)] selectively retrieved 106 of the 108 members of the steroid receptor superfamily in the data bank (Wurtz et al 1996). The conserved residues of the signature sequence are mostly hydrophobic and are thought to be a major stabilizing component of the folded LBD. This sequence also constitutes the putative hsp90 binding site for many of the steroid receptors (see Section II.B.1.).

IV. Is the LBD an Independently Active Domain?

The above examples of modular activity and conserved 3° structure of receptor LBDs, along with the general belief that most separately folding domains are independently active (Frankel and Kim 1991), have led many to suspect that an isolated LBD would be independently active, at least with regard to ligand binding. The first indication that this might not be so was that the boundaries of the exons encoding the LBD usually did not correspond to what was thought to constitute the ends of the LBD (Ponglikitmongkol et al 1988, Encio and Detera-Wadleigh 1991, Misrahi et al 1993).

A. LBD Stability Can Require Non-LBD Sequences

For at least the GR, the stability of the LBD depends on the method of preparation, and initially requires non-LBD sequences. *In vitro* translation of the rGR cDNA encoding amino acids 547–795 gave a species that bound Dex with 1/300th the affinity of full-length GR (Rusconi and Yamamoto 1987). In contrast, the same fragment fused to dihyrdofolate reductase had essentially wild-type affinity, and the cDNA for a larger fragment (537–795) by itself did not yield a stable protein after transient transfection into COS-7 cells (Xu et al 1996b). However, trypsin cleavage of intact, full-length receptors afforded an even smaller 16 kDa fragment, containing Cys-656 (Xu et al 1996b), that bound steroids with the same relative specificity as the full length receptor and

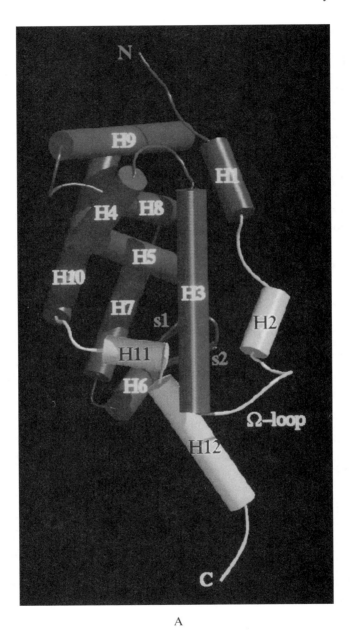

A

Figure 2A.—X-ray structures of LBD of (A) ligand-free hRXRα and (B) ligand-bound hRARγ (reprinted from Wurtz et al 1996).

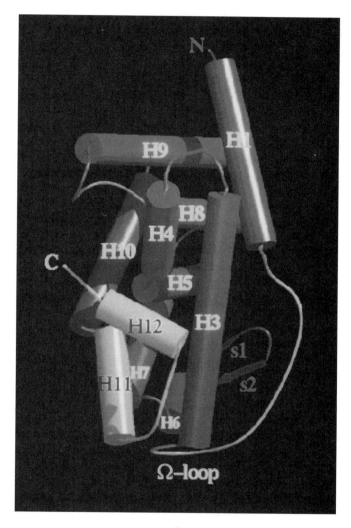

B

Figure 2B.

only a 23-fold reduction in affinity (Simons et al 1989). Although ligand binding to this 16 kDa rGR fragment may involve the noncovalent association of another tryptic fragment of the GR LBD, as has been observed for porcine ER (Thole and Jungblut 1994), it is clear that the rGR sequence of 550–795 is sufficient for steroid binding activity when fused to other proteins, but that it is insufficient in isolation for the expression and maintenance of a stable, active protein.

It is not known whether a similar situation exists for the other receptors, partially because such precise assignments of the LBD boundaries have not been made. However, it is worth noting that thrombin cleavage of the GST chimeras with hRXRα LBD (residues 226–462, which correspond to 545–795 of rGR [see Figure 3]) (Cheng et al 1994) and with hRARα LBD (domains E and F) (Dallery et al 1993) yielded functionally unstable proteins.

B. Possible Examples of Independently Active LBDs

If one makes the unproven assumption that the amino terminal end of all LBDs is in the middle of helix 1, as for rGR (Figure 3), the question then becomes whether or not truncated receptors extending from this point to the C–terminus alone will give a functionally active LBD, as defined by ligand binding. Every functionally active LBD that was found in the literature contained additional amino terminal sequences in the form of receptor and/or other proteins such as GST and polyhistidine tails. Examples exist for AR (Simental et al 1991, Nemoto et al 1992, Cooper et al 1996), PR (Eul et al 1989), TR (Lin et al 1991, Wagner et al 1995), RAR (Dallery et al 1993, Lupisella et al 1995, Rochel et al 1997), and RXR (Leng et al 1995, Bourguet et al 1995b). It will be interesting to see if stable, functional LBD domains can be expressed without additional upstream sequences. In the meantime, these results are of major significance for the preparation of samples for x-ray analysis, for studies of the interactions of receptor LBDs with other proteins, and for an understanding of the folding of newly synthesized receptors.

The best candidates for a receptor LBD that is independently active are those of ER and PR. A mER sequence (positions 313–559, which correspond to 544–786 of rGR) plus three leader amino acids, could be overexpressed in *E. coli* to give an SDS pure protein that bound steroid with wild-type affinity after dilution in buffer (Coffer et al 1996). An even smaller PR LBD construct (residues 1–4/688–933), which contained a four-amino-acid leader sequence in front of the equivalent of 551–795 of rGR, bound progesterone with only a 6–7 fold reduction in affinity (Tetel et al 1997). Another candidate is the hRARγ fragment of 178–423 (analogous to 539–782 of rGR) plus a leader sequence of six histidines that possessed wild-type affinity after overproduction in *E. coli* (Rochel et al 1997).

V. Specificity of Ligand Binding

A longstanding problem among the steroid receptors is that very few ligands exhibit absolute specificity. Most ligands bind to more than one receptor

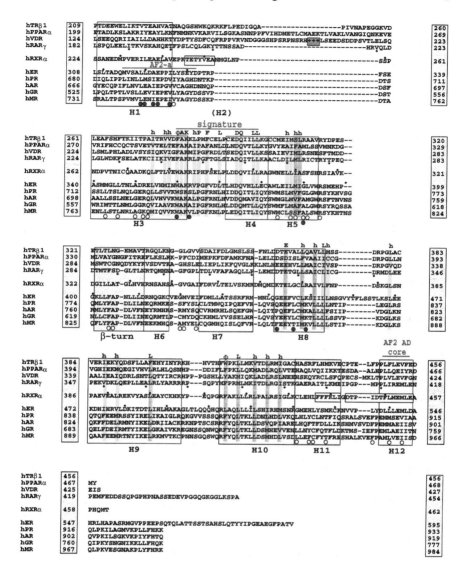

Figure 3.—Sequence alignment C-terminal sequences of selected selected receptors. For hVDR, the box with +++ indicates an insertion corresponding to amino acids 175–206. Conserved residues are indicated above the alignment by the single-letter code for invariant or highly conserved amino acids (h, hydrophobic; φ, aromatic). Highly conserved residues of the hydrophobic core are highlighted by light shading (in H3–H4 and H8–H9 regions corresponding to hTRβ residues of 275–308 and 373–390); residues involved in the interaction of H10 with H5, H8, and H9 are indicated by darker shading (among hTRβ1 residues 309–371 of H5 and H8, excluding the B–turn region, and 398–426 of H9–H10). Open circles below the alignment denote residues within 4.5 Å of the ligand in hRARγ, filled circles represent amino acids involved in key contacts between H1 and the LBD core. (adapted from Wurtz et al 1996).

(Ojasoo et al 1988). This promiscuity is not unexpected in view of homologies of > 50% among the LBDs of these receptors (Evans 1988), and it is much less of a problem among the nuclear receptors. Conversely, an examination of the determinants for the selective binding to AR and PR of steroids varying only in the 17β substituent disclosed not only that different sequences were most relevant (helices 10–12 of hPR for progesterone binding versus helices 6–7, and especially amino acids 788–791, of hAR [see footnote 4] for testosterone binding) but also that chimeric receptors with altered ligand specificity were no longer transcriptionally active (Vivat et al 1997). These results suggest that the LBD is a conformationally malleable structure, as opposed to a rigid structure with fixed ligand-receptor contact points. Consistent with this conclusion is the finding from alanine scanning mutations of the 515–535 region of hER that the receptor residues important for hormone binding vary with the structure of the ligand (Ekena et al 1997). For each receptor, attention initially focused on the specificity of agonist binding, but antagonists are known for all of the receptors except cPR and TR. The lack of antiprogestins for cPR has been found to stem from a species-specific point mutation (Benhamou et al 1992).

A. Point Mutations of LBDs

Volumes of information have resulted from studies of LBD point mutants. However, a constant proviso of mutational analyses is that changes in the folding, stability, and/or abundance of the mutant receptors (Byravan et al 1991, Powers et al 1993, Xu et al 1996b) can complicate one's interpretation of the data or even lead to erroneous conclusions. One of the earliest examples of this was for the hER, for which the original cDNA clone contained a mutation (G400V) that imparted temperature-sensitive steroid binding (Tora et al 1989).

Most mutations of receptor LBDs have led to decreased binding or loss of binding. Some compilations can be found for point mutations of AR (Quigley et al 1995), GR (Simons 1994), TR (Refetoff et al 1993), and VDR (Whitfield et al 1996). Initial fears to the contrary notwithstanding, not all mutations are detrimental. Reports have also appeared concerning ligand binding cores of receptors (Simons et al 1989, Ostrowski et al 1995, Wurtz et al 1996) that provide most of the affinity and binding selectivity. As seen in Figures 2 and 3, those sequences that make most of the contacts with bound ligand are located

---

[4] All references to the hAR will use the longer form (1–919) (Lubahn et al 1988), as opposed to a form that is 9 amino acids shorter (Veldscholte et al 1990). An even shorter, 87 kDa "A-form" of hAR has been described but is not considered here, as its abundance is ≤10% of the larger, more common 110 kDa form (Wilson and McPhaul 1994).

in the ~130 amino acids of helices 2–8, which might be capable of independent existence if it is conformationally stable.

It is beyond the scope of this review to cover all of the point mutations in the literature, but a few examples indicate the variety of modified properties that are possible. Cys-575 instead of Gly is largely responsible for the lack of binding and biological activity of RU 486 (RU 38,486) with cPR. Substitution of the comparable glycine residue by cysteine eliminated RU 486 binding, but not agonist binding, in hPR (with G722C) and destroyed all steroid binding in hGR (with G567C) (Benhamou et al 1992). Thus, precise ligand binding specificity may be achievable by genetic engineering. Mutation of Cys-656 to glycine in rGR created a "super" glucocorticoid receptor with increased steroid binding affinity, specificity, and transcriptional potency (Chakraborti et al 1991), which provides a precedent for augmenting the properties of native receptor proteins. The point mutant I747T in hGR LBD decreased the cell-free binding of Dex and cortisol at 0°C by ≤2-fold, with no change in whole-cell binding affinity of Dex at 37°C, while the bioactivity dose-response curve was right shifted by 100 for Dex. This right-shift was not due to altered interactions between AF-1 and AF-2 or to nuclear translocation (Roux et al 1996). Conversely, the Y537A mutant of hER produced a left shift in the dose-response curve without a change in affinity (Weis et al 1996). These results are particularly interesting because such dissociations of steroid binding and transactivation are not easily explained by the current models of steroid receptor action (Simons et al 1992). Finally, the preparation of mutants that both selectively blocked the binding of agonist steroids and perhaps increased the transcriptional activity of partial antagonists (Lanz and Rusconi 1994, Mahfoudi et al 1995) has provided a promising avenue to explore the mechanism of action of agonists, as compared to antagonists. An excellent, and expanding, online source for information about the steroid receptors and the properties of various point mutants is *http://nrr.georgetown.edu/NRR/NRR.html* (Martinez et al 1997).

B. Role of Cysteines in LBD Structure and Function

Cysteine residues can play a major role in protein structure because of their proclivity to form disulfide bonds reversibly. Their nucleophilicity and weak hydrogen bonding capacity make them important components of enzymes and binding proteins as well. While the formation of intramolecular disulfides will block steroid binding (Chakraborti et al 1990), no disulfide bonds were observed in any of the LBD x-ray structures (Bourguet et al 1995b, Renaud et al 1995, Wagner et al 1995), as was initially concluded from a chemical analysis of the native rGR LBD (Opoku and Simons 1994). However, the high concentration of cysteine residues in those helices surrounding bound ligand (i.e., helices 3–8 and 11) suggests that cysteines are intimately involved in the binding process. Thus, it is not surprising that several cysteines have been affinity labeled in GR (Simons et al 1987, Carlstedt-Duke et al 1988, Stromstedt et al 1990) and ER (Harlow et al 1989, Reese et al 1992). Thiol-reagents generally block ligand binding to receptors, although the effects of one of the smallest thiol-specific reagents (methyl methanethiolsulfonate or MMTS [Simons 1987]) varies with the receptor (Miller and Simons 1988, Dallery et al 1993, Souque et al 1996, but not Harlow et al 1989, Takagi et al 1990). Arsenite is a particularly useful reagent that selectively reacts at submicromolar concentrations with dithiols that are closely positioned in space. This specificity provided a means whereby 0.1 μM concentrations of arsenite blocked ligand binding only to GR, and not to any other of the steroid (Chakraborti et al 1990, Lopez et al 1990) or nuclear (Takagi et al 1990, Dallery et al 1993) receptors. Mutation of cysteine residues can have a major effect on ligand binding capacity and specificity (Chakraborti et al 1991, Byravan et al 1991). Nevertheless, most of the cysteine residues do not appear to be required for steroid binding (Chakraborti et al 1991, Reese et al 1992, Simons 1994, Neff et al 1994, Lamour et al 1996, Wolfgang et al 1997 but not Nakajima et al 1996), and their role may be more structural than functional (Hegy et al 1996).

C. Mutation Data That Support the LBD Model Derived from X-Ray Structures

One of the many attractive features of the 3-D model of the LBD for all steroid receptors, which was constructed from the x-ray data for hRXR and hRAR (Figures 2 and 3), is that it nicely accounted for the role of cysteine residues and the phenotypes of many published point mutants (Wurtz et al 1996). In the intervening months, several reports have appeared that bolster this general model, in which helices 5–7 and 11–12 contact the bound ligand, and helix 12 is pivotal for transactivation (see Section VIII.D.2.). With regard to ligand binding, the two naturally occurring hot spots for mutations of hTRβ have been localized to positions 310–350 (helices 5–7) and 438–454 (helices 11–12) (Refetoff et al 1993). Neutral, acidic, or basic point mutations of hTRβ His-435 (= 430 in 1–456 numbering) at the start of helix 11 caused the loss of T3 binding and gave dominant negative mutants with regard to transactivation (Nomura et al 1996). Four mutations of hER (G521A, H524A, L525A, and M528A), which both decreased the cell-free affinity for steroid at 0°C and increased the EC_{50} for transactivation, could be organized such that they all were on the same face of helix 11 that contacted bound ligand in the structures of RAR or TR LBDs (Ekena et al 1996). Interestingly, a different mutation of 524 (H524Q) in the Val-400 variant of hER did not affect the affinity of estradiol but still shifted the dose-response curve for transactivation to the right (Whelan and Miller 1996).

Helix 12 participates in both ligand binding and transactivation. Currently it is not possible to predict which activity will be affected by a given mutation. Thus, several mutations of helix 12 of hTRβ caused reduced binding and transactivation, or had no effect. However, two mutations (L454A and E457A in the receptor of 461 amino acids) dissociated the two properties to give transcriptionally inactive receptors with nearly wild-type ligand binding affinity (Tone et al 1994). Amino acids 405–419 of hRARα ($=$ 407–421 of hRARγ and overlaps helix 12), and especially Met-406 and Ile-410, were identified as being required for high affinity binding of 9-cis retinoic acid, but not of all trans-retinoic acid (Tate and Grippo 1995). The mutation L540Q in hER eliminated the activity of estrogens while maintaining or increasing the activity of antiestrogens, but not to more than 25–30% of that seen for estradiol with the wild-type receptor (Montano et al 1996). The proximity of helices 3 and 4 to 12, suggesting that they might contribute to the "transactivation surface" of the receptor, led to the discovery that K366 ($=$ K362 of hER), but not R367, was required for mER transactivation activity (Henttu et al 1997).

D. Unresolved Questions About the Model for Receptor LBDs

Not surprisingly, every piece of experimental data does not yet fit the model of Figure 3. It is likely that some minor refinements will be required for those receptors for which no x-ray structure yet exists, especially with GR, for which a detailed model has been advanced (Wurtz et al 1996). Thus, it is not obvious how the deletion of 5 or 14 amino acids from the C-terminus of the rGR would cause either a 20-fold loss (Rusconi and Yamamoto 1987) or a complete loss (Zhang et al 1996a) of binding affinity respectively, when none of the amino acids is even part of the structure of the LBD (Wurtz et al 1996). Furthermore, there was no obvious sequence requirement, as replacement of the 14 C-terminal amino acids of mGR by the highly divergent sequence of mAR had little effect on steroid binding or biological activity (Zhang et al 1996a). Similarly, the C-terminal 14 amino acids of cPR are not included in the model structure, but their reactivity with the monoclonal antibody C-262 was blocked by steroid binding (Weigel et al 1992). Finally, isoform specific differences in biological activity and binding affinity were noted for among mRARα and β mutants (Scafonas et al 1997). Thus, even very similar receptors may possess subtle, unpredictable structural modifications that contribute to ligand and/or transcription selectivity.

The placement of the steroid structure of Dex in the predicted binding cavity (Wurtz et al 1996) seems to need revision, since the electrophilic affinity labeling substituent of dexamethasone 21-mesylate is currently at the opposite end of the binding cavity from the affinity labeled Cys-638 (Cys-656 in rGR [Simons et al 1987]). More problematic is the fact that Met-604 and Cys-

736 (hGR numbering) are both photoaffinity labeled by triamcinolone acetonide or R5020 in rat or human GR (Carlstedt-Duke et al 1988, Stromstedt et al 1990), but are at opposite ends of the binding cavity. Only Cys-736 is close to the photoactivated diene-one of the steroid A-ring in the current model (Wurtz et al 1996). Notwithstanding, the current model represents a tremendous advance in our attempt to understand the functioning of receptor LBDs.

VI. COVALENT MODIFICATIONS OF THE LBD

A. Phosphorylation

The ability of protein phosphorylation/dephosphorylation to control transcription factor activity is well documented (Hunter and Karin 1992) and provides an attractive paradigm for the regulation of receptor activity. It seems that all of the receptors can be phosphorylated, predominantly on serines (see Chapter 9 for more detailed discussion). However, of the steroid receptors, only ER, which is also a nuclear receptor, has been reported to be phosphorylated in the LBD (Orti et al 1992), and this point is proving to be controversial. Phosphorylation of the hER LBD (Migliaccio et al 1989, Arnold et al 1997) has been detected by an anti-phosphotyrosine antibody that did not occur with the Y537F mutant (Arnold et al 1997). Nevertheless, no phosphorylated amino acids were detected in the porcine, human, or mouse ER LBD by sequencing (Bokenkamp et al 1994, Thole and Jungblut 1996) or mass spectral analysis (Seielstad et al 1995, Coffer et al 1996). This discrepancy may be resolved if it is found that the phosphorylated Y537 is very labile and decomposes during purification and/or analysis.

Among the other nuclear receptors, phosphorylation of VDR after exposure to vitamin D_3 was limited to Ser-205 (= Ser-208 in receptor of 427 amino acids) (Darwish et al 1993, Hilliard et al 1994). Phosphorylation of mRARα1 by PKA both *in vitro* and *in vivo* occurs only at Ser-369, but *in vivo* phosphorylation also occurs at other sites and does not occur at Ser-369 unless PKA is cotransfected or forskolin is added (Rochette-Egly et al 1995).

Demonstrations of an effect of receptor phosphorylation on receptor activity have generally proved to be elusive. ER phosphorylation was reported to increase the binding capacity, but not affinity, of the receptor (Migliaccio et al 1989, Castoria et al 1993, Arnold et al 1997) and to mediate ER dimerization (Arnold and Notides 1995) and DNA binding (Arnold et al 1995). Unfortunately, the significance of these findings is currently uncertain in view of the more recent inability to observe ER LBD phosphorylation (see above) and the observations that Y537 is not required for either dimerization and DNA binding (White et al 1997) or biological activity (Weis et al 1996).

Perhaps the function of Y537 phosphorylation is to stabilize the ER. Phosphorylation of Y537 may also facilitate the acquisition of transcriptional activity, as replacement of this residue with charged, or less hydrophobic, amino acids in mER provided mutant receptors with both increased steroid-independent transcriptional activity and association with coactivator proteins in mammalian cells (Weis et al 1996, White et al 1997). Phosphorylation of hVDR by casein kinase II at Ser-208 (= 205) is not required for transactivation but does permit enhanced transactivation (Jurutka et al 1996).

One of the problems encountered in understanding the role of phosphorylation in receptor function is that when phosphorylated residues have been mutated to a non-phosphorylatable amino acid, there usually has been no demonstration that alternative phosphorylation has not occurred. Mutation of the only phosphorylated residue of hVDR (Ser-205 [= Ser-208]) afforded a mutant receptor (S205A) with unaltered receptor activity, possibly because a new site had been found to be compensatorily phosphorylated (Hilliard et al 1994). On the other hand, steroid-induced phosphorylation of a mutant mGR containing alanine substitution at seven of the eight known phosphorylation sites did not occur at any new positions (Hu et al 1997).

B. Glycosylation of Receptor LBD and Other Possible Modifications

Glycosylation, to afford residues such as O-linked N-acetylglucosamine (O-GlcNAc), has been found among some transcription factors (Chou et al 1995) and has been proposed to provide regulatory control on protein activities (Kearse and Hart 1991; see Haltiwanger et al 1997 for review). Thus, glycosylation might also modify receptor activity. However, no evidence for significant amounts of glycosylation has been found for any of the receptors (Yen and Simons 1991, Bokenkamp et al 1994, Thole and Jungblut 1996, Seielstad et al 1995, Coffer et al 1996). Small amounts (5–10%) of calf or human ER have been found to contain O-GlcNAc on Thr-575, as determined by chromatography on wheat germ agglutinin columns of receptors affinity labeled with [^{3}H]tamoxifen aziridine. However, the inability to detect O-GlcNAc residues in non-radioactively-labeled receptors (Jiang and Hart 1997) is consistent with the failure of others to detect this modification of ER (Bokenkamp et al 1994, Thole and Jungblut 1996, Seielstad et al 1995, Coffer et al 1996).

To the extent that ligand binding activity is seen after expression in *E. coli* of many of the receptors, it can be concluded that eukaryotic-specific post-translational modifications are not required for steroid binding to rAR (Nemoto et al 1992), hER (Wooge et al 1992), cPR (Eul et al 1989), hRARα (Lamour et al 1996), hRXRα (Cheng et al 1994), or hTRβ1 (Lin et al 1990). Whether the inability hitherto to obtain steroid binding from GR (Nemoto et al 1990; Bayly and Simons, unpublished results) and MR

(Caamano et al 1993, Jalaguier et al 1996) that have been overexpressed in *E. coli* implies unique posttranslation modifications or, more likely, specific eukaryotic auxiliary proteins such as hsp90, remains to be determined.

VII. CONFORMATIONAL CHANGES IN LBD

A. Ligand-Induced Changes Detected by Proteases

The conformational changes described above as a result of ligand binding to receptors (see Figure 2) appear to have been first detected by a protease digestion analysis of GR. Major changes in the trypsin digestion pattern of rGR LBD which were observed as a consequence of steroid binding (Simons et al 1989), were not further altered by activation of the receptor-steroid complex (Reichman et al 1984). This method was extended and popularized by Allen et al in studies of PR and RAR (Allan et al 1992) and has been used to show a ligand-induced conformational change in the LBD for AR (Zeng et al 1994, Kuil et al 1995), ER (Fritsch et al 1993, Beekman et al 1993), MR (Trapp and Holsboer 1995), TR (Leng et al 1993, Toney et al 1993), RXR (Leid 1994, Leng et al 1995), VDR (Peleg et al 1995), and PPAR (Elbrecht et al 1996). This protease digestion assay has recently been refined with porcine ER and a panel of 9 proteases to probe the surface accessibility of receptor LBDs (Thole et al 1995).

These conformational changes are consistent with the experimentally observed two-step model for steroid binding (Pratt et al 1975) and with the postulates that steroid binding is the rate-limiting step in steroid hormone action (Munck 1984) and that a conformational reorganization of the LBD is a prerequisite for transcriptional activation (Beekman et al 1993, Renaud et al 1995, Wagner et al 1995). These proposals have received mechanistic support from the x-ray data of RAR LBD with steroid versus RXR LBD without steroid, which suggest a "flip" of the Ω loop between helices 2 and 3 (Figure 2) and a reorientation of helix 12 (Renaud et al 1995, Wurtz et al 1996). The existence of these two states in the same receptor has been supported by the observation of a ligand-induced dissociation of hRARβ from a monoclonal antibody against amino acids 207–222 of the Ω loop (Driscoll et al 1996).

It has not yet been established that the x-ray structures and protease digestion assays are detecting the same conformational changes. Cleavage in the flexible Ω loop has been observed only among the nuclear receptor subfamily with porcine ER (Thole et al 1995) and hRARα (Lamour et al 1996). Conversely, the ligand-induced conformational changes may not be sufficient to account for the dramatic increases in protease resistance of ligand-bound LBDs in general. Alternately, most of the differences in protease digestion could stem from associated proteins.

B. Ligand-Induced Changes Detected by Other Methods

The monoclonal antibody C-262 to the carboxyl terminal 14 amino acids of cPR recognized steroid-free, but not steroid-bound, receptors (Weigel et al 1992). Curiously, this region is C-terminal to the structures elucidated by x-ray analysis (Figure 3). Thus, the steroid-induced inaccessibility of antibody C262 may reflect interactions with nonreceptor sequences. A total of six antibodies has been employed to probe domains D and E of hRARγ. Three antibodies had no effect on ligand-binding affinity but did not recognize ligand bound receptors, thus supporting an LBD conformational change that occurs after the rate limiting step of ligand binding (Driscoll et al 1996). Estrogen and antiestrogen binding caused a decrease in receptor surface hydrophobicity (as determined by an aqueous two-phase partitioning system) that was localized to the LBD (Fritsch et al 1992a). CD spectroscopy of purified hER LBD (300–551 with an S305E mutation) was unable to detect significant changes in overall secondary structures due to the binding of either agonist or antagonist (Brandt and Vickery 1997).

C. Agonist-Versus Antagonist-Specific Conformational Changes in the LBD

The determinants of agonist versus antagonist activity for receptor-ligand complexes are of major mechanistic and clinical interest. While structure-activity relationships for the ligands have yielded only marginal advances, recent studies of the protease digestion patterns of receptors bound by agonists as opposed to antagonists have afforded promising results. Following the initial discovery that hPR lacking the C-terminal 42 or 54 amino acids displayed full agonist activity with the antiprogestin RU 486, but was inactive with and did not bind agonists (Vegeto et al 1992), it was found that trypsin digestion of antagonist-bound hPR-A or -B afforded a more resistant species of 27 kDa that was missing at least the C-terminal 14 amino acids (Allan et al 1992). It was therefore proposed that these changes in digestion are both specific for agonists versus antagonists and localized to the carboxyl terminal end of the LBD. This region of the LBD contains a transactivation sequence called AF2-AD (see Figure 3 and Section VIII.D.2.). Thus, not only would antagonists uniquely yield a smaller protease-resistant fragment, but the protease-detected conformational reorganization could account for the inactivity of antisteroids as well (Allan et al 1992, Vegeto et al 1992). These conclusions have been nicely supported by subsequent studies showing that (1) all antiprogestins examined were agonists with C-terminal truncated PRs (Wang et al 1994), (2) the monoclonal antibody C-262 to sequences beyond helix 12 of cPR could immunoprecipitate agonist-bound but not antagonist-bound receptors (Weigel et al 1992), (3) the major functional distinction between agonist- and antagonist-bound hPR was due to virtually the same C-terminal segment (917–928) with intrinsic repressor activity (Xu et al 1996a), and (4) unique trypsin cleavage patterns were seen for each of three types of PR-ligand complexes (agonist, antagonist, and mixed agonist) (Wagner et al 1996). A similar agonist-specific induced conformational change appears to occur in the hRAR LBD (Durand et al 1994, Keidel et al 1994). It should be noted that several lines of evidence with PR (Weigel et al 1992, Xu et al 1996a) and ER (Montano et al 1995) implicate the region beyond helix 12 (Figure 3), which is not included in the current three-dimensional model (Figure 2), as a determinant of agonist versus antagonist activity. Whether this region binds other factors or modifies the surface of the ligand-bound LBD will be the subject of future studies.

D. Ligand-Induced Conformational Changes in the LBD that Are Not Antagonist Specific

There is not a uniformity of opinion concerning the above model for antagonist binding and action, even with PR and RAR. One report with a different selection of PR ligands and 23 proteases (including trypsin) noted no correlation between protease fragment size and agonist versus antagonist activity of the complexes (Allan et al 1996). Furthermore, C-terminal truncations of hPR-B by 12 and 32 amino acids were completely inactive with the antiprogestin RU 486 (Lanz and Rusconi 1994). While the same constructs as described by Vegeto et al with 42 or 54 residue deletions of hPR-B (Vegeto et al 1992) were not examined, the 12 and 32 amino acid truncations of Lanz and Rusconi did remove most or all of the 12 amino acid region that has subsequently been shown to be sufficient to repress RU 486 agonist activity (Xu et al 1996a), and therefore would be expected to permit agonist activity with RU 486. These seemingly contradictory results could be due to the additional nonreceptor amino acids at the C-terminus of the various constructs. However, this would argue that the role of the C-terminal tail in determining agonist versus antagonist activity is not as straightforward as initially proposed.

A detailed study of trypsin digestion of hRARα complexes established that the size variation seen with the trypsin-resistant fragments of agonist versus antagonist complexes was due to different N termini (Keidel et al 1994), with R217 being the probable site of trypsin cleavage of antagonist-specific complexes (Lamour et al 1996). Elastase and chymotrypsin digestion of various hRARα complexes also gave smaller fragments with antisteroids. Here the antagonist fragments contained shorter N-terminal and longer C-terminal sequences (Keidel et al 1994). C-terminal digestion occurred within the F domain, probably at Arg-432 with trypsin, so that both the C-terminus of the LBD and the region corresponding to the PR repressor sequence (Xu et al 1996a) would be intact in all fragments. Therefore, antiretinoid binding to at least hRARα does give smaller protease-resistant fragments, but not as a result of preferential digestion of a C-terminal sequence.

Results with all of the other steroid receptors also are not compatible with a common mechanism of antisteroid action that can be detected simply by the presence of a smaller proteolytic fragment of antagonist-bound proteins. Trypsin digestion of hER bound by estradiol did produce a larger (32 kDa) species than that seen with various antagonist complexes (McDonnell et al 1995). However, the significance of these observations was not immediately evident since antiestrogens like 4-OH tamoxifen, which displayed partial agonist activity, afforded less of the "agonist-specific" 32 kDa tryptic fragment than did ICI 164,384, which was a more consistent antagonist and has been classified as a pure antiestrogen. In fact, as was noted for hRARα (Keidel et al 1994), the size differences could derive from deletions of the F domain of ER, which appears to play a crucial role in the expression of the agonist versus antagonist activity of ER ligands (Montano et al 1995). Antagonist-specific loss of activity was observed in chimeras of FLP recombinase and hER lacking ER hinge region sequences (Nichols et al 1997). This suggests that sequences amino terminal to the LBD also participate in the expression of antisteroid activity.

C-terminal deletions of 28, 53, or 55 residues from rGR, all of which are sufficient to remove the AF2-AD sequence, afforded mutants that were biologically inactive with the antiglucocorticoid RU 486 and bound only trace amounts of the antagonist (Lanz and Rusconi 1994). These results have been confirmed with mouse and rat GRs lacking 14 or 42 (Zhang et al 1996a) and 29 (Modarress et al 1997) amino acids respectively. It had been proposed earlier that steroid binding to GR effects changes mostly at the carboxyl terminus of the LBD (Simons et al 1989). Recent data have refined the assignment of protease digestion sites and determined that various cleavage patterns are not unique for agonists or antagonists (Modarress et al 1997). Similarly, others have found that trypsin digestion of hGR gave a 28 kDa fragment with all steroids and a smaller 25 kDa fragment when bound by either the agonist fluocinolone or the antagonist RU 486 (Guido et al 1996).

Digestion of hAR complexes with subtilisin gave smaller fragments with some, but not all, antiandrogens (Zeng et al 1994). Subsequent investigations of the trypsin digestion patterns established that the binding of the antiandrogen RU 486 does result in C-terminal cleavage of hAR. However, a similar cleavage was not observed with any of the other four antagonists examined (Kuil et al 1995). Deletion of the C-terminal 12 residues containing most of the highly homologous sequence of hPR required for antagonist activity (Xu et al 1996a) prevented the binding of androgens and four of the five antiandrogens examined (Kuil et al 1995). This same 12-amino-acid truncation of hAR also did not display any agonist activity with RU 486 (Lanz and Rusconi 1994), the one antiandrogen that would still bind.

Finally, the fragments produced by digestion of hMR complexes with three different proteases were not specific to agonists or antagonists (Trapp and Holsboer 1995). Thus, the involvement of C-terminal receptor sequences in the expression of antagonist activity seems to vary with the receptor and may be modified by the ligand, by other amino acids that might be added to the truncated receptor, and possibly by the reporter and cell used. Further studies of this topic should be very enlightening.

E. Conformational Changes in the LBD Induced by DNA Binding

Limited information exists about changes in the LBD in response to receptor binding to DNA. The effect of DNA binding on the protease digestion patterns of receptors is difficult to assess without a concomitant quantitation of the DNA-bound receptors and their stability. Notwithstanding, trypsin digestion of ligand-free mRXR$\alpha \pm$ hRARα was unaffected by added βRARE oligonucleotide under conditions where DNA binding was observed. The presence of 9-cis-retinoic acid had no effect on the digestion of RXR $\pm$ DNA but appeared to decrease the magnitude of ligand-induced conformational changes in the DNA-bound heterodimers (Minucci et al 1997). Antibody to the C-terminal 8 amino acids of mGR was observed to block the specific DNA binding of GR in a gel shift assay (Rowan and IP 1996). Binding to the estrogen response element (ERE) of the vitellogenin gene was reported to increase the rate of estradiol dissociation from rER by 2-fold (Fritsch et al 1992b), and a similar response was observed for rGR binding to the glucocorticoid response element (GRE) of the rat tyrosine aminotransferase gene (Modarress KJ, and Simons SS, Jr., unpublished results).

VIII. Other Activities of the LBD

A. Molybdate and Modulator Stabilization

The discovery that molybdate stabilizes the steroid receptors in the non-DNA-binding form (Nielsen et al 1977) opened many doors for subsequent investigations of receptor action. While it was originally thought that molybdate interacted with a thiol-containing region of the LBD (Dalman et al 1991a), further studies with receptors containing cysteines modified by methyl methanethiolsulfonate discounted this model (Modarress et al 1994) and greatly strengthened the present hypothesis that molybdate binds to hsp90 (Pratt et al 1993, see Pratt and Toft 1997 for comprehensive review). In view of the role of hsp90 (see Section II.B.), which does not associate with the nuclear receptors except perhaps ER, it is not surprising that molybdate has been utilized mostly with the steroid receptors.

A functionally related, but chemically different, endogenous compound that keeps GR (Bodine and Litwack 1988) and MR (Schulman et al 1992), but not ER or AR (Celiker et al 1993), in the non-DNA-binding state consists of two incompletely characterized substances called "modulator" with an unusual phosphoglyceride structure (Bodine and Litwack 1990). Studies with exogenously added peptides, one with an intramolecular disulfide, have led to the proposal that the modulator binding site in rGR is amino acids 596–614 (Bodine et al 1995).

B. Nuclear Localization of Receptors

Ligand-regulated nuclear localization is unique to the steroid receptors, but the mechanism is poorly understood. Recent results argue that translocation may actually result from the resetting of an initial dynamic equilibrium in which cytoplasmic localization was more favored (Guiochon-Mantel et al 1991, Madan and DeFranco 1993, see Defranco et al 1995 for review, and a detailed discussion in Chapter 2). Several nuclear localization sequences (NLS) have been located, with one residing in the LBD (Picard and Yama-moto 1987, Guiochon-Mantel et al 1989). However, while this site is essential for ligand-regulated nuclear translocation of GR and PR, but not ER (Chambraud et al 1990), it has not been shown to have independent translocation activity. Therefore, the action of the LBD nuclear translocation site has been suggested to be indirect, possibly as the result of the association of some molecule with the LBD that inhibits the other NLS (Urda et al 1989, Miyashita et al 1993, Scherrer et al 1993). For the PR, however, this mechanism has been discounted (Ylikomi et al 1992).

C. Dimerization

Most receptors dimerize upon binding to a hormone response element (HRE) in a manner that involves the d box of the DNA binding domain. Several receptors have been found to dimerize in solution before binding to an HRE. Depending on the receptor, homodimers and/or heterodimers have been observed (see Chapter 5 for more detailed discussion, especially for the DNA binding of dimers and the kinetics of formation of DNA-bound dimers).

1. Homodimerization in Solution

Full-length ER (Fawell et al 1990, Coffer et al 1996), PR (Guiochon-Mantel et al 1989, DeMarzo et al 1991), and RXRα (Kersten et al 1995a, Bourguet et al 1995b) (Noy N, Gronemeyer H, et al, submitted) formed homodimers in solution. The hER LBD appears to have an increased propensity towards dimerization that occurs in the absence of ligand at elevated temperatures, conditions under which the full-length hER remained as a monomer (Salomonsson et al 1994). Dissociation of hER LBD (residues 300–551 with an S305E mutation) dimers was retarded by bound estrogen or antiestrogen (Brandt and Vickery 1997). In contrast, the isolated hPR LBD (amino acids 688–933) steroid dimerized either with intact PR-A in a coimmunoprecipitation assay or with an His_6-tagged version of itself in a metal affinity chromatography assay only when more amino terminal sequences were present (Tetel et al 1997). The DE domain of hRXRα was monomeric in solution (Cheng et al 1994). Larger portions of the ligand-free mRXRα (C–E domains) existed as monomers, dimers, and tetramers, with tetramers predominating at receptor concentrations above 70 nM (Kersten et al 1995a) but these portions of mRXRα dissociated largely to dimers with added ligand (Kersten et al 1995b). Ligand-free and bound hVDR (Sone et al 1991, Cheskis and Freedman 1994), and ligand-bound hRARγ (Renaud et al 1995), did not form homodimers in solution. The ligand-occupied rTRα1 LBD existed as a monomer (Wagner et al 1995), apparently because of a ligand-mediated dissociation (Yen et al 1992) of the dimeric, full-length receptor (Holloway et al 1990).

2. Heterodimerization in Solution

RXR heterodimerizes with multiple receptors (see also Chapter 5) such as RAR, TR and VDR (Kliewer et al 1992a, Zhang et al 1992), PPAR (Kliewer et al 1992b), and COUP (Berrodin et al 1992). TRα and RARβ also heterodimerize with multiple proteins (each other, RXRα, RXRβ, COUP-TF, and an endogenous liver nuclear protein) (Berrodin et al 1992). COUP-TF heterodimerizes with the LBDs of RXR, TR, and RAR (Leng et al 1996). Thus, a tremendous amount of combinatory diversity is possible among these transcription factors.

A series of heptad repeats, distinct from leucine zippers (Rosen et al 1993), was originally identified as mediating the dimerization between TRs and RARs, but not between the steroid receptors (Forman et al 1989, Forman and Samuels 1990). Further work disclosed the most important element as being the ninth heptad (Leid et al 1995), which is encompassed by helix 10 (Figure 3). The x-ray structure of the dimeric hRXRα revealed that helix 10, with some contribution from helix 9, formed the main dimer interface. However, the previously determined crucial residues of the ninth heptad were found not to constitute the dimer interface (Bourguet et al 1995b, Wurtz et al 1996).

Interestingly, different amino acids were important for homo- versus heterodimerization (Marks et al 1992). Mutations of the first or last hydrophobic amino acid in the ninth heptad had no effect on homodimerization but eliminated heterodimerization of ligand-free TR, RAR, and RXR (Au-Fliegner et al 1993). Lysines in helix 9 of hRARα were the only residues involved in heterodimerization with hRXRα, whereas histidines uniquely participated in RAR homodimerization along with Val-361 (Rachez et al 1996).

When ligand-free heterodimers exist, an obvious question is whether the binding of ligand to each subunit occurs independently or not. So far, this question has been examined only for binding to RXR associated with various partners (see also Chapter 5). The nature of the heterodimer did influence ligand binding to RXR that was bound to DNA, but not to RXR that was free in solution (Kurokawa et al 1994, Forman et al 1995b). However, no effect of RAR-RXR heterodimer formation, either free in solution or bound to DNA, was reported by researchers using a potentially more accurate optical method to follow 9-cis retinoic acid binding to RXR (Kersten et al 1996). This issue is of major relevance for additional levels of control of receptor action, and further studies with all of the heteromeric complexes should yield interesting results. For example, the antiretinoid LG100754 is an antagonist for hRXRα homodimers but an agonist for RXR/PPAR or RXR/RAR (but not RXR/ TR or RXR/VDR) heterodimers (Lala et al 1996), apparently because of a "phantom ligand effect" on the associated ligand-free receptor (Schulman et al 1997). Nonetheless, RAR and RXR agonists synergized in the binding to, and transcriptional activity of, RXR-RAR heterodimers in a manner that required the activation activity at the C-terminus of the LBD of both receptors (Botling et al 1997, Minucci et al 1997) (see also Section VIII.D).

3. DNA-Bound Homo-and Heterodimers

Given the separation of the dominant sequences for dimerization (helix 10) and transactivation (helix 12, see Section VIII.D.) and the ability of point mutations to specifically affect homo- versus heterodimerization (see preceding section), it is not surprising that differential effects have also been seen for DNA binding and transactivation. The R429Q mutation in helix 10 of hTRβ1 LBD prevented homodimerization without altering ligand binding affinity or the induction of genes, but did block the repression of some genes (Flynn et al 1994). Disruption of the ninth heptad in TRα resulted in the loss of ligand-free homodimer binding to all TREs, but not of heterodimer binding with RXRα, thus leading to the proposal that different portions of the ninth heptad are important for homo- and heterodimer binding to DNA (Nagaya et al 1996, Yang et al 1996). This hypothesis has been expanded with the demonstration that the larger, 40-amino-acid region of helices 9 and 10 constituted a transferable sequence, called the I box, which is critical for the formation of ligand-free RXR-RAR and RXR-TR heterodimers as well as COUP-TF homodimers. However, the specificity of DNA binding of these dimers is further modified by the dimerization domain in the DNA binding domain. RXR sequences outside of the I box were required, in order to obtain ligand-free heterodimers with VDR and PPAR (Perlmann et al 1996) (see also Chapter 5).

The same ninth heptad (Forman and Samuels 1990) that is crucial for heterodimerization (see above section) is largely responsible, but not sufficient, for the homodimerization of DNA-bound ER. Thus, a 22-amino-acid peptide (501–522) of mER (= 497–518 of hER) restored the dimerization activity to defective mutants of mER (Lees et al 1990). The function of this region in the other steroid receptors appears limited, despite good sequence homology, as the comparable sequence of AR in an ER chimera was unable to replace the original ER sequence in yielding good dimer binding to DNA in a gel shift assay (White et al 1991).

Among the steroid receptors, the LBD was not required for the dimerization of DNA-bound proteins, at least for rGR (Chalepakis et al 1990), but was necessary for optimal dimer formation (Segard-Maurel et al 1996). These results are consistent with an interaction between the N- and C-terminal domains. More convincing support for such interactions came from studies of the geometry of dimerization of functional hAR and hER in mammalian two-hybrid assays with the N- or C-terminal halves of the receptors present either as chimeras or as simple receptor fragments. The results of three studies were essentially the same, although the two ER chimeras contained a common overlap of 66 amino acids in helices 1–4. Agonists, but not antagonists, promoted the association of the amino- and carboxyl-terminal regions to induce gene transcription, thereby suggesting that an agonist-induced interaction of the LBD with the N-terminal domains takes place, which may or may not require additional intermediary proteins (Langley et al 1995, Kraus et al 1995, Doesburg et al 1997). An association of LBDs was not observed with any receptor (Langley et al 1995, Kraus et al 1995, Tetel et al 1997) except with high expression levels of AR LBD chimera (Doesburg et al 1997). The data in one study of the AR were interpreted as resulting from the formation of an intermolecular, antiparallel dimer (Langley et al 1995), while an intramolecular association of the two ends of the receptor was favored in a more recent study of AR (Doesburg et al 1997) and for ER (Kraus et al 1995). However, in view of the inability of AR sequences to substitute for ER LBD dimerization sequences in the formation of DNA bound dimers (White et al 1991), some difference in geometry of the agonist-induced dimers of AR versus ER might be expected.

Heterodimerization was originally thought to be the province of the nuclear receptor subfamily, excluding ER. Recent evidence, however, suggests that GR and MR, and perhaps others, can heterodimerize upon binding to DNA (Trapp et al 1994), although, the heterodimerization of the steroid receptors is thought to occur through the DBD as opposed to the LBD (Liu et al 1995).

D. Transcriptional Activation

Two transcriptional activation functions (TAF-1 and −2) were initially localized to the amino terminal (A and B) and the carboxyl terminal (D–F) domains of receptors, respectively (Webster et al 1989, Bocquel et al 1989) (see Chapter 6 for a more detailed discussion with particular emphasis on the mechanism of transactivation). With time, the designation of these domains changed to AF-1 and -2, to distinguish them from the TBP associated factors (TAFs). It should be realized, however, that not all receptors seem to have both AF-1 (Sone et al 1991) and AF-2 (Simental et al 1991, Doesburg et al 1997) domains. The understanding of transcriptional activation also increased such that a variety of activities in the LBD AF-2 region emerged. The core of AF-2 activity was found in a 15-amino-acid region of receptors (Danielian et al 1992) that was eventually localized to helix 12 (Figure 3) (Wurtz et al 1996). A less-well characterized component of AF-2, called AF2-a, exists in the amino-terminal end of the LBD (Pierrat et al 1994, Milhon et al 1994, Norris et al 1997) (see also Figure 3). Many of the new activities involved associations with other proteins in cell-free systems, or artificial whole cell assays such as the two-hybrid assays, but with little direct evidence that the interactions detected were relevant to the whole-cell functioning of receptors. Furthermore, many of the interactions have not been shown to be influenced by added ligand or mutations in the AF-2 domain, which leaves their biological significance in doubt. Notwithstanding, it is likely that these interactions will prove to be highly informative in our attempts to reconstruct how steroid receptors regulate the transcription of responsive genes.

1. Interaction with the RNA Polymerase II Transcription Complex

a. Interaction with TBP and TFIID

TFIID is a complex of TBP and TBP-associated factors (TAFs) and is the first transcription factor to bind to the TATA box in the assembly of the transcription initiation complex with RNA polymerase II. It seems that not all TAFs are present in the same complex with TBP and that different TAFs interact with different transcription factors (Brou et al 1993, Jacq et al 1994). Recent data suggest that a major role of transcription factors such as the receptors is to increase the binding of TFIID and/or TBP to promoter elements (Chatterjee and Struhl 1995, Klages and Strubin 1995). TBP has been reported to interact directly with the carboxyl terminal 19 amino acids of hRXRα (helix 12), and mutations in this region that prevented transactivation also blocked RXR LBD ligand-dependent association with TBP, although the cell-free interactions were ligand-independent (Schulman et al 1995). A novel RXR antagonist failed to promote a functional coupling of hRXRα LBD with TBP in a yeast two-hybrid assay (Lala et al 1996). TBP could bind to the AF-2 domain in helix 12 of hER and potentiate estrogen-induced, but not basal, transcription (Sadovsky et al 1995). The LBD of hTRα interacted with C-terminal domain of TBP in cell-free systems and is thought to repress basal transcription by interacting with TBP/TFIIA or

TBP/TFIIA/TFIIB on DNA in a manner that is disrupted by thyroid hormone binding (Fondell et al 1996), although others failed to observe any interaction with TBP in similar cell-free systems (Schulman et al 1995).

The relationships of receptor LBDs with $TAF_{II}110$ are currently ambiguous. $hTR\alpha$ and $hRXR\alpha$ interacted with Drosophila $TAF_{II}110$ ($dTAF_{II}110$) in a ligand-dependent fashion in a yeast two hybrid assay (Schulman et al 1995). An $RXR\alpha$ antagonist caused an even stronger association with $TAF_{II}110$ in a yeast two-hybrid assay than did the agonist 9-cis-retinoic acid (Lala et al 1996). In contrast, the specific interactions of $TR\alpha1$, and $TR\beta$, with $dTAF_{II}110$ (but not several other TAFs) were (1) inhibited by added steroid *in vitro* in the presence or absence of DNA binding, (2) unaffected by ligand in a yeast two-hybrid assay, and (3) increased by thyroid hormone in HeLa cells (Petty et al 1996).

Human $TAF_{II}30$ was present in a distinct TFIID complex, interacted with amino acids 300–330 of hER, and was required for the transcriptional activity of ER AF-2 (Jacq et al 1994). The ability of human $TAF_{II}28$ to promote the transcriptional activity of RXR AF-2 was found not to require direct contacts of RXR and $TAF_{II}28$, but rather to be correlated with the ability of $TAF_{II}28$ to interact with TBP (May et al 1996).

b. Interaction with TFIIB

TFIIB has been reported to interact with the hER LBD (Ing et al 1992). The carboxyl terminal region of hVDR (positions 123–311) was found to be required for association with TFIIB in cell-free (Blanco et al 1995) and yeast two-hybrid systems (MacDonald et al 1995). Added TFIIB modified VDR activity in a ligand-dependent manner, but the nature of the response (increased or decreased) depended on the cell line, suggesting the involvement of a third factor (Blanco et al 1995). There is agreement that hTR can interact with TFIIB, but several of the details are different, which may be specific for the α or β isoforms. Added thyroid hormone was reported to decrease the binding of $hTR\beta$ LBD to the N-terminus of TFIIB (Baniahmad et al 1993), while the $hTR\alpha$ LBD was found to associate with the C-terminal region of TFIIB and to be unaffected by ligand (Fondell et al 1996). The repression of basal activity by ligand-free $hTR\beta$, which results from C-terminal sequences (Tong et al 1995, Fondell et al 1996), has been reproduced in a cell-free transcription assay and is thought to result from the interaction of $TR\beta$ homodimers, or TR/RXR heterodimers, with TFIIB-TBP to prevent further transcription complex assembly. Ligand binding to TR is proposed to reverse repression by reducing TFIIB-TBP interactions and promoting transcription complex assembly (Tong et al 1995). However, the ability of transcriptionally inactive $hTR\beta$ mutants to display the same ligand-induced reduction of association with both TBP and TFIIB suggests that these interactions are not determining factors of TR activity (Tone et al 1994).

c. Interactions in Mammalian Cells Versus Yeast

Despite the tremendous power and utility of conducting experiments in yeast, there are sufficient differences from mammalian cells that one must be wary of overinterpretation. As has already been mentioned, TRα1 interactions with dTAF$_{II}$110 were affected by added steroid *in vitro* and in mammalian cells, but not in yeast (Petty et al 1996). These types of observations may result from species differences among some of the TAF$_{II}$s (May et al 1996). Furthermore, factors coupling hVDR AF-2 to the transcriptional machinery were different in yeast versus mammalian cells (Jin et al 1996), and polymerase complex assembly onto the TATA box is thought to involve a preassembled complex in yeast, as opposed to a step-wise assembly in mammalian cells (Wu et al 1996, Zawel and Reinberg 1993 and references in Schwerk et al 1995).

2. Transactivation Activity

Endeavors to understand receptor control of gene transcription have been complicated by the observations that the response to a given receptor-ligand complex is not constant but can vary with the promoter (Berry et al 1990, Nagpal et al 1992, Tzukerman et al 1994, Guido et al 1996), composition (Durand et al 1994, Guido et al 1996) or spacing (Kraus et al 1994) of the response element, gene (Mercier et al 1986, Wasner et al 1988, Turcotte et al 1990), cell (Berry et al 1990, Tzukerman et al 1994, Guido et al 1996), cell density (Oshima and Simons 1992a), cis-acting elements (Oshima and Simons 1992b), and agents such as Br-cAMP (Beck et al 1993, Fujimoto and Katzenellenbogen 1994, Chauchereau et al 1994) and dopamine (Power et al 1991). Furthermore, studies with hER suggest that the AF-1 and -2 activation domains do not function equally with agonists and antagonists. Thus, AF-2 is sufficient (along with the DNA binding domain) for a response with estradiol, but receptors must contain both AF-1 and -2 for antiestrogens to display partial agonist activity (Berry et al 1990).

Numerous studies (Roux et al 1996, Schulman et al 1996, and references therein) have confirmed the initial observation that a crucial component of AF-2 is what is now called AF2-AD in helix 12 (Danielian et al 1992). Autonomous activity of this sequence from mER was not observed (Danielian et al 1992) and seems to require other sequences, such as the C-terminus of helix 3 containing Lys-366 (= K362 of hER) (Henttu et al 1997). The AF2-AD of hER was also inactive when substituted for the homologous sequence of hRARα (Tate et al 1996). However, independent activity could be demonstrated for the homologous sequence of other receptors, as seen by the ability of the mRARα, or mRXRα, AF2-AD to convey activation properties to the DNA binding domain fragment of GAL4 or ER respectively (Durand et al 1994). Similar results were obtained with the AF2-AD of RXRβ (Leng et al 1995), although results with TR indicated that adjacent amino acids can be of major significance. Thus, transferable activity was observed with the C-terminal 35, but not 12, amino acids of cTRα (Barettino et al 1994), and with the C-terminal 17, but not 49, amino acids of hTRβ1 (Baniahmad et al 1995a). Two other regions of activation activity have been identified in the hTRβ1 LBD: a weak activation domain at 207–214 and a strong activation domain at 339–368 (1–456 numbering) (Baniahmad et al 1995a). A transactivation sequence labeled as AF2-a in Figure 3, which is adjacent to the weak activation domain of hTRβ1, has been identified in mGR by virtue of clustered point mutations that reduced or eliminated transcriptional potency (Milhon et al 1994) and in C-terminal truncations of hER (Pierrat et al 1994). However, it should be noted that the AF2-a sequence of hER was active only in yeast (Pierrat et al 1994).

For many receptors, but not all (see Section III.C.), the AF2-AD can be removed without compromising ligand binding activity. In most cases, mutations within the region also eliminated transactivation activity without perturbing ligand binding (Damm et al 1993, Tzukerman et al 1994, Tone et al 1994, Leng et al 1995, Schulman et al 1996). One unresolved discrepancy concerns the Met-Leu and Ile-Ile pairs in helix 12 of GR (Figure 3). It was initially reported that the two double mutants of mGR (M758A/L759A and I762A/I763A) retained steroid binding activity after *in vitro* translation, but not transcriptional capacity (Danielian et al 1992). Overexpression of a comparable rGR mutant (M770A/L771A) using vaccinia virus in HeLa cells yielded receptors with a 3-fold reduced affinity (Schmitt and Stunnenberg 1993), while neither rGR mutant (M770A/L771A or I774A/I775A) retained either transcriptional or binding activity with the agonist Dex after transient transfection in CV-1 cells or *in vitro* translation, respectively. These receptors, however, did bind the antiglucocorticoid RU 486 to give agonist activity in intact cells (Lanz et al 1994, Lanz and Rusconi 1994). A resolution of the differences between these mutations in rat and mouse GR will be required in order to determine whether the transcriptional activation properties of agonists and antagonists can be selectively disrupted without affecting ligand binding. It should be noted that the comparable mutations of mER have recently been reported to bind estrogens and antiestrogens, but to display biological activity only with antagonists (Mahfoudi et al 1995).

The x-ray data (Figure 2) have led to the proposal that the acquisition of transcriptional activity, which follows ligand binding to receptors and the previously discussed conformational changes (see Section VII), entails a repositioning of helix 12 to generate a surface or surfaces that could interact with the transcriptional machinery and/or intermediary factors, such as coactivators and cointegrators (Wurtz et al 1996). The nature of the activation sequence is not known, but it is different from the acidic activator sequences that have been characterized in the AF-1 region of RAR and GR (Folkers et al 1995). The ligand-induced reorientation of the carboxyl terminal tail could concomitantly afford transcriptional activation by causing the dissociation of corepressors such as SMRT (see Section VIII.H.2.), which binds to RAR and TR but dissociates with added steroid (Chen and Evans 1995, Baniahmad et al 1995a). It should also be noted that even the AF2-AD sequence may have structural functions. Analyses by circular dichroism of truncated hTRβ1 proteins suggested that five amino acids (EVFED) of an amphipathic alpha-helix in helix 12 are required to maintain the tertiary structure of the LBD (Bhat et al 1995).

3. Biological Activity of Receptors Lacking AF2-AD

As was discussed above (see Sections VII.C. and D.), it was initially proposed from studies of PR that steroid receptors with C-terminal deletions of AF2-AD would be inactive with agonists but would elicit agonist activity when bound by antagonists. It now appears that this behavior is not general for the steroid receptors and may not be consistently observed with PR. The more common result of deletion of the AF2-AD sequence from receptors is the production of dominant negative receptors. This consequence has been observed for hGR (Hollenberg et al 1985, Bamberger et al 1995, Oakley et al 1996), ER (Ince et al 1993), mRXRβ (Leng et al 1995), and RAR (Damm et al 1993, Durand et al 1994). In many cases, these truncated receptors were also inactive when bound by antagonists (Lanz and Rusconi 1994, Zhang et al 1996a, Modarress et al 1997), or did not bind antisteroids (Kuil et al 1995). Studies of TRs indicated that receptor dimerization is required for dominant negative inhibition and that most dominant negative mutants contain altered amino acids on either side of the dimerization domain (Nagaya and Jameson 1993). This is consistent with the presence of transcription activation domains on either side of the dimerization domain of helices 9 and 10 (Baniahmad et al 1995a). Somewhat surprisingly, maximal activation by RXR heterodimers was found to require an intact AF2-AD sequence in both heterodimer partners even when only one was ligand-bound (Botling et al 1997).

4. Transcriptional Activation in Mammalian Cells Versus Yeast

Several differences need to be considered when using yeast systems as models for eukaryotic transcriptional transactivation by steroid receptors. As mentioned above, the transcriptional complex is thought to be preformed in yeast (Wu et al 1996, and references in Schwerk et al 1995) and assembled on DNA in mammalian cells (Zawel and Reinberg 1993). Another difference that has yet to be resolved is why the activity of several ligands for the steroid receptors is different in yeast versus mammalian cells. Dex is a very potent agonist in mammalian cells but has little activity in yeast, apparently because the affinity of Dex for rGR expressed in yeast is 1/1000 that of GR expressed in mammalian cells (Picard et al 1990a). Deoxycorticosterone (DOC) is used as a full agonist in yeast (Kimura et al 1995, Hong et al 1996) but is usually an antiglucocorticoid in mammalian cells (Ojasoo et al 1988) (Simons et al, data not shown). The antiestrogens 4-HO-Tamoxifen and ICI 164,384 can be partial (Lyttle et al 1992) or full agonists (Wrenn and Katzenellenbogen 1993) in yeast and did not inhibit the activity of estradiol even in hyper-permeable yeast (Lyttle et al 1992). Ro 41–5253 has been described as a pure RARα-selective antiretinoid with < 10% agonist activity in COS cells (Apfel et al 1992), but it displayed 20–30% activity in yeast (Joyeux et al 1997). Furthermore, the activity of random point mutants of hER (Wrenn and Katzenellenbogen 1993) and rGR (Schena et al 1989) can be different in yeast and mammalian cells. Similarly, studies with truncated hER revealed that amino acids 303–338 (corresponding to helices 1 and 2) were transcriptionally active in yeast but not in animal cells (Pierrat et al 1994). Clearly, additional studies aimed at understanding the differences between yeast and mammalian cells will be very informative with regard to elucidating the mechanism of steroid receptor action.

E. Synergism

The details of synergism are still poorly understood. While studies on the induction from single and tandem GREs by truncated hGRs indicated that the AF-1 sequences were the sole determinants of synergism (Wright and Gustafsson 1991), other experiments in which the GR DBD was replaced by the GAL4 DNA binding domain argued that C-terminal hGR sequences (488–777) were capable of synergism in the absence of co-operative DNA binding (Baniahmad et al 1991). Synergism between hRARα and hRXRα (1) has been found to require a functional RXRα LBD, (2) was not seen with a mutant RXRα missing the AF2-AD that still bound ligand, and (3) was maintained in the absence of the AB domain of RXRα (Chen et al 1996a). Thus, at least in some situations, it appears that the LBD, and probably the AF2-AD, can participate in synergism.

F. Repression

The term "repression" has been applied in the literature to (1) the ability of ligand-free receptors to suppress basal level expression, (2) the blockage of activity of other receptors (often called dominant negative activity), and (3) the ability of other sequences of the same receptor to prevent the expression of agonistic ligand activity. Repression is to be distinguished from the reduction of expression of selected genes that is affected by the binding of receptor-agonist ligand complexes (see also Chapter 6).

Suppression of basal level expression was first documented for TR and ascribed to the τ_i domain, a region of homology for all steroid receptors that covers amino acids 282–321 of hTRβ1 (Forman and Samuels 1990), and that subsequently was found to correspond to the "signature region" (Wurtz et al 1996) (see Figure 3). A transferable silencing domain was found in the C-terminal halves of rTRα (120–410) (Holloway et al 1990, Baniahmad et al 1992), hER (254–595) (Holloway et al 1990), and hRARα (143–403) (Baniahmad et al 1992). This suppression was thought to result from a destabilization of the transcription initiation complex, as opposed to competitive binding with other transcription factors (Baniahmad et al 1992). Other indirect data suggest that an additional suppressive sequence of hTRβ1 could lie between 408 and 421 (numbering of 456 amino acid TR), which is the region of the ninth heptad. Specifically, C-terminal 35 amino acids (Barettino et al 1994), but not the C-terminal 49 (Baniahmad et al 1995a), were observed to possess independent transcriptional activity. Whether the inactivity of the longer segment resulted from dimerization, from suppression by adjacent sequences, or simply from improper folding remains to be established.

While steroid-induced down-regulation of gene expression probably occurs via a different mechanism, an hTRβ1 point mutation (R429Q of the 461 amino acid receptor) that blocked the TR response of genes that are down-regulated, but not of those that are induced, has recently been described in a patient with abnormal pituitary TSH secretion (PRTH) (Flynn et al 1994). This mutation occurs in helix 11 between AF2-AD and the dimerization domain of the ninth heptad and may offer a unique opportunity to explore the mechanism of steroid-regulated induction versus down-regulation.

The amino acids C-terminal to helix 12 were not resolved in the x-ray structures of the LBDs and thus are not modeled in the general 3-D structure (Figure 3). However, recent data with hPR-B suggest that this region is functionally very significant. In pursuing earlier observations regarding the importance of the carboxyl terminal 42 amino acids in the binding and activity of progestins and antiprogestins (see Section VII.C.), researchers found that a 12-amino-acid sequence of hPR-B (917–928) that is distal to helix 12 was necessary and sufficient to prevent the expression of any agonist activity upon binding an antiprogestin. Furthermore, both mutations in this sequence and overexpression of the peptide imparted activity to otherwise inactive receptor-antisteroid complexes. This last observation is consistent with the idea that prevention of induction by antisteroid is due to the interaction with some corepressor that appears to be different from NCoR and SMRT (Xu et al 1996a) (see Section VIII.H.2.). In view of the moderate homology of this 12-amino-acid sequence among the steroid receptors, it will be very interesting to see whether these properties will be maintained among other receptors.

Repression of receptor activity can also be produced by dominant negative receptors. The most common mechanism for this is heterodimerization with inactive receptors that usually are lacking the AF2-AD domain and do not bind steroid (Koenig et al 1989, Fuqua et al 1992, Damm et al 1993, Bamberger et al 1995, Matsui and Sashihara 1995, Piedrafita et al 1995,

Oakley et al 1996) (see also Section II.C.). As was recently pointed out in studies with hRXRα and hTRα, these data argue that the activating and repressing activities of receptors are dissociable (Schulman et al 1996). However, the separation does not appear to be absolute and can depend on the heterodimerizing partner. mRXR missing the C-terminal 21 residues was a dominant suppressor of RXR, but the activity with other receptors (TR, RAR) could be negative or positive (Leng et al 1995). An interesting dominant negative receptor of TR activation was found in the RXR isoform δ from zebra fish. This mutant contains a 14-amino-acid insert at the beginning of helix 7, which is in the middle of the region homologous to the second strong activation region of TR (Baniahmad et al 1995a), but it does not disrupt the major activation domain AF2-AD, the τ_i domain, or the ninth heptad repeat that is crucial for dimerization (Jones et al 1995).

1. Repression in Mammalian Cells Versus Yeast

There is a major unresolved difference between mammalian cells and yeast with regard to repression. No repression by ligand-free TR has been observed in yeast. In fact, ligand-free TR activates gene expression in yeast as opposed to the repression noted in mammalian cells (Privalsky et al 1990, Lee et al 1994).

G. Squelching

Squelching is a well-known phenomenon that is thought to arise from the competition of limiting transcriptional cofactors (Ptashne 1988, Levine and Manley 1989). Abundant data exist on the interactions of cofactors with receptor LBDs, some of which are supported by effects of these factors on squelching (see next section), but few studies have focused on squelching. The inhibition of hER-induced transcription in MCF-7 cells by transfected hPR-B required only the AF-2 of ER (domains DEF) (Chalbos and Galtier 1994). hTRα or hTRβ inhibited transactivation by hPR-B in transiently transfected CV-1 cells, and vice versa, but only in presence of ligand. Removal of the carboxyl terminal 6 amino acids of TRβ, which are part of AF2-AD, prevented the squelching by TR, presumably by disrupting a sequence that interacts with a soluble transcription factor. In support of this interpretation, ligand-bound TR LBD (amino acids 145–456 = part of domain C to C-terminus) attenuated the cell-free transactivation by PR (Zhang et al 1996b).

H. Interaction with Coactivators and Corepressors and Cointegrators

Over the last few years, numerous putative "adapter" proteins, or cofactors, have been found to associate with receptors, and have been advanced as mediating the action of receptors (for review, see Horwitz et al 1996, see also

Chapter 6). In most cases, though, the significance of these interactions, not to mention their physiological relevance, has been debated (for review, see Triezenberg 1995, Horwitz et al 1996) because they have been observed either in non-native contexts (e.g., cell-free conditions or yeast versus mammalian cells), with protein fragments, or with very high concentrations of artificial constructs in cells where receptor and/or factor are no longer limiting components. Furthermore, the designation of a protein as a coactivator, or corepressor, is often based on a steroid-dependent association with the receptor, as opposed to a demonstration of altered transactivation. It should be realized, however, that even ligand-regulated interactions between a receptor and a selected factor are not unambiguous evidence for a direct association of the receptor LBD. For example, thyroid hormone has been reported to increase the interaction of the hematopoietic bZip protein p45/NF-E2 with rTRα, and rTRβ, DBD without any detectable association of p45/NF-E2 with TR LBD (Cheng et al 1997). Notwithstanding the uncertainty of the location of the interacting site, a high affinity association of two proteins is often a good predictor of biologically relevant interactions. Thus such studies do alert investigators to the existence of previously unsuspected components (Ptashne 1988, Levine and Manley 1989, Szapary et al 1996), and have identified many proteins that warrant increased investigation. Another nontrivial problem is simply keeping track of all of the cofactors and determining whether or not a given protein is the same as a previously reported factor (Horwitz et al 1996). It should be noted that almost all of these cofactors have been discovered by interactions with receptor sequences that extend upstream of what is probably the minimum LBD sequence. However, the existing evidence suggests that these extra hinge region, or D domain, residues are not required for association with the receptors. Yet other cofactors might be expected to exist, since most receptors contain a second target in the AF-1 activation domain of the amino terminal half of the receptor.

A common theme for all of these cofactors is that their association with receptors is almost always modulated by ligand binding. This phenomenon is taken as further evidence that the association is directly with the receptor LBD. It also is an excellent functional demonstration of the ligand-induced conformational changes that were initially detected by protease digestion studies (see Section VII).

1. Coactivators

Perhaps the best documented coactivator is hSRC-1 (Onate et al 1995), which is a homolog of the longer mouse ERAP160 (or p160) (Halachmi et al 1994) that is unrelated to RIP160 (Cavailles et al 1994). SRC-1 was initially reported to interact with D/E domains of hPR in a yeast two-hybrid assay (Onate et al 1995). This association of SRC-1 required agonists and did not occur with steroid-free or antagonist-bound PR, although it has not yet been shown that SRC-1 interactions are independent of hinge region sequences or that they involve direct contacts with the PR LBD. Cotransfected SRC-1 enhanced the transactivation of agonist-bound receptors (progesterone, estrogen, glucocorticoid, T3, and RXR receptors, but not CREB) in HeLa cells. Furthermore, cotransfected SRC-1 overcame the squelching of PR by added ER, while an N-terminal truncated form of SRC-1 acted as a dominant negative inhibitor of PR and TRβ induction in Lmtk$^-$ cells (Onate et al 1995). Mutations at Y537 of hER produced two constitutively active receptors (Y537A and Y537S) that bound SRC-1 in the absence of steroid in direct proportion to their constitutive activity (Weis et al 1996, White et al 1997), although the binding and biological activity also required the presence of K362 of helix 3 (White et al 1997, Henttu et al 1997). The ability of SRC-1 to augment hER transactivation did not require an intact AF2-AD domain (Smith et al 1997). Finally, SRC-1 appeared to augment the transcriptionally productive interactions between AF1 and AF2 domains of hER (McInerney et al 1996). Thus, SRC-1 appears to be a *bona fide* co-activator of many steroid receptors.

Three proteins that are closely related to, and members of the same family as, the human SRC-1 are ERAP160, GRIP1, and TIF2. ERAP160 (or p160) (Halachmi et al 1994) appears to be the murine equivalent of SRC-1 (Kamei et al 1996, Hanstein et al 1996). GRIP1 (Hong et al 1996) is a partial mouse clone that is different from ERAP160 but homologous to the human protein TIF2, another 160 kDa protein (Voegel et al 1996). ERAP160 associated in cell-free systems with the carboxyl half ($\sim$288–596) of hER, but not with ER truncated at 534 and missing the AF2-AD, in the presence of estradiol but not with antiestrogens. ERAP160 also bound to RARβ and RXRα. The interaction of ERAP160 correlated with, and may be required for, transcriptional activation by ER (Halachmi et al 1994). The partial clone GRIP1, identified by interaction with mGR D/E domain sequences 513–783 (= 506–777 of hGR) in a yeast two-hybrid screen, was equally functional in assays with the LBDs of hER (amino acids 274–595) and hAR (amino acids 644–919). GRIP1 was found to have independent transactivation activity and increased the transcriptional activity of receptor LBD chimeras in yeast, although squelching was observed in mammalian cells (Hong et al 1996). Consistent with these observations, the full-length human homolog of GRIP1, TIF2, bound directly to the ER, RARα, RXRα, and TRα LBDs in a manner that required bound agonists and that was eliminated by almost all mutations within AF2-AD that prevented transactivation. TIF2 possessed endogenous transactivation activity, relieved the self-squelching of hER in HeLa cells, and augmented the activity of AR and PR chimeras, but had no effect on the activation by cotransfected GR, RAR, RXR, VDR, or TR chimeras (Voegel et al 1996). Thus, there may be receptor-specific coactivators.

Three potential coactivators (RIP160, RIP140, and RIP80) were detected by far-Western blotting with a GST/mER LBD (residues 313–599 = 309–595 of hER) chimera. These three proteins displayed a steroid-dependent association with the LBD of mER that was not seen with antiestrogens (Cavailles et al 1994), and they also interacted with RAR, RXR, and TR (L'Horset et al 1996). The cloned 127 kDa RIP140 had intrinsic transactivating activity (L'Horset et al 1996), and it bound receptors in a manner that correlated with the transcriptional activity of AF2-AD mutations, although the increase of ER, RAR, and TR transcriptional activity by overexpressed RIP140 in mammalian or yeast cells was never more than 2-fold (Cavailles et al 1995, L'Horset et al 1996, Henttu et al 1997) and was repressed by higher concentrations of RIP140 (Henttu et al 1997). In a different yeast strain (YPH250), overexpressed RIP140 caused a larger (4–10-fold) increase in transactivation by a chimeric hRARα, or hER, and a slight ($\leq$factor of 2) left shift in the dose-response curves, but not with high concentrations of transfected hER (Joyeux et al 1997).

A two-hybrid system succeeded in isolating TIF1, which interacts with the DE domains of several receptors (RXRα, RARα, VDR, PR, and ER). More detailed studies indicated that amino acids 263–467 of mRXRα were sufficient for TIF1 association, which was prevented by mutations in the AF2-AD sequence. Finally, cotransfection of a C-terminal clone of TIF1 (amino acids 396–1017) appeared to cause a left shift in the dose-response curve for RXR (α or γ) regulated gene expression in yeast (Le Douarin et al 1995). Interestingly, while SRC-1, TIF-1, TIF-2, and RIP140 all interacted with mER in the presence of agonists, the binding of only RIP140 was maintained in the K336A mutation of mER, which indicates that there are distinct interaction domains for some of these mER-associated factors (Henttu et al 1997).

An interaction trap assay in yeast was used to isolate 15 different human proteins, called Trip1–15, all of which displayed T3-regulated behavior with a fusion protein of LexA and amino acids 169–461 of rTRβ1 (Lee et al 1995a). One of these, Trip1, was cloned and found to be highly homologous with, and functionally equivalent to, the yeast transcriptional mediator Sug1. Trip1 displayed ligand-dependent association with TR and RXR, but not with GR (Lee et al 1995b). Others identified the mouse equivalent of Trip1 (mSUG1), which is homologous to yeast SUG1 and identical to the human p45 protein of the 26S proteosome (vom Baur et al 1996). Thus, the effects of SUG1 could be at the level of protein turnover. Other mechanisms are also possible with the recent demonstration that mSUG1 possessed 3'-5' DNA helicase activity (Fraser et al 1997). The interactions in yeast between various receptors (ER, TRα, VDR, RXRα, and RARα) and either mSUG1 or TIF1 appeared to be different, even though the same receptor sequence (AF2-AD) was involved (vom Baur et al 1996).

Less well characterized putative coactivators include the large form (13S) of E1A, which augments the AF-2 activity of RARβ, possibly by forming a bridge between RARβ and TBP (Folkers and van der Saag 1995). Yeast SPT6 increased hER transcriptional activity ~3-fold in a steroid dependent manner, and also increased hGR activity. No attempt has yet been made to show that the association of SPT6 with amino acids 179–595 of hER does not require the non-LBD sequences (Baniahmad et al 1995b). Similarly, a yeast two-hybrid assay was used to identify a 70 kDa protein (ARA70) from a human brain library that binds to the CDE domain fragment of AR only in the presence of agonist steroids, and is 99% homologous with the RET-fused gene protein. ARA70 caused an agonist-dependent, 10-fold increase in the maximal induction by AR, but 2-fold increase with ER, GR, or PR (Yeh and Chang 1996). The η-isoform of the human 14-3-3 proteins was isolated by its glucocorticoid agonist-dependent interaction with the hGR DE domains (residues 485–777) in a yeast two-hybrid assay. When cotransfected into COS-7 cells, 14–3–3η caused increased total transactivation by hGR and perhaps a 2-fold left shift in the dose-response curve (Wakui et al 1997). Far-Western blotting of GH3 and HeLa cell nuclear extracts with a fusion protein of GST and the DE domain of mRXRβ (150–410) fusion protein revealed 4–6 potential coactivators (Suen and Chin 1995).

2. Corepressors

Several corepressors have been described (Horwitz et al 1996), but most appear to interact predominantly with receptor sequences just amino terminal of the LBD. One of the first corepressors described was the 270 kDa mouse protein NCoR. NCoR binds to ligand-free TRβ1 and RARα, either in solution or bound to DNA, but not to RXR or steroid-bound receptors when complexed with DNA. NCoR was identified by its ability to be retained on a column containing immobilized TR/RXR heterodimer-DNA complexes. The immobilized protein was visualized on SDS gels by far-Western blotting with ^{32}P-labeled GST fusion protein containing the C-terminal region of TRβ1 (amino acids 169–456) (Horlein et al 1995). It is proposed that NCoR binding is required for the repressive effects of ligand-free TRα (and RARα), and that ligand binding causes both the dissociation of NCoR and the subsequent association of the coactivators RIP140 and RIP160 (Horlein et al 1995, Kurokawa et al 1995). It should be noted that the binding of NCoR to RARα was inhibited by agonist ligands only when RXR/RAR heterodimers were bound to a direct repeat RARE with a spacing of 5 nucleotides (Kurokawa et al 1995). NCoR was found to associate with TRβ1 that was free in solution, via sequences (CoR box) in the hinge region that overlap the start of the LBD (positions 206–235 of 1–456 TRβ1) (Horlein et al 1995, Kurokawa et al 1995), and mutations in the CoR box that eliminated NCoR association also decreased the repressive activity of ligand-free receptors (Horlein et al 1995). However, additional sequences seem to participate, since NCoR failed to associate with VDR, which possesses a highly homologous CoR box (Horlein et al 1995). NCoRI, a human homolog to the mouse NCoR but lacking the two amino terminal repressor regions, was isolated by a yeast two-hybrid screen from a human placental library. NCoRI bound to the ligand-free, but not ligand-bound, LBD of hTRβ1 (amino acids 204–461) either in solution or bound to DNA, and competitively inhibited the repressive activity of NCoR with TRα1 and TRβ1, but not TRβ2 (Hollenberg et al 1996). The partial clone RIP13 is identical to the C-terminal region of NCoR. Two full-length isoforms that arise from alternative splicing, RIP13a and RIP13Δ1, were identified that diverge from NCoR in the N-terminal region but also interact with the hinge and LBD sequences of hTRβ and hRARα in yeast and mammalian two-hybrid assays (Seol et al 1996b). Recent results indicate that repression is considerably more complex as overexpression of NCoR can actually increase transactivation (Söderström et al 1997), presumably by squelching (Szapary et al 1996), in which case excess NCoR would immobilize other components required for repression, such as Sin3 and the histone deacetylase RPD3 (Heinzel et al 1997).

A related corepressor is a human 168 kDa protein (SMRT) that was isolated in a yeast two-hybrid assay from its functional association with the hRXRα LBD (residues 228–462). Thus, the region of receptor with which SMRT interacts may be wholly contained within the LBD. However, the association of SMRT in yeast was much stronger with hRARα or with TR than with hRXRα, and this association was decreased with added ligand for RAR and TR, either free in solution or bound to DNA, while it was increased for RXR. Thus, the interactions with RXR may be somewhat different from those with RAR or TR. SMRT did not interact with GR. SMRT possessed independent repressor activity, and overexpression was able to restore the inhibitory activity of GAL-TR or -RAR fusion proteins that had been "activated" by cotransfection of v-ErbA. Thus, SMRT appears to function in intact cells as a corepressor (Chen and Evans 1995). Interestingly, the RXR-specific ligand LG10054 is an agonist for RXR/RAR heterodimers but did not cause the dissociation of SMRT (Lala et al 1996). In a more sensitive whole-cell assay in CV-1 cells, LG100754 did inhibit the interaction of SMRT with the complex of RAR LBD and RXR LBD (Schulman et al 1997). Interestingly, this same assay did not detect an interaction between hRXRα LBD and SMRT, which was how SMRT was originally discovered. Consequently, additional mechanisms of activation may be possible in the presence of SMRT.

One of the 20 potential clones of proteins interacting with amino acids 120–408 of cTRα in a yeast two-hybrid assay was called TRAC-1, which functionally associated with several other receptors (RARβ, RXRα, and PPARα). The binding of TRAC-1 to cTRα was inhibited by ligand, but was also prevented by a point mutation (P144R) in the hinge region, thereby raising questions about the binding site in cTRα. A second TRAC, TRAC-2, was found to be an N-terminal extension of TRAC-1 and seems to be identical to SMRT (Sande and Privalsky 1996). The activity of TRAC-1 appears to be limited to that of a competitive inhibitor of TRAC-2/SMRT (Sande and Privalsky 1996), and thus TRAC-1 is functionally homologous to NCoRI (Hollenberg et al 1996). A 266-amino-acid protein (TRUP) interacted with residues 168–259 (hinge and amino terminal end of LBD) of hTRβ in a yeast two-hybrid assay. Overexpression of TRUP inhibited transactivation by TR and RAR (but not ER or RXR) by inhibiting DNA binding ± ligand (Burris et al 1995).

3. Cointegrators

CBP (CREB binding protein) and p300 are two closely related proteins with such similar activity that they are usually considered as functional homologs (Lundblad et al 1995). CBP interacted with the LBD of numerous receptors (121–410 of rTRα1, 143–462 of hRARα, 227–463 of mRXRγ, and 251–595 of hER) in a yeast two-hybrid assay in a ligand-dependent manner, and was coimmunoprecipitated with anti-GR antibody (Kamei et al 1996). Furthermore, LBD mutations that rendered the receptors transcriptionally inactive eliminated the ability to functionally interact with CBP in the yeast assay. CBP also associated with ERAP160/SRC-1, leading to the proposal that steroid receptors activate transcription via a complex of CBP/p300, ERAP160, and perhaps other cofactors (Kamei et al 1996). Further studies revealed that the interaction site with CBP/p300 could be restricted to amino acids 288–595 of hER, but that CBP/p300 did not directly contact hER. Instead, it seemed that CBP/p300 binding was mediated by ERAP160/SRC-1 (Hanstein et al 1996). Several of these features were confirmed in an independent study which also demonstrated that microinjection of anti-CBP antibody into the nuclei of mammalian fibroblasts prevented gene induction by receptor-ligand complexes (Chakravarti et al 1996).

4. Mammalian Cells Versus Yeast

No obvious homolog to the mammalian proteins p300 and CBP have yet been found in yeast (Ogryzko et al 1996). This suggests that other proteins, or processes, may be functioning in yeast. Interestingly, GRIP1 was found to augment GR action in yeast, but to have no effect (Voegel et al 1996) or to cause squelching (Hong et al 1996) in mammalian cells. Conversely, added ligand strongly inhibited the interaction of RIP13 with hTRβ or hRARα in mammalian cells, but was without effect in yeast (Seol et al 1996b). These results could simply reflect cell-specific variation in co-factor concentrations. Alternatively, the different responses may provide clues regarding other examples of differential activity of receptors in mammalian cells versus yeast.

I. Interaction with Other Proteins

Far-Western blotting revealed interactions of mRXRβ LBD with 4–6 nuclear proteins in GH3 and HeLa cells (Suen and Chin 1995). Multiple ^{35}S-labeled proteins were found to associate with a GST/hER LBD (amino acids 288–595) fusion protein in an agonist-specific manner. The most prominent were 100, 90, and 30 kDa proteins (Hanstein et al 1996), in addition to the previously identified ERAP160 and ERAP140 (Halachmi et al 1994, Cavailles et al 1994). Numerous proteins have been found to interact with GR, but very few have been shown to bind directly to the LBD (see Simons 1994 for review).

IX. Future Directions

The LBD of steroid nuclear receptors constitutes the binding site for the cognate ligands and is therefore absolutely required for translating the structural information of the ligand into the induction or repression of responsive genes. The role of the LBD is further evident in interpreting the structures of ligands that can display either agonist or antagonist activity, depending on the environment. Given the pharmacological importance of understanding the determinants of agonist versus antagonist activity, this will continue to be a major research effort. One avenue that should prove very rewarding is solving the x-ray structure of other receptor LBDs, because all of the structures obtained to date have been for the members of the subfamily of nuclear receptors. In addition, an analysis at this level may be required to understand the relationships between ligand structure and activity. However, it is unlikely that x-ray data alone will suffice because the binding and activity of ligand can be influenced by other portions of the same receptor, by other receptors that can form heterodimers, and by nonreceptor proteins. This last category of associated nonreceptor molecules is ever expanding to encompass proteins with numerous functions, from protein folding to DNA binding, to transcriptional activation, to protein degradation. Currently, a very large number of proteins have been found to interact solely with the LBD of a given steroid receptor, and they are likely to participate integrally in the transmission of the information conveyed to the receptor after the binding of ligand. It is almost certain, however, that each protein is not simultaneously present in a single receptor-containing complex. Thus, one of the major challenges of the future will be to determine (1) which associations are relevant, (2) whether or not there is a temporal sequence for protein-receptor interactions, and (3) how these new complexes influence receptor-regulated gene transcription.

ACKNOWLEDGMENTS

I thank numerous people for communicating their recent studies prior to publication: (Jan Carlstedt-Duke [Huddinge], Dean Edwards [Denver], Hinrich Gronemeyer [Strasbourg], Benita Katzenellenbogen [Urbana], Malcom Parker [London], William Pratt [Ann Arbor], and David Toft [Rochester]). I am also grateful to Jean-Marie Wurtz (Strasbourg) for the preparation of Figures 2 and 3, and to Vera Nikodem (NIDDK/NIH) and members of the Steroid Hormones Section for constructive criticism of the manuscript.

REFERENCES

Allan GF, Leng X, Tsai SY, Weigel NL, Edwards DP, Tsai M-J, and O'Malley BW (1992): Hormone and antihormone induce distinct conformational changes which are central to steroid receptor activation. *J. Biol. Chem.* 267: 19513–19520.

Allan GF, Lombardi E, Haynes-Johnson D, Palmer S, Kiddoe M, Kraft P, Campen C, Rybczynski P, Combs DW, and Phillips A (1996): Induction of a novel conformation in the progesterone receptor by ZK299 involves a defined region of the carboxyl-terminal tail. *Mol. Endo.* 10: 1206–1213.

Alnemri ES, and Litwack G (1993): The steroid binding domain influences intracellular solubility of the baculovirus overexpressed glucocorticoid and mineralocorticoid receptors. *Biochemistry* 32: 5387–5393.

Apfel C, Bauer F, Crettaz M, Forni L, Kamber M, Kaufmann F, LeMotte P, Pirson W, and Klaus M (1992): A retinoic acid receptor alpha antagonist selectively counteracts retinoic acid effects. *Proc. Natl. Acad. Sci. USA* 89: 7129–7133.

Arnold SF, and Notides AC (1995): An antiestrogen: a phosphotyrosyl peptide that blocks dimerization of the human estrogen receptor. *Proc. Natl. Acad. Sci. USA* 92: 7475–7479.

Arnold SF, Vorojeikina DP, and Notides AC (1995): Phosphorylation of tyrosine 537 on the human estrogen receptor is required for binding to an estrogen response element. *J. Biol. Chem.* 270: 30205–30212.

Arnold SF, Klotz DM, Collins BM, Vonier PM, Guillette, Jr. LJ, and McLachlan JA (1996): Synergistic activation of estrogen receptor with combinations of environmental chemicals. *Science* 272: 1489–1492.

Arnold SF, Melamed M, Vorojeikina DP, Notides AC, and Sasson S (1997): Estradiol-Binding mechanism and binding capacity of the human estrogen receptor is regulated by tyrosine phosphorylation. *Mol. Endo.* 11: 48–53.

Ashby J, Lefevre PA, Odum J, Harris CA, Routledge EJ, and Sumpter JP (1997): Synergy between synthetic oestrogens? [scientific correspondence]. *Nature* 385: 494.

Au-Fliegner M, Helmer E, Casanova J, Raaka BM, and Samuels HH (1993): The conserved ninth C-terminal heptad in thyroid hormone and retinoic acid receptors mediates diverse responses by affecting heterodimer but not homodimer formation. *Mol. Cell. Biol.* 13: 5725–5737.

Bamberger CM, Bamberger AM, de Castro M, and Chrousos GP (1995): Glucocorticoid receptor beta, a potential endogenous inhibitor of glucocorticoid action in humans. *J Clin Invest* 95: 2435–2441.

Baniahmad A, Kohne AC, and Renkawitz R (1992): A transferable silencing domain is present in the thyroid hormone receptor, in the v-erbA oncogene product and in the retinoic acid receptor. *EMBO J.* 11: 1015–1023.

Baniahmad A, Ha I, Reinberg D, Tsai S, Tsai M-J, and O'Malley BW (1993): Interaction of human thyroid hormone receptor beta with transcription factor TFIIB may mediate target gene derepression and activation by thyroid hormone. *Proc. Natl. Acad. Sci. USA* 90: 8832–8836.

Baniahmad A, Leng X, Burris TP, Tsai SY, Tsai M-J, and O'Malley BW (1995a): The tau4 activation domain of the thyroid hormone receptor is required for release of a putative corepressor(s) necessary for transcriptional silencing. *Mol. Cell. Biol.* 15: 76–86.

Baniahmad C, Muller M, Altschmied J, and Renkawitz R (1991): Co-operative binding of the glucocorticoid receptor DNA binding domain is one of at least two mechanisms for synergism. *J. Mol. Biol.* 222: 155–165.

Baniahmad C, Nawaz Z, Baniahmad A, Gleeson MAG, Tsai M-J, and O'Malley BW (1995b): Enhancement of human estrogen receptor activity by SPT6: a potential coactivator. *Mol. Endo.* 9: 34–43.

Barettino D, Vivanco Ruiz MM, and Stunnenberg HG (1994): Characterization of the ligand-dependent transactivation domain of thyroid hormone receptor. *EMBO J.* 13: 3039–3049.

Beck CA, Weigel NL, Moyer ML, Nordeen SK, and Edwards DP (1993): The progesterone antagonist RU486 acquires agonist activity upon stimulation of cAMP signaling pathways. *Proc. Natl. Acad. Sci. USA* 90: 4441–4445.

Becker–Andre M, Wiesenberg I, Schaeren–Wiemers N, Andre E, Missbach M, Saurat J–H, and Carlberg C (1994): Pineal gland hormone melatonin binds and activates an orphan of the nuclear receptor superfamily. *J. Biol. Chem.* 269: 28531–28534.

Beekman JM, Allan GF, Tsai SY, Tsai M-T, and O'Malley BW (1993): Transcriptional activation by the estrogen receptor requires a conformational change in the ligand binding domain. *Mol. Endo.* 7: 1266–1274.

Benhamou B, Garcia T, Lerouge T, Vergezac A, Gofflo D, Bigogne C, Chambon P, and Gronemeyer H (1992): A single amino acid that determines the sensitivity of progesterone receptors to RU486. *Science* 255: 206–209.

Berrodin TJ, Marks MS, Ozato K, Linney E, and Lazar MA (1992): Heterodimerization among thyroid hormone receptor, retinoic acid receptor, retinoid X receptor, chicken ovalbumin upstream promoter transcription factor, and an endogenous liver protein. T3, associated proteins, T3 receptor and retinoic acid receptor both heterodimerize with multiple proteins (each other, RXR alpha, RXR beta, COUP-TF, and a liver nuclear protein). *Mol. Endo.* 6: 1468–1478.

Berry M, Metzger D, and Chambon P (1990): Role of the two activating domains of the oestrogen receptor in the cell-type and promoter-context dependent agonistic activity of the anti-oestrogen 4-hydroxytamoxifen. *EMBO J.* 9: 2811–2818.

Bhat MK, McPhie P, Ting Y-T, Zhu X-G, and Cheng S-Y (1995): Structure of the carboxy-terminal region of tyroid hormone nuclear receptors and its possible role in hormone-dependent intermolecular interactions. *Biochemistry* 34: 10591–10599.

Blanco JCG, Wang I-M, Tsai SY, Tsai MJ, O'Malley BW, Jurutka PW, Haussler MR, and Ozato K (1995): Transcription factor TFIIB and the vitamin D receptor cooperatively activate ligand-dependant transcription. *Proc. Natl. Acad. Sci. USA* 92: 1535–1539.

Bocquel MT, Kumar V, Stricker C, Chambon P, and Gronemeyer H (1989): The contribution of the N- and C-terminal regions of steroid receptors to activation of transcription is both receptor and cell-specific. *Nucleic Acids Res.* 17: 2581–2595.

Bodine PV, and Litwack G (1988): Evidence that the modulator of the glucocorticoid-receptor complex is the endogenous molybdate factor. *Proc. Natl. Acad. Sci. USA* 85: 1462–1466.

Bodine PV, and Litwack G (1990): Purification and characterization of two novel phosphoglycerides that modulate the glucocorticoid-receptor complex. *J. Biol. Chem.* 265: 9544–9554.

Bodine PVN, Alnemri ES, and Litwack G (1995): Synthetic peptides derived from the steroid binding domain block modulator and molybdate action toward the rat glucocorticoid receptor. *Receptor* 5: 117–122.

Bokenkamp D, Jungblut PW, and Thole HH (1994): The C-terminal half of the porcine estradiol receptor contains no post-translational modification: determination of the primary structure. *Mol. Cell. Endo* 104: 163–172.

Botling J, Castro DS, Öberg F, Nilsson K, and Perlmann T (1997): Retinoic acid receptor/retinoid X receptor heterodimers can be activated through both subunits providing a basis for synergistic transactivation and cellular differentiation. *J. Biol. Chem.* 272: 9443–9449.

Bouhoute A, and Leclercq G (1995): Modulation of estradiol and DNA binding to estrogen receptor upon association with calmodulin. *Biochem. Biophys. Res. Commun.* 208: 748–755.

Bourguet W, Ruff M, Bonnier D, Granger F, Boeglin M, Chambon P, Moras D, and Gronemeyer H (1995a): Purification, functional characterization, and crystallization of the ligand binding domain of the retinoid X receptor. *Protein Express. Purif.* 6: 604–608.

Bourguet W, Ruff M, Chambon P, Gronemeyer H, and Moras D (1995b): Crystal structure of the ligand-binding domain of the human nuclear receptor RXR-alpha. *Nature* 375: 377–382.

Brand N, Petkovich M, Krust A, Chambon P, de The H, Marchio A, Tiollais P, and Dejean A (1988): Identification of a second human retinoic acid receptor. *Nature* 332: 850–853.

Brandt ME, and Vickery LE (1997): Cooperativity and dimerization of recombinant human estrogen receptor hormone-binding domain. *J. Biol. Chem.* 272: 4843–4849.

Bresnick EH, Dalman FC, Sanchez ER, and Pratt WB (1989): Evidence that the 90-kDa heat shock protein is necessary for the steroid binding conformation of the L cell glucocorticoid receptor. *J. Biol. Chem.* 264: 4992–4997.

Brou C, Wu J, Ali S, Scheer E, Lang C, Davidson , Chambon P, and Tora L (1993): Different TBP-associated factors are required for mediating the stimulation of transcription in vitro by the acidic transactivator GAL-VP16 and the two nonacidic activation functions of the estrogen receptor. *Nucleic Acids Res.* 21: 5–12.

Burris TP, Nawaz Z, Tsai M-J, and O'Malley BA (1995): Nuclear hormone receptor-associated protein that inhibits transactivation by the thyroid hormone and retinoic acid receptors. *Proc. Natl. Acad. Sci. USA* 92: 9525–9529.

Byravan S, Milhon J, Rabindran SK, Olinger B, Garabedian MJ, Danielsen M, and

Stallcup MR (1991): Two point mutations in the hormone binding domain of the receptor that dramatically reduce its function. *Mol. Endocrinol.* 5: 752–758.

Caamano CA, Morano MI, Patel PD, Watson SJ, and Akil H (1993): A bacterially expressed mineralocorticoid receptor is associated in vitro with 90-kilodalton heat shock protein and shows typical hormone- and DNA-binding characteristics. *Biochemistry* 32: 8589–8595.

Cadepond F, Schweizer-Groyer G, Segard-Maurel I, Jibard N, Hollenberg SM, Giguere V, Evans RM, and Baulieu E-E (1991): Heat shock protein 90 as a critical factor in maintaining glucocorticosteroid receptor in a nonfunctional state. *J. Biol. Chem.* 266: 5834–5841.

Caplan AJ, Langley E, Wilson EM, and Vidal J (1995): Hormone-dependent transactivation by the human androgen receptor is regulated by a dnaJ protein. *J. Biol. Chem.* 270: 5251–5257.

Carlstedt-Duke J, Stromstedt P-E, Persson B, Cederlund E, Gustafsson J-A, and Jornvall H (1988): Identification of hormone-interacting amino acid residues within the steroid-binding domain of the glucocorticoid receptor in relation to other steroid hormone receptors. *J. Biol. Chem.* 263: 6842–6846.

Castoria G, Migliaccio A, Green S, Domenico DM, Chambon P, and Auricchio F (1993): Properties of a purified estradiol-dependent calf uterus tyrosine kinase. *Biochemistry* 32: 1740–1750.

Cavailles V, Dauvois S, Danielian PS, and Parker MG (1994): Interaction of proteins with transcriptionally active estrogen receptors. *Proc. Natl. Acad. Sci. USA.* 91: 10009–10013.

Cavailles V, Dauvois S, L'Horset F, Lopez G, Hoare S, Kushner PJ, and Parker MG (1995): Nuclear factor RIP140 modulates transcriptional activation by the estrogen receptor. *EMBO J.* 14: 3741–3751.

Celiker MY, Haas A, Saunders D, and Litwack G (1993): Specific regulation of male rat liver cytosolic estrogen receptor by the modulator of the glucocorticoid receptor. *Biochem. Biophys. Res. Commun.* 195: 151–157.

Chakraborti PK, Hoeck W, Groner B, and Simons, Jr. SS (1990): Localization of the vicinal dithiols involved in steroid binding to the rat glucocorticoid receptor. *Endocrinology* 127: 2530–2539.

Chakraborti PK, Garabedian MJ, Yamamoto KR, and Simons, Jr. SS (1991): Creation of "super" glucocorticoid receptors by point mutations in the steroid binding domain. *J. Biol. Chem.* 266: 22075–22078.

Chakravarti D, LaMorte VJ, Nelson MC, Nakajima T, Schulman IG, Juguilon H, Montminy M, and Evans RM (1996): Role of CBP/P300 in nuclear receptor signalling. *Nature* 383: 99–103.

Chalbos D, and Galtier F (1994): Differential effect of forms A and B of human progesterone receptor on estradiol-dependent transcription. *J. Biol. Chem.* 269: 23007–23012.

Chalepakis G, Schauer M, Cao X, and Beato M (1990): Efficient binding of glucocorticoid receptor to its responsive element requires a dimer and a DNA flanking sequence. *DNA Cell Biol.* 9: 355–368.

Chambraud B, Berry M, Redeuilh G, Chambon P, and Baulieu E-E (1990): Several regions of human estrogen receptor are involved in the formation of receptor-heat shock protein 90 complexes. *J. Biol. Chem.* 265: 20686–20691.

Chang C, Whelan CT, Popovich TC, Kokontis J, and Liao S (1989): Fusion proteins containing androgen receptor sequences and their use in the production of poly- and monoclonal anti-androgen receptor antibodies. *Endocrinology* 123: 1097–1099.

Chatterjee S, and Struhl K (1995): Connecting a promoter-bound protein to TBP bypasses the need for a transcriptional activation domain. *Nature* 374: 820–822.

Chauchereau A, Cohen-Solal K, Jolivet A, Bailly A, and Milgrom E (1994): Phosophorylation sites in ligand-induced and ligand-independent activation of the progesterone receptor. *Biochemistry* 33: 13295–13303.

Chen J-Y, Clifford J, Zusi C, Starrett J, Tortolani D, Ostrowski J, Reczek PR, Chambon P, and Gronemeyer H (1996a): Two distinct actions of retinoid-receptor ligands. *Nature* 382: 819–822.

Chen JD, and Evans RM (1995): A transcriptional co-repressor that interacts with nuclear hormone receptors. *Nature* 377: 454–457.

Chen M-S, Silverstein AM, Pratt WB, and Chinkers M (1996b): The tetratricopeptide repeat domain of protein phosphatase 5 mediates binding to glucocorticoid receptor heterocomplexes and acts as a dominant negative mutant. *J. Biol. Chem.* 271: 32315–32320.

Cheng L, Norris AW, Tate BF, Rosenberger M, Grippo JF, and Li E (1994): Characterization of the ligand binding domain of human retinoid X receptor alpha expressed in Escherichia coli. *J. Biol. Chem.* 269: 18662–18667.

Cheng X, Reginato MJ, Andrews NC, and Lazar MA (1997): The transcriptional integrator CREB-binding protein mediates positive cross talk between nuclear hormone receptors and the hematopoietic bZip protein p45/NF-E2. *Mol. Cell. Biol.* 17: 1407–1416.

Cheskis B, and Freedman LP (1994): Ligand modulates the conversion of DNA-bound vitamin D3 receptor (VDR) homodimers into VDR-retinoid X receptor heterodimers. *Mol. Cell. Biol.* 14: 3329–3338.

Chou TY, Dang CV, and Hart GW (1995): Glycosylation of the c-Myc transactivation domain. *Proc Natl Acad Sci U S A* 92: 4417–4421.

Coffer A, Cavailles V, Knowles P, and Pappin D (1996): Biochemical characterization and novel isolation of pure estrogen receptor hormone-binding domain. *J. Steroid Biochem. Mol. Biol.* 58: 467–477.

Conneely OM, Kettelberger DM, Tsai M-J, Schrader WT, and O'Malley BW (1989): The chicken progesterone receptor A and B isoforms are products of an alternate translation initiation event. *J. Biol. Chem.* 264: 14062–14064.

Cooper B, Gruber JA, and McPhaul MJ (1996): Hormone-binding and solubility properties of fusion proteins containing the ligand-binding domain of the human androgen receptor. *J. Steroid Biochem. Molec. Biol.* 57: 251–257.

Dallery N, Sablonniere B, Grillier I, Formstecher P, and Dautrevaux M (1993): Purification and functional characterization of the ligand-binding domain from the retinoic acid receptor alpha: Evidence that sulfhydryl groups are involved in ligand-receptor interactions. *Biochemistry* 32: 12428–12436.

Dalman FC, Koenig RJ, Perdew GH, Massa E, and Pratt WB (1990): In contrast to the glucocorticoid receptor, the thyroid hormone receptor is translated in the DNA binding state and is not associated with hsp90. *J. Biol. Chem.* 265: 3615–3618.

Dalman FC, Scherrer LC, Taylor LP, Akil H, and Pratt WB (1991a): Localization of the 90-kDa heat shock protein-binding site within the hormone-binding domain of the glucocorticoid receptor by peptide competition. *J. Biol. Chem.* 266: 3482–3490.

Dalman FC, Sturzenbecker LJ, Levin AA, Lucas DA, Perdew GH, Petkovitch M, Chambon P, Grippo JF, and Pratt WB (1991b): Retinoic acid receptor belongs to a subclass of nuclear receptors that do not form "docking" complexes with hsp90. *Biochemistry* 30: 5605–5608.

Damm K, Heyman RA, Umesono K, and Evans RM (1993): Functional inhibition of retinoic acid response by dominant negative retinoic acid receptor mutants. *Proc. Natl. Acad. Sci. USA* 90: 2989–2993.

Danielian PS, White R, Lees JA, and Parker MG (1992): Identification of a conserved region required for hormone dependent transcriptional activation by steroid hormone receptors. *EMBO J.* 11: 1025–1033.

Darwish HM, Burmester JK, Moss VE, and DeLuca HF (1993): Phosphorylation is involved in transcriptional activation by the 1,25-dihydroxyvitamin D3 receptor. *Biochimica Biophysica Acta* 1167: 29–36.

DeFranco DB, Qi M, Borror KC, Garabedian MJ, and Brautigan DL (1991): Protein phosphatase types 1 and/or 2A regulate nucleocytoplasmic shuttling of glucocorticoid receptors. *Mol. Endocrinol.* 5: 1215–1228.

Defranco DB, Madan AP, Tang Y, Chandran UR, Xiao N, and Yang J (1995): Nucleocytoplasmic shuttling of steroid receptors. *Vitam. Horm.* 51: 315–338.

DeMarzo AM, Beck CA, Onate SA, and Edwards DP (1991): Dimerization of mammalian progesterone receptors occurs in the absence of DNA and is related to the release of the 90-kDa heat shock protein. *Proc. Natl. Acad. Sci. USA* 88: 72–76.

Dittmar KD, Hutchison KA, Owens-Grillo JK, and Pratt WB (1996): Reconstitution of the steroid receptor hsp90 heterocomplex assembly system of rabbit reticulocyte lysate. *J. Biol. Chem.* 271: 12833–12839.

Doesburg P, Kuil CW, Berrevoets CA, Steketee K, Faber PW, Mulder E, Brinkmann AO, and Trapman J (1997): Functional in vivo interaction between the amino-terminal, transactivation domain and the ligand binding domain of the androgen receptor. *Biochemistry* 36: 1052–1064.

Drewes T, Clairmont A, Klein-Hitpass L, and Ryffel GU (1994): Estrogen-inducible derivatives of hepatocyte nuclear factor-4, hepatocyte nuclear factor-3 and liver factor B1 are differently affected by pure and partial antiestrogens. *Eur. J. Biochem.* 225: 441–448.

Driscoll JE, Seachord CL, Lupisella JA, Darveau RP, and Reczek PR (1996): Ligand-induced conformational changes in the human retinoic acid receptor beta detected using monoclonal antibodies. *J. Biol. Chem.* 271: 22969–22975.

Duina AA, Chang H-CJ, Marsh JA, Lindquist S, and Gaber RF (1996): A cyclophilin function in hsp90-dependent signal transduction. *Science* 274: 1713–1715.

Durand B, Saunders M, Gaudon C, Roy B, Losson R, and Chambon P (1994): Activation function 2 (AF-2) of retinoic acid receptor and 9-cis retinoic acid receptor: presence of a conserved autonomous constitutive activating domain and influence of the nature of the response element on AF-2 activity. *EMBO J.* 13: 5370–5382.

Ekena K, Weis KE, Katzenellenbogen JA, and Katzenellenbogen BS (1996): Identification of amino acids in the hormone binding domain of the human estrogen receptor important in estrogen binding. *J. Biol. Chem.* 271: 20053–20059.

Ekena K, Weis KE, Katzenellenbogen JA, and Katzenellenbogen BS (1997): Different residues of the human estrogen receptor are involved in the recognition of structurally diverse estrogens and antiestrogens. *J. Biol. Chem.* 272: 5069–5075.

Elbrecht A, Chen Y, Cullinan CA, Hayes N, Leibowitz MD, Moller DE, and Berger J (1996): Molecular cloning, expression and characterization of human peroxisome proliferator activated receptors gamma 1 and gamma 2. *Biochem Biophys Res Commun* 224: 431–437.

Encio IJ, and Detera-Wadleigh SD (1991): The genomic structure of the human glucocorticoid receptor. *J. Biol. Chem.* 266: 7182–7188.

Enmark E, and Gustafsson J-A (1996): Orphan nuclear receptors—the first eight years. *Mol. Endo.* 10: 1293–1307.

Estes PA, Suba EJ, Lawler-Heavner J, Elashry-Stowers D, Wei LL, Toft DO, Sullivan WP, Horwitz KB, and Edwards DP (1987): Immunologic analysis of human breast cancer progesterone receptors. 1. Immunoaffinity purification of transformed receptors and production of monoclonal antibodies. *Biochemistry* 26: 6250–6262.

Eul J, Meyer ME, Tora L, Bocquel MT, Quirin-Stricker C, Chambon P, and Gronemeyer H (1989): Expression of active hormone and DNA-binding domains of the chicken progesterone receptor in E. coli. *EMBO J.* 8: 83–90.

Evans RM (1988): The steroid and thyroid hormone receptor superfamily. *Science* 240: 889–895.

Fabris D, Zaia J, Hathout Y, and Fenselau C (1996): Retention of thiol protons in two classes of protein zinc ion coordination centers. *J. Am. Chem. Soc.* 118: 12242–12243.

Fang Y, Fliss AE, Robins DM, and Caplan AJ (1996): Hsp90 regulates androgen receptor hormone binding affinity in vivo. *J. Biol. Chem.* 271: 28697–28702.

Fankhauser CP, Briand P-A, Picard D (1994): The hormone binding domain of the mineralocorticoid receptor can regulate heterologous activities in cis. *Biochem. Biophys. Res. Comm.* 200: 195–201.

Fawell SE, Lees JA, White R, and Parker MG (1990): Characterization and colocalization of steroid binding and dimerization activities in the mouse estrogen receptor. *Cell* 60: 953–962.

Fields S, and Song O.-k. (1989): A novel genetic system to detect protein-protein interactions. *Nature* 340: 245–246.

Flynn TR, Hollenberg AN, Cohen O, Menke JB, Usala SJ, Tollin S, Hegarty MK, and Wondisford FE (1994): A novel C-terminal domain in the thyroid hormone receptor selectively mediates thyroid hormone inhibition. *J. Biol. Chem.* 269: 32713–32716.

Folkers GE, and van der Saag PT (1995): Adenovirus E1A functions as a cofactor for retinoic acid receptor beta (RARbeta) through direct interaction with RARbeta. *Mol. Cell. Biol.* 15: 5868–5878.

Folkers GE, van Heerde EC, and van der Saag PT (1995): Activation function 1 of retinoic acid receptor beta2 is an acidic activator resembling VP16. *J. Biol. Chem.* 270: 23552–23559.

Fondell JD, Brunel F, Hisatake K, and Roeder RG (1996): Unliganded thyroid hormone receptor can target TATA-binding protein for transcriptional repression. *Mol. Cell. Biol.* 16: 281–287.

Forman BM, Yang C, Au M, Casanova J, Ghysdael J, and Samuels HH (1989): A domain containing leucine-zipper-like motifs mediates novel in vivo interactions between the thyroid hormone and retinoic acid receptors. *Mol. Endo.* 3: 1610–1626.

Forman BM, and Samuels HH (1990): Interactions among a subfamily of nuclear hormone receptors: The regulatory zipper model. *Mol. Endocrinology* 4: 1293–1301.

Forman BM, Goode E, Chen J, Oro AE, Bradley DJ, Perlmann T, Noonan DJ, Burka LT, McMorris T, Lamph WW, Evans RM, and Weinberger C (1995a): Identification of a nuclear receptor that is activated by farnesol metabolites. *Cell* 81: 687–693.

Forman BM, Umesono K, Chen J, and Evans RM (1995b): Unique response pathways are established by allosteric interactions among nuclear hormone receptors. *Cell* 81: 541–550.

Frankel AD, and Kim PS (1991): Modular structure of transcription factors: Implications for gene regulation. *Cell* 65: 717–719.

Fraser RA, Rossignol M, Heard DJ, Egly J-M, and Chambon P (1997): SUG1, a putative transcriptional mediator and subunit of the PA700 proteasome regulatory complex, is a DNA helicase. *J. Biol. Chem.* 272: 7122–7126.

Freedman LP, Luisi BF, Korszun ZR, Basavapa R, Sigler PB, and Yamamoto KR (1988): The function and structure of the metal coordination sites within the glucocorticoid receptor DNA binding domain. *Nature* 334: 543–546.

Fritsch M, Leary CM, Furlow JD, Ahrens H, Schuh TJ, Mueller GC, and Gorski JA (1992a): Ligand-induced conformational change in the estrogen receptor is localized in the steroid binding domain. *Biochemistry* 31: 5303–5311.

Fritsch M, Welch RD, Murdoch FE, Anderson I, and Gorski J (1992b): DNA allosterically modulates the steroid binding domain of the estrogen receptor. *J. Biol. Chem.* 267: 1823–1828.

Fritsch M, Anderson I, and Gorski J (1993): Structural characterization of the trypsinized estrogen receptor. *Biochemistry* 32: 14000–14008.

Fujimoto N, and Katzenellenbogen BS (1994): Alteration in the agonist/antagonist balance of antiestrogens by activation of protein kinase A signaling pathways in breast cancer cells: antiestrogen selectivity and promoter dependence. *Mol. Endo.* 8: 296–304.

Fuqua SAW, Fitzgerald SD, Allred DC, Elledge RM, Nawaz Z, McDonnell DP, O'Malley BW, Greene GL, and McGuire WL (1992): Inhibition of estrogen receptor action by a naturally occurring variant in human breast tumors. *Cancer Res.* 52: 483–486.

Gaitan D, DeBold CR, Turney MK, Zhou P, Orth DN, and Kovacs WJ (1995): Glucocorticoid receptor structure and function in an adrenocorticotropin-secreting small cell lung cancer. *Mol. Endo.* 9: 1193–1201.

Garabedian MJ, and Yamamoto KR (1992): Genetic dissection of the signaling domain of a mammalian steroid receptor in yeast. *Mol. Biol. Cell* 3: 1245–1257.

Giambiagi N, and Pasqualini JR (1990): Interaction of three monoclonal antibodies with the nonactivated and activated forms of the estrogen receptor. *Endocrinology* 126: 1403–1409.

Giguere V, Ong ES, Segui P, and Evans RM (1987): Identification of a receptor for the morphogen retinoic acid. *Nature* 330: 624–629.

Green S, and Chambon P (1987): Oestradiol induction of a glucocorticoid-responsive gene by a chimaeric receptor. *Nature* 325: 75–78.

Greene GL, Sobel NB, King WJ, and Jensen EV (1984): Immunochemical studies of estrogen receptors. *J. Steroid Biochem.* 20: 51–56.

Guido EC, Delorme EO, Clemm DL, Stein RB, Rosen J, and Miner JN (1996):

Determinants of promoter-specific activity by glucocorticoid receptor. *Mol. Endo.* 10: 1178–1190.

Guiochon-Mantel A, Loosfelt H, Lescop P, Sar S, Atger M, Perrot-Applanat M, and Milgrom E (1989): Mechanisms of nuclear localization of the progesterone receptor: Evidence for interaction between monomers. *Cell* 57: 1147–1154.

Guiochon-Mantel A, Lescop P, Christin-Martre S, Loosfelt H, Perrot-Applanat M, and Milgrom E (1991): Nucleocytoplasmic shuttling of the progesterone receptor. *EMBO J.* 10: 3851–3859.

Halachmi S, Marden E, Martin G, MacKay H, Abbondanza C, and Brown M (1994): Estrogen receptor-associated proteins: Possible mediators of hormone-induced transcription. *Science* 264: 1455–1458.

Haltiwanger RS, Busby S, Grove K, Li S, Mason D, Medina L, Moloney D, Philipsberg G, and Scartozzi R (1997): O-Glycosylation of nuclear and cytoplasmic proteins: Regulation analogous to phosphorylation? *Biochem. Biophys. Res. Commun.* 231: 237–242.

Hanstein B, Eckner R, DiRenzo J, Halachmi S, Liu H, Searcy B, Kurokawa R, and Brown M (1996): p300 is a component of an estrogen receptor coactivator complex. *Proc. Natl. Acad. Sci. USA* 93: 11540–11545.

Harlow KW, Smith DN, Katzenellenbogen JA, Greene GL, and Katzenellenbogen BS (1989): Identification of cysteine 530 as the covalent attachment site of an affinity-labeling estrogen (ketononestrol aziridine) and antiestrogen (tamoxifen aziridine) in the human estrogen receptor. *J. Biol. Chem.* 264: 17476–17485.

Hegy GB, Shackleton CHL, Carlquist M, Bonn T, Engstrom O, Sjoholm P, and Witkowska HE (1996): Carboxymethylation of the human estrogen receptor ligand-binding domain-estradiol complex: HPLC/ESMS peptide mapping shows that cysteine 447 does not react with iodoacetic acid. *Steroids* 61: 367–373.

Heinzel T, Lavinsky RM, Mullen T-M, Söderström M, Laherty CD, Torchia J, Yang W-M, Brard G, Ngo SD, Davie JR, Seto E, Eisenman RN, Rose DW, Glass CK, and Rosenfeld MG (1997): A complex containing N-CoR, mSin3 and histone deacetylase mediates transcriptional repression. *Nature* 387: 43–48.

Henttu PMA, Kalkhoven E, and Parker MG (1997): AF-2 activity and recruitment of the coactivator SRC-1 to the estrogen receptor depend on a lysine residue conserved in nuclear receptors. *Mol. Cell. Biol.* in press.

Hilliard GMIV, Cook RG, Weigel NL, and Pike JW (1994): 1,25-dihydroxyvitamin D3 modulates phosphorylation of serine 205 in the human vitamin D receptor: Site-directed mutagenesis of this residue promotes alternative phosphorylation. *Biochemistry* 33: 4300–4311.

Hollenberg AN, Monden T, Madura JP, Lee K, and Wondisford FE (1996): Function of nuclear co-repressor protein on thyroid hormone response elements is regulated by the receptor A/B domain. *J. Biol. Chem* 271: 28516–28520.

Hollenberg SM, Weinberger C, Ong ES, Cerelli G, Oro A, Lebo R, Thompson EB, Rosenfeld MG, and Evans RM (1985): Primary structure and expression of a functional human glucocorticoid receptor cDNA. *Nature* 318: 635–641.

Holloway JM, Glass CK, Adler S, Nelson CA, and Rosenfeld MG (1990): The C'-terminal interaction domain of the thyroid hormone receptor confers the ability of the DNA site to dictate positive or negative transcriptional activity. *Proc. Natl. Acad. Sci.* 87: 8160–8164.

Hong H, Kohli K, Trivedi A, Johnson DL, and Stallcup MR (1996): GRIP1, a novel mouse protein that serves as a transcriptional coactivator in yeast for the hormone binding domains of steroid receptors. *Proc. Natl. Acad. Sci. USA* 93: 4948–4952.

Horlein AJ, Naar AM, Heinzel T, Torchia J, Gloss B, Kurokawa R, Ryan A, Kamei Y, Soderstrom M, Glass CK, and Rosenfeld MG (1995): Ligand-independent repression by the thyroid hormone receptor mediated by a nuclear receptor co-repressor. *Nature* 377: 397–404.

Horwitz KB, Jackson TA, Bain DL, Richer JK, Takimoto GS, and Tung L (1996): Nuclear receptor coactivators and corepressors. *Mol. Endo.* 10: 1167–1177.

Hu J-M, Bodwell JE, and Munck A (1997): Control by basal phosphorylation of cell cycle-dependent, hormone-induced glucocorticoid receptor hyperphosphorylation. *Mol. Endocrinol.* 11: 305–311.

Hunter T, and Karin M (1992): The regulation of transcription by phosphorylation. *Cell* 70: 375–387.

Ince BA, Zhuang Y, Wrenn CK, Shapiro DJ, and Katzenellenbogen BS (1993): Powerful dominant negative mutants of the human estrogen receptor. *J. Biol. Chem.* 268: 14026–14032.

Ing NH, Beekman JM, Tsai SY, Tsai MJ, and O'Malley BW (1992): Members of the steroid hormone receptor superfamily interact with TFIIB (S300-II). *J. Biol. Chem.* 267: 17617–17623.

Israel DI, and Kaufman RJ (1993): Dexamethasone negatively regulates the activity of a chimeric dihydrofolate reductase/glucocorticoid receptor protein. *Proc. Natl. Acad. Sci. USA* 90: 4290–4294.

Issemann I, and Green S (1990): Activation of a member of the steroid hormone receptor superfamily by peroxisome proliferators. *Nature* 347: 645–650.

Jacq X, Brou C, Lutz Y, Davidson I, Chambon P, and Tora L (1994): Human TAF 30 is present in a distinct TFIID complex and is required for transcriptional activation by the estrogen receptor. *Cell* 79: 107–117.

Jalaguier S, Mesnier D, Leger JJ, and Auzou G (1996): Putative steroid binding domain of the human mineralocorticoid receptor, expressed in E. coli in the presence of heat shock proteins, shows typical native receptor characteristics. *J. Steroid Biochem. Molec. Biol.* 57: 43–50.

Jiang M-S, and Hart GW (1997): A subpopulation of estrogen receptors are modified by O-linked N-acetylglucosamine. *J. Biol. Chem.* 272: 2421–2428.

Jin CH, Kerner SA, Hong MH, and Pike JW (1996): Transcriptional activation and dimerization functions in the human vitamin D receptor. *Mol. Endo.* 10: 945–957.

Jones BB, Ohno CK, Allenby G, Boffa MB, Levin AA, Grippo JF, and Petkovich M (1995): New retinoid X receptor subtypes in zebra fish (Danio rerio) differentially modulate transcription and do not bind 9-cis retinoic acid. *Mol. Cell. Biol.* 15: 5226–5234.

Joyeux A, Cavaillè s V, Balaguer P, and Nicolas JC (1997): RIP 140 enhances nuclear receptor-dependent transcription in yeast. *Mol. Endo.* 11: 193–202.

Jurutka PW, Hsieh J-C, Nakajima S, Haussler CA, Whitfield K, and Haussler MR (1996): Human vitamin D receptor phosphorylation by casein kinase II at Ser-208 potentiates transcriptional activation. *Proc. Natl. Acad. Sci. USA* 93: 3519–3524.

Kamei Y, Xu L, Heinzel T, Torchia J, Kurokawa R, Gloss B, Lin S-C, Heyman RA,

Rose DW, Glass CK, and Rosenfeld MG (1996): A CBP integrator complex mediates transcriptional activation and AP-1 inhibition by nuclear receptors. *Cell* 85,: 403–414.

Kastner P, Krust A, Turcotte B, Stropp U, Tora L, Gronemeyer H, and Chambon P (1990): Two distinct estrogen-regulated promoters generate transcripts encoding the two functionally different human progesterone receptor forms A and B. *EMBO J.* 9: 1603–1614.

Kearse KP, and Hart GW (1991): Lymphocyte activation induces rapid changes in nuclear and cytoplasmic glycoproteins. *Proc Natl Acad Sci U S A* 88: 1701–1705.

Keidel S, LeMotte P, and Apfel C (1994): Different agonist- and antagonist-induced conformational changes in retinoic acid receptors analyzed by protease mapping. *Mol. Cell. Biol.* 14: 287–298.

Kersten S, Kelleher D, Chambon P, Gronemeyer H, and Noy N (1995a): Retinoid X receptor alpha forms tetramers in solution. *Proc. Natl. Acad. Sci. USA* 92: 8645–8649.

Kersten S, Pan L, Chambon P, Gronemeyer H, and Noy N (1995b): Role of ligand in retinoid signaling. 9-cis-retinoic acid modulates the oligomeric state of the retinoid X receptor. *Biochemistry* 34: 13717–13721.

Kersten S, Dawson MI, Lewis BA, and Noy N (1996): Individual subunits of heterodimers comprised of retinoic acid and retinoid X receptors interact with their ligands independently. *Biochemistry* 35: 3816–3824.

Kimura Y, Yahara I, and Lindquist S (1995): Role of the protein chaperone YDJ1 in establishing hsp90-mediated signal transduction pathways. *Science* 268: 1362–1365.

Klages N, and Strubin M (1995): Stimulation of RNA polymerase II transcription initiation by recruitment of TBP in vivo. *Nature* 374: 822–823.

Kliewer SA, Umesono K, Mangelsdorf DJ, and Evans RM (1992a): Retinoid X receptor interacts with nuclear receptors in retinoic acid, thyroid hormone and vitamin D3 signalling. *Nature* 355: 446–449.

Kliewer SA, Umesono K, Noonan DJ, Heyman RA, and Evans RM (1992b): Convergence of 9-cis retinoic acid and peroxisome proliferator signalling pathways through heterodimer formation of their receptors. *Nature* 358: 771–774.

Koelle MR, Talbot WS, Segraves WA, Bender MT, Cherbas P, and Hogness DS (1991): The Drosophila EcR gene encodes an ecdysone receptor, a new member of the steroid receptor superfamily. *Cell* 67: 59–77.

Koenig RJ, Warne RL, Brent GA, Harney JW, Larsen PR, and Moore DD (1988): Isolation of a cDNA clone encoding a biologically active thyroid hormone receptor. *Proc. Natl. Acad. Sci. USA* 85: 5031–5035.

Koenig RJ, Lazar MA, Hodin RA, Brent GA, Larsen PR, Chin WW, and Moore DD (1989): Inhibition of thyroid hormone action by a non-hormone binding c-erbA protein generated by alternative mRNA splicing. *Nature* 337: 659–661.

Kraus WL, Montano MM, and Katzenellenbogen BS (1994): Identification of multiple, widely spaced estrogen-responsive regions in the rat progesterone receptor gene. *Mol. Endo.* 8: 952–969.

Kraus WL, McInerney EM, and Katzenellenbogen BS (1995): Ligand-dependent, transcriptionally productive association of the amino- and carboxyl-terminal regions of a steroid hormone nuclear receptor. *Proc. Natl. Acad. Sci. USA* 92: 12314–12318.

Krust A, Kastner P, Petkovich M, Zelent A, and Chambon P (1989): A third human retinoic acid receptor, hRAR-gamma. *Proc. Natl. Acad. Sci. USA* 86: 5310–5314.

Kuil CW, Berrevoets CA, and Mulder E (1995): Ligand-induced conformational alterations of the androgen receptor analyzed by limited trypsinization. *J. Biol. Chem.* 270: 27569–27576.

Kuiper GGJM, Enmark E, Pelto-Huikko M, Nilsson S, and Gustafsson J-A (1996): Cloning of a novel estrogen receptor expressed in rat prostate and ovary. *Proc. Natl. Acad. Sci. USA* 93: 5925–5930.

Kuiper GGJM, Carlsson B, Grandien K, Enmark E, Häggblad J, Nilsson S, and Gustafsson J-A (1997): Comparison of the ligand binding specificity and transcript tissue distribution of estrogen receptors a and b. *Endocrinology* 138: 863–870.

Kurokawa R, DiRenzo J, Boehm M, Sugarman J, Gloss B, Rosenfeld MG, Heyman RA, and Glass CK (1994): Regulation of retinoid signalling by receptor polarity and allosteric control of ligand binding. *Nature* 371: 528–531.

Kurokawa R, Soderstrom M, Horlein A, Halachmi S, Brown M, Rosenfeld M, and Glass C (1995): Polarity-specific activities of retinoic acid receptors determined by a co-repressor. *Nature* 377: 451–454.

L'Horset F, Dauvois S, Heery DM, Cavailles V, and Parker MG (1996): RIP-140 interacts with multiple nuclear receptors by means of two distinct sites. *Mol. Cell. Biol.* 16: 6029–6036.

Lala DS, Mukherjee R, Schulman IG, Koch SS, Dardashti LJ, Nadzan AM, Croston GE, Evans RM, and Heyman RA (1996): Activation of specific RXR heterodimers by an antagonist of RXR homodimers. *Nature* 383: 450–453.

Lamour FPY, Lardelli P, and Apfel CM (1996): Analysis of the ligand-binding domain of human retinoic acid receptor and by site-directed mutagenesis. *Mol. Cell. Biol.* 16: 5386–5392.

Langley E, Zhou Z-x, and Wilson EM (1995): Evidence for an anti-parallel orientation of the ligand-activated human androgen receptor dimer. *J. Biol. Chem.* 270: 29983–29990.

Lanz RB, Hug M, Gola M, Tallone T, Wieland S, and Rusconi S (1994): Active, interactive, and inactive steroid receptor mutants. *Steroids* 59: 148–152.

Lanz RB, and Rusconi S (1994): A conserved carboxy-terminal subdomain is important for ligand interpretation and transactivation by nuclear receptors. *Endocrinology* 135: 2183–2194.

Laudet V, Hanni C, Coll J, Catzeflis F, and Stehelin D (1992): Evolution of the nuclear receptor gene superfamily. *EMBO J.* 11: 1003–1013.

Lazar MA (1993): Thyroid hormone receptors: multiple forms, multiple possibilities. *Endocr Rev* 14: 184–193.

Le Douarin B, Zechel C, Garnier J-M, Lutz Y, Tora L, Pierrat B, Heery D, Gronemeyer H, Chambon P, and Losson R (1995): The N-terminal part of TIF1, a putative mediator of the ligand-dependent activation function (AF-2) of nuclear receptors, is fused to B-raf in the oncogenic protein T18. *EMBO J.* 14: 2020–2033.

Lee HS, Aumais J, and White JH (1996): Hormone-dependent transactivation by estrogen receptor chimeras that do not interact with hsp90. *J. Biol. Chem* 271: 25727–25730.

Lee JW, Moore DD, and Heyman RA (1994): A chimeric thyroid hormone receptor constitutively bound to DNA requires retinoid X receptor for hormone-dependent transcriptional activation in yeast. *Mol. Endo.* 8: 1245–1252.

Lee JW, Choi H-S, Gyuris J, Brent R, and Moore DD (1995a): Two classes of proteins dependent on either the presence or absence of thyroid hormone for interaction with the thyroid hormone receptor. *Mol. Endocrinol.* 9: 243–254.

Lee JW, Ryan F, Swaffield JC, Johnston SA, and Moore DD (1995b): Interaction of thyroid-hormone receptor with a conserved transcriptional mediator. *Nature* 374: 91–94.

Lees JA, Fawell SE, White R, and Parker MG (1990): A 22-amino-acid peptide restores DNA-binding activity to dimerization-defective mutants of the estrogen receptor. *Mol Cell. Biol.* 10: 5529–5531.

Leid M (1994): Ligand-induced alteration of the protease sensitivity of retinoid X receptor alpha. *J. Biol. Chem.* 269: 14175–14181.

Leid M, Kastner P, Lyons R, Nakshatri H, Saunders M, Zacharewski T, Chen J-Y, Staub A, Garnier J-M, Mader S, and Chambon P (1995): Purification, cloning, and RXR identity of the HeLa cell factor with which RAR or TR heterodimerizes to bind target sequences efficiently. *Cell* 68: 377–395.

Leng X, Tsai SY, O'Malley BW, and Tsai M-J (1993): Ligand-dependent conformational changes in thyroid hormone and retinoic acid receptors are potentially enhanced by heterodimerization with retinoic X receptor. *J. Steroid Biochem. Molec. Biol.* 46: 643–661.

Leng X, Blanco J, Tsai SY, Ozato K, O'Malley BW, and Tsai M-J (1995): Mouse retinoid X receptor contains a separable ligand-binding and transactivation domain in its E region. *Mol. Cell. Biol.* 15: 255–263.

Leng X, Cooney AJ, Tsai SY, and Tsai M-J (1996): Molecular mechanisms of COUP-TF-mediated transcriptional repression: evidence for transrepression and active repression. *Mol. Cell. Biol.* 16: 2332–2340.

Levine M, and Manley JL (1989): Transcriptional repression of eukaryotic promoters. *Cell* 59: 405–408.

Lin K-H, Fukuda T, and Cheng S-y (1990): Hormone and DNA binding activity of a purified human thyroid hormone nuclear receptor expressed in Escherichia coli. *J. Biol. Chem.* 265: 5161–5165.

Lin K-H, Parkinson C, McPhie P, and Cheng S-Y (1991): An essential role of domain D in the hormone-binding activity of human beta 1 thyroid hormone nuclear receptor. *Mol. Endocrinology* 5: 485–492.

Littlewood TD, Hancock DC, Danielian PS, Parker MG, and Evan GI (1995): A modified oestrogen receptor ligand-binding domain as an improved switch for the regulation of heterologous proteins. *Nucleic Acids Res.* 23: 1686–1690.

Liu W, Wang J, Sauter NK, and Pearce D (1995): Steroid receptor heterodimerization demonstrated in vitro and in vivo. *Proc. Natl. Acad. Sci. USA* 92: 12480–12484.

Logie C, and Stewart F (1995): Ligand-regulated site-specific recombination. *Proc. Natl. Acad. Sci. USA* 92: 5940–5944.

Lopez S, Miyashita Y, and Simons, Jr. SS (1990): Structurally based, selective interaction of arsenite with steroid receptors. *J. Biol. Chem.* 265: 16039–16042.

Lubahn DB, Joseph DR, Sar M, Tan J-A, Higgs HN, Larson RE, French FS, and Wilson EM (1988): The human androgen receptor: Complementary deoxyribonucleic acid cloning, sequence analysis and gene expression in prostate. *Mol. Endo.* 2: 1265–1275.

Lundblad JR, Kwok RP, Laurance ME, Harter ML, and Goodman RH (1995): Adenoviral E1A-associated protein p300 as a functional homologue of the transcriptional co-activator CBP. *Nature* 374: 85–88.

Lupisella JA, Driscoll JE, Metzler WJ, and Reczek PR (1995): The ligand binding domain of the human retinoic acid receptor gamma is predominantly alpha-helical with a Trp residue in the ligand binding site. *J. Biol. Chem.* 270: 24884–24890.

Lyttle CR, Damian-Matsumura P, Juul H, and Butt TR (1992): Human estrogen receptor regulation in a yeast model system and studies on receptor agonists and antagonists. *J. Steroid Biochem. Molec. Biol.* 42: 677–685.

MacDonald PN, Sherman DR, Dowd DR, Jefcoat SCJ, and DeLisle RK (1995): The vitamin D receptor interacts with general transcription factor IIB. *J. Biol. Chem.* 270: 4748–4752.

Madan AP, and DeFranco DB (1993): Bidirectional transport of glucocorticoid receptors across the nuclear envelope. *Proc. Natl. Acad. Sci. USA* 90: 3588–3592.

Mahfoudi A, Roulet E, Dauvois S, Parker MG, and Wahli W (1995): Specific mutations in the estrogen receptor change the properties of antiestrogens to full agonoists. *Proc. Natl. Acad. Sci. USA* 92: 4206–4210.

Maksymowych AB, Hsu T-C, and Litwack G (1992): A novel, highly conserved structural motif is present in all members of the steroid receptor superfamily. *Receptor* 2: 225–240.

Mangelsdorf DJ, and Evans RM (1995): The RXR heterodimers and orphan receptors. *Cell* 83: 841–850.

Marivoet S, Van Dijck P, Verhoeven G, and Heyns W (1992): Interaction of the 90-KDa heat shock protein with native and in vitro translated androgen receptor and receptor fragments. *Mol. Cell. Endo.* 88: 165–174.

Marks MS, Hallenbeck PL, Nagata T, Segars JH, Appella E, Nikodem VM, and Ozato K (1992): H-2RIIBP (RXRbeta) heterodimerization provides a mechanism for combinatorial diversity in the regulation of retinoic acid and thyroid hormone responsive genes. *EMBO J.* 11: 1419–1435.

Martin PM, Berthois Y, and Jensen EV (1988): Binding of antiestrogens exposes an occult antigenic determinant in the human estrogen receptor. *Proc. Natl. Acad. Sci. USA* 85: 2533–2537.

Martinez E, Moore DD, Keller E, Pearce D, Robinson V, MacDonald PN, Simons, Jr. SS, Sanchez E, and Danielsen M (1997): The nuclear receptor resource project. *Nucl. Acids Res.* 25: 163–165.

Matsui T, and Sashihara S (1995): Tissue-specific distribution of a novel C-terminal truncation retinoic acid receptor mutant which acts as a negative repressor in a promotor- and cell-type-specific manner. *Mol. Cell. Biol.* 15: 1961–1967.

May M, Mengus G, Lavigne A-C, Chambon P, and Davidson I (1996): Human TAF28 promotes transcriptional stimulation by activation function 2 of the retinoid X receptors. *EMBO J.* 15: 3093–3104.

McDonnell DP, Clemm DL, Hermann T, Goldman ME, and Pike JW (1995): Analysis of estrogen receptor function in vitro reveals three distinct classes of antiestrogens. *Mol. Endo.* 9: 659–669.

McInerney EM, Tsai M-J, O'Malley BW, and Katzenellenbogen BS (1996): Analysis of estrogen receptor transcriptional enhancement by a nuclear hormone receptor coactivator. *Proc. Natl. Acad. Sci. USA* 93: 10069–10073.

Mercier L, Miller PA, and Simons, Jr. SS (1986): Antiglucocorticoid steroids have increased agonist activity in those hepatoma cell lines that are more sensitive to glucocorticoids. *J. Steroid Biochem.* 25: 11–20.

Migliaccio A, Domenico MD, Green S, de Falco A, Kajtaniak EL, Blasi F, Chambon P, and Auricchio F (1989): Phosphorylation on tyrosine of in vitro synthesized human estrogen receptor activates its hormone binding. *Mol. Endo.* 3: 1061–1069.

Milhon J, Kohli K, and Stallcup MR (1994): Genetic analysis of the N-terminal end of the glucocorticoid receptor hormone binding domain. *J. Steroid Biochem. Molec. Biol.* 51: 11–19.

Miller NR, and Simons, Jr. SS (1988): Steroid binding to hepatoma tissue culture cell glucocorticoid receptors involves at least two sulfhydryl groups. *J. Biol. Chem.* 263: 15217–15225.

Minucci S, Leid M, Toyama R, Saint-Jeannet J-P, Peterson VJ, Horn V, Ishmael JE, Bhattacharyya N, Dey A, Dawid IB, and Ozato K (1997): Retinoid X receptor(RXR) within the RXR-retinoic acid receptor heterodimer binds its ligand and enhances retinoid-dependent gene expression. *Mol. Cell. Biol.* 17: 644–655.

Misrahi M, Venencie P-Y, Saugier-Veber P, Sar S, Dessen P, and Milgrom E (1993): Structure of the human progesterone receptor gene. *Biochim. Biophys. Acta* 1216: 289–292.

Miyashita Y, Miller M, Yen PM, Harmon JM, Hanover JA, and Simons, Jr. SS (1993): Glucocorticoid receptor binding to rat liver nuclei occurs without nuclear transport. *J. Steroid Biochem. Molec. Biol.* 46: 309–320.

Modarress KJ, Cavanaugh AH, Chakraborti PK, and Simons, Jr. SS (1994): Metal oxyanion stabilization of the rat glucocorticoid receptor is independent of thiols. *J. Biol. Chem.* 269: 25621–25628.

Modarress KJ, Opoku J, Xu M, Sarlis NJ, and Simons, Jr. SS (1997): Steroid-induced conformational changes at ends of the hormone binding domain in the rat glucocorticoid receptor are independent of agonist vs. antagonist activity. *J. Biol. Chem.* in press.

Montano MM, Muller V, Trobaugh A, and Katzenellenbogen BS (1995): The carboxy-terminal F domain of the human estrogen receptor: Role in the transcriptional activity of the receptor and the effectiveness of antiestrogens as estrogen antagonists. *Mol. Endo.* 9: 814–825.

Montano MM, Ekena K, Krueger KD, Keller AL, and Katzenellenbogen BS (1996): Human estrogen receptor ligand activity inversion mutants: Receptors that interpret antiestrogens as estrogens and estrogens as antiestrogens and discriminate among different antiestrogens. *Mol. Endo.* 10: 230–242.

Mosselman S, Polman J, and Dijkema R (1996): ER beta: identification and characterization of a novel human estrogen receptor. *FEBS Lett* 392: 49–53.

Munck A, and Holbrook NJ (1984): Glucocorticoid-receptor complexes in rat thymus cells. Rapid kinetic behavior and a cyclic model. *J. Biol. Chem.* 259: 820–831.

Murray MB, Zilz ND, McCreary NL, MacDonald MJ, and Towle HC (1988): Isolation and characterization of rat cDNA clones for two distinct thyroid hormone receptors. *J. Biol. Chem.* 263: 12770–12777.

Nagaya T, and Jameson JL (1993): Thyroid hormone receptor dimerization is required for dominant negative inhibition by mutations that cause thyroid hormone resistance. *J. Biol. Chem.* 268: 15766–15771.

Nagaya T, Nomura Y, Fujieda M, and Seo H (1996): Heterodimerization preferences of thyroid hormone receptor alpha isoforms. *Biochem. Biophys. Res. Commun.* 226: 426–430.

Nagpal S, Saunders M, Kastner P, Durand B, Nakshatri H, and Chambon P (1992): Promoter context- and response element-dependent specificity of the transcriptional activation and modulating functions of retinoic acid receptors. *Cell* 70: 1007–1019.

Nakajima S, Hsieh J-C, Jurutka PW, Galligan MA, Haussler CA, Whitfield GK, and Haussler MR (1996): Examination of the potential functional role of conserved cysteine residues in the hormone binding domain of the human 1,25-dihydroxyvitamin D3 receptor. *J. Biol. Chem.* 271: 5143–5149.

Neff S, Sadowski C, and Miksicek RJ (1994): Mutational analysis of cysteine residues within the hormone-binding domain of the human estrogen receptor identifies mutants that are defective in both DNA-binding and subcellular distribution. *Mol. Endo.* 8: 1215–1223.

Nemoto T, Ohara-Nemoto Y, Denis M, and Gustafsson J (1990): The transformed glucocorticoid receptor has a lower steroid-binding affinity than the nontransformed receptor. *Biochemistry* 29: 1880–1886.

Nemoto T, Ohara-Nemoto Y, and Ota M (1992): Association of the 90-kDa heat shock protein does not affect the ligand-binding ability of androgen receptor. *J. Steroid Biochem. Molec. Biol.* 42: 803–812.

Nemoto T, Ohara-Nemoto Y, Shimazaki S, and Ota M (1994): Dimerization characteristics of the DNA- and steroid-binding domains of the androgen receptor. *J. Steroid Biochem. Molec. Biol.* 50: 225–233.

Nichols M, Rientjes JMJ, Logie C, and Stewart AF (1997): FLP recombinase/estrogen receptor fusion proteins require the receptor D domain for responsiveness of antagonists, but not agonists. *Mol. Endocrinol.* 11: 950–961.

Nielsen CJ, Sando JJ, Vogel WM, and Pratt WB (1977): Glucocorticoid receptor inactivation under cell-free conditions. *J. Biol. Chem.* 252: 7568–7578.

Nishikawa J-I, Kitaura M, Imagawa M, and Nishihara T (1995): Vitamin D receptor contains multiple dimerization interfaces that are functionally different. *Nucleic Acids Res.* 23: 606–611.

Nomura Y, Nagaya T, Tsukaguchi H, Takamatsu J, and Seo H (1996): Amino acid substitutions of thyroid hormone receptor-beta codon 435 with resistance to thyroid hormone selectively alter homodimer formation. *Endocrinology* 137: 4082–4086.

Norris JD, Fan D, Kerner SA, and McDonell DP (1997): Identification of a third autonomous activation domain within the human estrogen receptor. *Mol. Endocrinol.* 11: 747–754.

Notides AC, Lerner N, and Hamilton DE (1981): Positive cooperativity of the estrogen receptor. *Proc. Natl. Acad. Sci. USA* 78: 4926–4930.

Oakley RH, Sar M, and Cidlowski JA (1996): The human glucocorticoid receptor beta isoform. *J. Biol. Chem.* 271: 9550–9559.

Ogryzko VV, Schiltz RL, Russanova V, Howard BH, and Nakatani Y (1996): The transcriptional coactivators p300 and CBP are histone acetyltransferases. *Cell* 87: 953–959.

Ojasoo T, Dore J-C, Gilbert J, and Raynaud J-P (1988): Binding of steroid to the progestin and glucocorticoid receptors analyzed by correspondence analysis. *J. Med. Chem.* 31: 1160–1169.

Onate SA, Tsai SY, Tsai M-J, and O'Malley BW (1995): Sequence and characterization of a coactivator for the steroid hormone receptor superfamily. *Science* 270: 1354–1357.

Opoku J, and Simons, Jr. SS (1994): Absence of intramolecular disulfides in the structure and function of native rat glucocorticoid receptors. *J. Biol. Chem.* 269: 503–510.

Orti E, Bodwell JE, and Munck A (1992): Phosphorylation of steroid hormone receptors. *Endocrine Reviews* 13: 105–128.

Oshima H, and Simons, Jr. SS (1992a): Modulation of glucocorticoid induction of tyrosine aminotransferase gene expression by variations in cell density. *Endocrinology* 130: 2106–2112.

Oshima H, and Simons, Jr. SS (1992b): Modulation of transcription factor activity by a distant steroid modulatory element. *Mol. Endocrinol.* 6: 416–428.

Ostrowski J, Hammer L, Roalsvig T, Pokornowski K, and Reczek PR (1995): The N-terminal portion of domain E of retinoic acid receptors alpha and beta is essential for the recognition of retinoic acid and various analogs. *Proc. Natl. Acad. Sci. USA* 92: 1812–1816.

Owens-Grillo JK, Hoffmann D, Hutchison KA, Yem AW, Deibel MRJ, Handschumacher RE, and Pratt WB (1995): The cyclosporin A-binding immumophilin CyP-40 and the FK506-binding immumophilin hsp56 bind to a common site on hsp90 and exist in independent cytosolic heterocomplexes with the untransformed glucocorticoid receptor. *J. Biol. Chem.* 270: 20479–20484.

Pakdel F, Reese JC, and Katzenellenbogen BS (1993): Identification of charged residues in N-terminal portion of the hormone-binding domain of the human estrogen receptor important in transcriptional activity of the receptor. *Mol. Endo.* 7: 1408–1417.

Peleg S, Sastry M, Collins ED, Bishop JE, and Norman AW (1995): Distinct conformational changes induced by 20-epi analogues of 1alpha, 25-dihydroxyvitamin D3 are associated with enhanced activation of the vitamin D receptor. *J. Biol. Chem.* 270: 10551–10558.

Perlmann T, Umesono K, Rangarajan PN, Forman BM, and Evans RM (1996): Two distinct dimerization interfaces differentially modulate target gene specificity of nuclear hormone receptors. *Mol. Endo.* 10: 958–966.

Petty KJ, Krimkevich YI, and Thomas D (1996): A TATA binding protein-associated factor functions as a coactivator for thyroid hormone receptors. *Mol. Endo.* 10: 1632–1645.

Picard D, and Yamamoto KR (1987): Two signals mediate hormone-dependent nuclear localization of the glucocorticoid receptor. *EMBO J.* 6: 3333–3340.

Picard D, Khursheed B, Garabedian MJ, Fortin MG, Lindquist S, and Yamamoto KR (1990a): Reduced levels of hsp90 compromise steroid receptor action in vivo. *Nature* 348: 166–168.

Picard D, Kumar V, Chambon P, and Yamamoto KR (1990b): Signal transduction by steroid hormones: Nuclear localization is differentially regulated in estrogen and glucocorticoid receptors. *Cell Regul* 1: 291–299.

Piedrafita FJ, Ortiz MA, and Pfahl M (1995): Thyroid hormone receptor-beta mutants associated with generalized resistance to thyroid hormone show defects in their ligand-sensitive repression function. *Mol. Endo.* 9: 1533–1548.

Pierrat B, Heery DM, Chambon P, and Losson R (1994): A highly conserved region in the hormone-binding domain of the human estrogen receptor functions as an efficient transactivation domain in yeast. *Gene* 143: 193–200.

Ponglikitmongkol M, Green S, and Chambon P (1988): Genomic organization of the human oestrogen receptor gene. *EMBO J.* 7: 3385–3388.

Power RF, Mani SK, Codina J, Conneely OM, and O'Malley BW (1991): Dopaminergic and ligand-independent activation of steroid hormone receptors. *Science* 254: 1636–1639.

Powers JH, Hillmann AG, Tang DC, and Harmon JM (1993): Cloning and expression of mutant glucocorticoid receptors from glucocorticoid-sensitive and resistant human leukemic cells. *Cancer Res.* 53: 4059–4065.

Pratt WB, Kaine JL, and Pratt DV (1975): The kinetics of glucocorticoid binding to the soluble specific binding protein of mouse fibroblasts. *J. Biol. Chem.* 250: 4584–4591.

Pratt WB (1993): The role of heat shock proteins in regulating the function, folding, and trafficking of the glucocorticoid receptor. *J. Biol. Chem.* 268: 21455–21458.

Pratt WB, Czar MJ, Stancato LF, and Owens JK (1993): The hsp56 immunophilin component of steroid receptor heterocomplexes: Could this be the elusive nuclear localization signal-binding protein? *J. Steroid Biochem. Molec. Biol.* 46: 269–279.

Pratt WB, and Toft DO (1997): Steroid receptor interactions with heat shock protein and immunophilin chaperones. *Endocrin. Rev.* in press.

Pratt WB (1990): Interaction of hsp90 with steroid receptors: Organizing some diverse observations and presenting the newest concepts. *Mol. Cell. Endocrinol.* 74: C69–76.

Privalsky ML, Sharif M, and Yamamoto KR (1990): The viral erbA oncogene protein, a constitutive repressor in animal cells, is a hormone-regulated activator in yeast. *Cell* 63: 1277–1286.

Ptashne M (1988): How eukaryotic transcriptional activators work. *Nature* 335: 683–689.

Quigley CA, De Bellis A, Marschke KB, el-Awady MK, Wilson EM, and French FS (1995): Androgen receptor defects: Historical, clinical, and molecular perspectives [published erratum appears in *Endocr Rev.* 1995 Aug;16(4):546]. *Endocr Rev* 16: 271–321.

Rachez C, Sautiere P, Formstecher P, and Lefebvre P (1996): Identification of amino acids critical for the DNA binding and dimerization properties of the human retinoic acid receptor alpha. Importance of lysine 360, lysine 365, and valine 361. *J. Biol. Chem.* 271: 17996–18006.

Ramamoorthy K, Wang F, Chen I-C, Safe S, Norris JD, McDonnell DP, Gaido KW, Bocchinfuso WP, and Korach KS (1997): Potency of combined estrogenic pesticides. [technical comment]. *Science* 275: 405.

Reese JC, Wooge CH, and Katzenellenbogen BS (1992): Identification of two cysteines closely positioned in the ligand-binding pocket of the human estrogen receptor: Roles in ligand binding and transcriptional activation. *Mol. Endo.* 6: 2160–2166.

Refetoff S, Weiss RE, and Ursala SJ (1993): The syndromes of resistance to thyroid hormone. *Endocr. Rev.* 14: 348–399.

Reichman ME, Foster CM, Eisen LP, Eisen HJ, Torain BF, and Simons, Jr. SS (1984):

Limited proteolysis of covalently labeled glucocorticoid receptors as a probe of receptor structure. *Biochemistry* 23: 5376–5384.

Renaud J-P, Rochel N, Ruff M, Vivat V, Chambon P, Gronemeyer H, and Moras D (1995): Crystal structure of the RAR-gramma ligand-binding domain bound to all-trans retinoic acid. *Nature* 378: 681–689.

Rochel N, Renaud J-P, Ruff M, Vivat V, Granger F, Bonnier D, Lerouge T, Chambon P, Gronemeyer H, and Moras D (1997): Purification of the human RARalpha ligand-binding domain and crystallization of its complex with all-trans retinoic acid. *Biochem. Biophys. Res. Commun.* 230: 293–296.

Rochette-Egly C, Oulad-Abdelghani M, Staub A, Pfister V, Scheuer I, Chambon P, and Gaub M-P (1995): Phosphorylation of the retinoic acid receptor-alpha by protein kinase A. *Mol. Endo.* 9: 860–871.

Rosen ED, Beninghof EG, and Koenig RJ (1993): Dimerization interfaces of thyroid hormone, retinoic acid, vitamin D, and retinoid X receptors. *J. Biol. Chem.* 268: 11534–11541.

Roux S, Terouanne B, Balaguer P, Jausons-Loffreda N, Pons M, Chambon P, Gronemeyer H, and Nicolas J-C (1996): Mutation of isoleucine 747 by a threonine alters the ligand responsiveness of the human glucocorticoid receptor. *Mol. Endo.* 10: 1214–1226.

Rowan BG, and Ip MM (1996): Differential binding of mutant glucocorticoid receptors to the glucocorticoid response element of the tyrosine aminotransferase gene. *J. Steroid Biochem. Molec. Biol.* 58: 147–162.

Ruh TS, and Baudendistel LJ (1977): Different nuclear binding sites for antiestrogen and estrogen receptor complexes. *Endocrinology* 100: 420–426.

Rusconi S, and Yamamoto KR (1987): Functional dissection of the hormone and DNA binding activities of the glucocorticoid receptor. *EMBO J.* 6: 1309–1315.

Sadovsky Y, Webb P, Lopez G, Baxter JD, Fitzpatrick PM, Gizang-Ginsberg E, Cavailles V, Parker MG, and Kushner PJ (1995): Transcriptional activators differ in their responses to overexpression of TATA-box-binding protein. *Mol. Cell. Biol.* 15: 1554–1563.

Salomonsson M, Haggblad J, O'Malley BW, and Sitbon GM (1994): The human estrogen receptor hormone binding domain dimerizes independently of ligand activation. *J. Steroid Biochem. Molec. Biol.* 48: 447–452.

Sande S, and Privalsky ML (1996): Identification of TRACs (T3 receptor-associating cofactors), a family of cofactors that associate with, and modulate the activity of, nuclear hormone receptors. *Mol. Endo.* 10: 813–825.

Scafonas A, Wolfgang CL, Gabriel JL, Soprano KJ, and Soprano DR (1997): Differential role of homologous positively charged amino acid residues for ligand binding in retinoic acid receptor α compared with retinoic acid receptor β. *J. Biol. Chem.* 272: 11244–11249.

Schena M, Freedman LP, and Yamamoto KR (1989): Mutations in the glucocorticoid receptor zinc finger region that distinguish interdigitated DNA binding and transcriptional enhancement activities. *Genes and Develop.* 3: 1590–1601.

Scherrer LC, Picard D, Massa E, Harmon JM, Simons, Jr. SS, Yamamoto KR, and Pratt WB (1993): Evidence that the hormone binding domain of steroid receptors confers hormonal control on chimeric proteins by determining their hormone-regulated binding to heat-shock protein 90. *Biochemistry* 32: 5381–5386.

Schmitt J, and Stunnenberg HG (1993): The glucocorticoid receptor hormone binding domain mediates transcriptional activation in vitro in the absence of ligand. *Nucleic Acids Res.* 21: 2673–2681.

Schowalter DB, Sullivan WP, Maihle NJ, Dobson ADW, Conneely OM, O'Malley BW, and Toft DO (1991): Characterization of progesterone receptor binding to the 90- and 70-kDa heat shock proteins. *J. Biol. Chem.* 266: 21165–21173.

Schulman G, Bodine PV, and Litwack G (1992): Modulators of the glucocorticoid receptor also regulate mineralocorticoid receptor function. *Biochemistry* 31: 1734–1741.

Schulman IG, Chakravarti D, Juguilon H, Romo A, and Evans RM (1995): Interactions between the retinoid X receptor and a conserved region of the TATA-binding protein mediate hormone-dependent transactivation. *Proc. Natl. Acad. Sci. USA* 92: 8288–8292.

Schulman IG, Juguilon H, and Evans RM (1996): Activation and repression by nuclear hormone receptors: Hormone modulates an equilibrium between active and repressive states. *Mol. Cell. Biol.* 16: 3807–3813.

Schulman IG, Li C, Schwabe JWR, and Evans RM (1997): The phantom ligand effect: Allosteric control of transcription by the retinoid X receptor. *Genes and Develop.* 11: 299–308.

Schwerk C, Klotzbucher M, Sachs M, Ulber V, and Klein-Hitpass L (1995): Identification of a transactivation function in the progesterone receptor that interacts with the TAFII110 subunit of the TFIID complex. *J. Biol. Chem.* 270: 21331–21338.

Segard-Maurel I, Rajkowski K, Jibard N, Schweizer-Groyer G, Baulieu E-E, and Cadepond F (1996): Glucocorticoid receptor dimerization investigated by analysis of receptor binding to glucocorticoid responsive elements using a monomer-dimer equilibrium model. *Biochemistry* 35: 1634–1642.

Seielstad DA, Carlson KE, Katzenellenbogen JA, Kushner PJ, and Greene GL (1995): Molecular characterization by mass spectrometry of the human estrogen receptor ligand-binding domain expressed in *Escherichia coli*. *Mol. Endo.* 9: 647–658.

Seol W, Choi H-S, and Moore DD (1996a): An orphan nuclear hormone receptor that lacks a DNA binding domain and heterodimerizes with other receptors. *Science* 272: 1336–1339.

Seol W, Mahon MJ, Lee Y-K, and Moore DD (1996b): Two receptor interacting domains in the nuclear hormone receptor corepressor RIP13/N-Cor. *Mol. Endo* 10: 1646–1655.

Simental JA, Sar M, Lane MV, French FS, and Wilson EM (1991): Transcriptional activation and nuclear targeting signals of the human androgen receptor. *J. Biol. Chem.* 266: 510–518.

Simons, Jr. SS (1987): Selective covalent labeling of cysteines in bovine serum albumin and in HTC cell glucocorticoid receptors by dexamethasone 21-mesylate. *J. Biol. Chem.* 262: 9669–9675.

Simons, Jr. SS, Pumphrey JG, Rudikoff S, and Eisen HJ (1987): Identification of cysteine-656 as the amino acid of HTC cell glucocorticoid receptors that is covalently labeled by dexamethasone 21-mesylate. *J. Biol. Chem.* 262: 9676–9680.

Simons, Jr. SS, Sistare FD, and Chakraborti PK (1989): Steroid binding activity is retained in a 16-kDa fragment of the steroid binding domain of rat glucocorticoid receptors. *J. Biol. Chem.* 264: 14493–14497.

Simons, Jr. SS, Oshima H, and Szapary D (1992): Higher levels of control: Modulation of steroid hormone-regulated gene transcription. *Mol. Endocrinol.* 6: 995–1002.

Simons, Jr. SS (1994): Function/activity of specific amino acids in glucocorticoid receptors. *Vitamins and Hormones* 48: 49–130.

Simons, Jr. SS (1996): Environmental estrogens: Can two "alrights" make a wrong? [comment]. *Science* 272: 1451.

Smith CL, Nawaz Z, and O'Malley BW (1997): Coactivator and corepressor regulation of the agonist/antagonist activity of the mixed antiestrogen, 4-hydroxytamoxifen. *Mol. Endocrinol.* 11: 657–666.

Smith DF, Whitesell L, Nair SC, Chen S, Prapapanich V, and Rimerman RA (1995): Progesterone receptor structure and function altered by geldanamycin, an hsp90-binding agent. *Mol. Cell. Biol.* 15: 6804–6812.

Smith DF (1993): Dynamics of heat shock protein 90-progesterone receptor binding and the disactivation loop model for steroid receptor complexes. *Mol. Endo.* 7: 1418–1429.

Söderström M, Vo A, Heinzel T, Lavinsky RM, Yang W-M, Seto E, Peterson DA, Rosenfeld MG, and Glass CK (1997): Differential effects of nuclear receptor corepressor (N-CoR) expression levels on retinoic acid receptor-mediated repression support the existence of dynamically regulated corepressor complexes. *Mol. Endocrinol.* 11: 682–692.

Sone T, Kerner S, and Pike JW (1991): Vitamin D receptor interaction with specific DNA. Association as a 1,25-dihyroxyvitamin D3-modulated heterodimer. *J. Biol. Chem.* 266: 23296–23305.

Souque A, Fagart J, Couette B, and Rafestin-Oblin M-E (1996): Sulfhydryl groups are involved in the binding of agonists and antagonists to the human mineralocorticoid receptor. *J. Steroid Biochem. Molec. Biol.* 57: 315–321.

Stancato LF, Silverstein AM, Gitler C, Groner B, and Pratt WB (1996): Use of the thiol-specific derivatizing agent N-iodoacetyl-3-[125I]iodotyrosine to demonstrate conformational differences between the unbound and hsp90-bound glucocorticoid receptor hormone binding domain. *J. Biol. Chem.* 271: 8831–8836.

Stromstedt P-E, Berkenstam A, Jornvall H, Gustafsson J-A, and Carlstedt-Duke J (1990): Radiosequence analysis of the human progestin receptor charged with [3H]promegestone. A comparison with the glucocorticoid receptor. *J. Biol. Chem.* 265: 12973–12977.

Suen C-S, and Chin WW (1995): A potential transcriptional adaptor(s) may be required in thyroid hormone-stimulated gene transcription in vitro. *Endocrinology* 136: 2776–2783.

Svec F, Teubner V, and Tate D (1989): Location of the second steroid-binding site on the glucocorticoid receptor. *Endocrinology* 125: 3103–3108.

Szapary D, Xu M, and Simons, Jr. SS (1996): Induction properties of a transiently transfected glucocorticoid-responsive gene vary with glucocorticoid receptor concentration. *J. Biol. Chem.* 271: 30576–30582.

Takagi S, Hummel BCW, and Walfish PG (1990): Thionamides and arsenite inhibit specific T3 binding to the hepatic nuclear receptor. *Biochem. Cell Biol.* 68: 616–621.

Tate BF, Allenby G, Janocha R, Kazmer S, Speck J, Sturzenbecker LJ, Abarzua P,

Levin AA, and Grippo JF (1994): Distinct binding determinants for 9-cis retinoic acid are located within AF-2 of retinoic acid receptor alpha. *Mol. Cell. Biol.* 14: 2323–2330.

Tate BF, and Grippo JF (1995): Mutagenesis of the ligand binding domain of the human retinoic acid receptor alpha identifies critical residues for 9-cis-retinoic acid binding. *J. Biol. Chem.* 270: 20258–20263.

Tate BF, Allenby G, Perez JR, Levin AA, and Grippo JF (1996): A systematic analysis of the AF-2 domain of human retinoic acid receptor alpha reveals amino acids critical for transcriptional activation and conformational integrity. *FASEB J.* 10: 1524–1531.

Tetel MJ, Jung S, Carbajo P, Ladtkow T, Skafar DF, and Edwards DP (1997): Hinge and amino-terminal sequences contribute to solution dimerization of human progesterone receptor. *Mol. Endo.* 11: 1114–1128.

Thole HH, and Jungblut PW (1994): The ligand-binding site of the estradiol receptor resides in a non-covalent complex of two consecutive peptides of 17 and 7 kDa. *Biochem. Biophys. Res. Commun.* 199: 826–833.

Thole HH, Maschler I, and Jungblut PW (1995): Surface mapping of the ligand-filled C-terminal half of the porcine estradiol receptor by restricted proteolysis. *Eur. J. Biochem.* 231: 510–516.

Thole HH, and Jungblut PW (1996): Completion of the amino acid sequence of the C-terminal half of the porcine estradiol receptor by Edman degradation: reconfirmation of the absence of O-linked sugars and phosphates. *Biochem. Biophys. Res. Commun.* 219: 227–230.

Tone Y, Collingwood TN, Adams M, and Chatterjee VK (1994): Functional analysis of a transactivation domain in the thyroid hormone beta receptor. *J. Biol. Chem.* 269: 31157–31161.

Toney JH, Wu L, Summerfield AE, Sanyal G, Forman BM, Zhu J, and Samuels HH (1993): Conformational changes in chicken thyroid hormone receptor alpha1 induced by binding to ligand or to DNA. *Biochemistry* 32: 2–6.

Tong G-X, Tanen MR, and Bagchi MK (1995): Ligand modulates the interaction of thyroid hormone receptor beta with the basal transcription machinery. *J. Biol. Chem.* 270: 10601–10611.

Tora L, Mullick A, Metzger D, Ponglikitmongkol M, Park I, and Chambon P (1989): The cloned human oestrogen receptor contains a mutation which alters its hormone binding properties. *EMBO J.* 8: 1981–1986.

Trapp T, Rupprecht R, Castren M, Reul JMHM, and Holsboer F (1994): Hetero-dimerization between mineralocorticoid and glucocorticoid receptor: A new principle of glucocorticoid action in the CNS. *Neuron* 13: 1457–1462.

Trapp T, and Holsboer F (1995): Ligand-induced conformational changes in the mineralocorticoid receptor analyzed by protease mapping. *Biochem. Biophys. Res. Comm.* 215: 286–291.

Tremblay GB, Tremblay A, Copeland NG, Gilbert DJ, Jenkins NA, Labrie F and Giguere V (1997): Cloning, chromosomal localization, and functional analysis of the murine estrogen receptor beta. *Mol. Endocrinol.* 11: 353–365.

Triezenberg SJ (1995): Structure and function of transcriptional activation domains. *Curr. Opin. Gen. Dev.* 5: 190–196.

Turcotte B, Meyer M-E, Bocquel M-T, Belanger L, and Chambon P (1990): Repression of the a-fetoprotein gene promoter by progesterone and chimeric receptors in the presence of hormones and antihormones. *Mol. Cell. Biol.* 10: 5002–5006.

Tzukerman MT, Esty A, Santiso-Mere D, Danielian P, Parker MG, Stein RB, Pike JW, and McDonnell DP (1994): Human estrogen receptor transactivational capacity is determined by both cellular and promoter context and mediated by two functionally distinct intramolecular regions. *Mol. Endo.* 8: 21–30.

Urda LA, Yen PM, Simons, Jr. SS, and Harmon JM (1989): Region-specific antiglucocorticoid receptor antibodies selectively recognize the activated form of the ligand-occupied receptor and inhibit the binding of activated complexes to deoxyribonucleic acid. *Mol. Endocrinol.* 3: 251–260.

Vegeto E, Allan GF, Schrader WT, Tsai M-J, McDonnell DP, and O'Malley BW (1992): The mechanism of RU486 antagonism is dependent on the conformation of the carboxy-terminal tail of the human progesterone receptor. *Cell* 69: 703–713.

Veldscholte J, Ris-Stalpers C, Kuiper GGJM, Jenster G, Berrevoets C, Claassen E, van Rooij HCJ, Trapman J, Brinkman AO, and Mulder E (1990): A mutation in the ligand binding domain of the androgen receptor of human LNCaP cells affects steroid binding characteristics and response to anti-androgens. *Biochem. Biophs. Res. Comm.* 173: 534–540.

Vivat V, Gofflo D, Garcia T, Wurtz J-M, Bourguet W, Philibert D, and Gronemeyer H (1997): Sequences in the ligand binding domains of the human androgen and progesterone receptors which determine their distinct ligand identities. *J. Mol. Endocrinol.* in press.

Voegel JJ, Heine MJS, Zechel C, Chambon P, and Gronemeyer H (1996): T1F2, a 160 kDa transcriptional mediatior for the ligand-dependent activation function AF-2 of nuclear receptors. *EMBO J.* 15: 3667–3675.

vom Baur E, Zechel C, Heery D, Heine MJS, Garnier JM, Vivat V, Le Douarin B, Gronemeyer H, Chambon P, and Losson R (1996): Differential ligand-dependent interactions between the AF-2 activating domain of nuclear receptors and the putative transcriptional intermediary factors mSUG1 and TIF1. *EMBO J.* 15: 110–124.

Wagner BL, Pollio G, Leonhardt S, Wani MC, Lee DY-W, Imhof MO, Edwards DP, Cook CE, and McDonnell DP (1996): 16alpha-substituted analogs of the antiprogestin RU486 induce a unique conformation in the human progesterone receptor resulting in mixed agonist activity. *Proc. Natl. Acad. Sci. USA* 93: 8739–8744.

Wagner RL, Apriletti JW, McGrath ME, West BL, Baxter JD, and Fletterick RJ (1995): A structural role for hormone in the thyroid hormone receptor. *Nature* 378: 690–697.

Wakui J, Wright APH, Gustafsson J-A, and Zilliacus J (1997): Interaction of the ligand-activated glucocorticoid receptor with the 14–3–3h protein. *J. Biol. Chem.* 272: 8153–8156.

Wang Y, O'Malley BWJ, Tsai SY, and O'Malley BW (1994): A regulatory system for use in gene transfer. *Proc. Natl. Acad. Sci. USA* 91: 8180–8184.

Wasner G, Oshima H, Thompson EB, and Simons, Jr. SS (1988): Unlinked regulation of the sensitivity of primary glucocorticoid-inducible responses in mouse mammary tumor virus infected Fu5–5 rat hepatoma cells. *Mol. Endocrinol.* 2: 1009–1017.

Webster NJG, Green S, Tasset D, Ponglikitmongkol M, and Chambon P (1989): The transcriptional activation function located in the hormone-binding domain of the human oestrogen receptor is not encoded in a single exon. *EMBO J.* 8: 1441–1446.

Weigel NL, Beck CA, Estes PA, Prendergast P, Altmann M, Christensen K, and Edwards DP (1992): Ligands induce conformational changes in the carboxyl-terminus of progesterone receptors which are detected by a site-directed antipeptide monoclonal antibody. *Mol. Endo.* 6: 1585–1597.

Weis KE, Ekena K, Thomas JA, Lazennec G, and Katzenellenbogen BS (1996): Constitutively active human estrogen receptors containing amino acid substitutions for tyrosine 537 in the receptor protein. *Mol. Endo.* 10: 1388–1398.

Whelan J, and Miller N (1996): Generation of estrogen receptor mutants with altered ligand specificity for use in establishing a regulatable gene expression system. *J. Steroid Biochem. Mol. Biol.* 58: 3–12.

White R, Fawell SE, and Parker MG (1991): Analysis of oestrogen receptor dimerization using chimeric proteins. *J. Steroid Biochem. Molec. Biol.* 40: 333–341.

White R, Sjö berg M, Kalkhoven E, and Parker MG (1997): Ligand independent activation of the oestrogen receptor by mutation of a conserved tyrosine. *EMBO J.* 16: 1427–1435.

Whitfield GK, Selznick SH, Haussler C, Hsieh J-C, Galligan MA, Jurutka PW, Thompson PD, Lee SM, Zerwekh JE, and Haussler MR (1996): Vitamin D receptors from patients with resistance to 1,25-dihydroxyvitamin D3: Point mutations confer reduced transactivation in response to ligand and impaired interaction with the retinoid X receptor heterodimeric partner. *Mol. Endo.* 10: 1617–1631.

Wilson CM, and McPhaul MJ (1994): A and B forms of the androgen receptor are present in human genital skin fibroblasts. *Proc. Natl. Acad. Sci. USA.* 91: 1234–1238.

Wolfgang CL, Zhang Z, Gabriel JL, Pieringer RA, Soprano KJ, and Soprano DR (1997): Identification of sulfhydryl-modified cysteine residues in the ligand binding pocket of retinoic acid receptor beta. *J Biol Chem* 272: 746–753.

Wooge CH, Nilsson GM, Heierson A, McDonnell DP, and Katzenellenbogen BS (1992): Structural requirements for high affinity ligand binding by estrogen receptors: A comparative analysis of truncated and full length estrogen receptors expressed in bacteria, yeast, and mammalian cells. *Mol. Endo.* 6: 861–869.

Wrenn CK, and Katzenellenbogen BS (1993): Structure-function analysis of the hormone binding domain of the human estrogen receptor by region-specific mutagenesis and phenotypic screening in yeast. *J. Biol. Chem.* 268: 24089–24098.

Wright APH, and Gustafsson J-A (1991): Mechanism of synergistic transcriptional transactivation by the human glucocorticoid receptor. *Proc. Natl. Acad. Sci. USA* 88: 8283–8287.

Wu Y, Reece RJ, and Ptashne M (1996): Quantitation of putative activator-target affinities predicts transcriptional activating potentials. *EMBO J.* 15: 3951–3963.

Wurtz J-M, Bourguet W, Renaud J-P, Vivat V, Chambon P, Moras D, and Gronemeyer H (1996): A canonical structure for the ligand-binding domain of nuclear receptors. *Nature Structural Biol.* 3: 87–94.

Xu J, Nawaz Z, Tsai SY, Tsai M-J, and O'Malley BW (1996a): The extreme C terminus of progesterone receptor contains a transcriptional repressor domain that functions through a putative corepressor. *Proc. Natl. Acad. Sci. USA* 93: 12195–12199.

Xu M, Chakraborti PK, Garabedian MJ, Yamamoto KR, and Simons, Jr. SS (1996b): Modular structure of glucocorticoid receptor domains is not equivalent to functional independence. Stability and activity of the steroid binding domain are controlled by sequences in separate domains. *J. Biol. Chem.* 271: 21430–21438.

Yang Y-Z, Burgos-Trinidad M, Wu Y, and Koeing RJ (1996): Thyroid hormone receptor variant alpha2. Role of the ninth heptad in DNA binding, heterodimerization with retinoid X receptors, and dominant negative activity. *J. Biol. Chem.* 271: 28235–28242.

Yeh S, and Chang C (1996): Cloning and characterization of a specific coactivator, ARA70, for the androgen receptor in human prostate cells. *Proc. Natl. Acad. Sci. USA* 93: 5517–5521.

Yen PM, and Simons, Jr. SS (1991): Evidence against posttranslational glycosylation of rat glucocorticoid receptors. *Receptor* 1: 191–205.

Yen PM, Darling DS, Carter RL, Forgione CM, Umeda PK, and Chin WW (1992): Triiodothyronine (T3) decreases binding to DNA by T3-receptor homodimers but not receptor-auxiliary protein heterodimers. *J. Biol. Chem.* 267: 3565–3568.

Ylikomi T, Bocquel MT, Berry M, Gronemeyer H, and Chambon P (1992): Cooperation of proto-signals for nuclear accumulation of estrogen and progesterone receptors. *EMBO J.* 11: 3681–3694.

Zanaria E, Muscatelli F, Bardoni B, Strom TM, Guioli S, Guo W, Lalli E, Moser C, Walker AP, McCabe ERB, Meitinger T, Monaco AP, Sassone-Corsi P, and Camerino G (1994): An unusual member of the nuclear hormone receptor superfamily responsible for X-linked adrenal hypoplasia congenita. *Nature* 372: 635–641.

Zawel L, and Reinberg D (1993): Initiation of transcription by RNA polymerase II: A multi-step process. *Prog Nucleic Acid Res Mol Biol* 44: 67–108.

Zeiner M, and Gehring U (1995): A protein that interacts with members of the nuclear hormone receptor family: Identification and cDNA cloning. *Proc. Natl. Acad. Sci. USA* 92: 11465–11469.

Zelent A, Krust A, Petkovich M, Kastner P, and Chambon P (1989): Cloning of murine alpha and beta retinoic acid receptors and a novel receptor gamma predominantly expressed in skin. *Nature* 339: 714–717.

Zeng Z, Allan GF, Thaller C, Cooney AJ, Tsai SY, O'Malley BW, and Tsai M-J (1994): Detection of potential ligands for nuclear receptors in cellular extracts. *Endocrinology* 135: 248–252.

Zhang S, Liang X, and Danielsen M (1996a): Role of the C terminus of the glucocorticoid receptor in hormone binding and agonist/antagonist discrimination. *Mol. Endocrinol.* 10: 24–34.

Zhang X, Jeyakumar M, and Bagchi MK (1996b): Ligand-dependent cross-talk between steroid and thyroid hormone receptors. Evidence for common transcriptional coactivator(s). *J. Biol. Chem.* 271: 14825–14833.

Zhang X-k, Hoffmann B, Tran PB-V, Graupner G, and Pfahl M (1992): Retinoid X receptor is an auxiliary protein for thyroid hormone and retinoic acid receptors. *Nature* 355: 441–446.

4

Structure and Function of the Steroid and Nuclear Receptor DNA Binding Domain

FRAYDOON RASTINEJAD

GENERAL OVERVIEW

The nuclear receptors form the largest known superfamily of genes encoding eukaryotic transcription factors. To date, over 150 related genes have been identified, all of which encode proteins with a common molecular architecture, shown in Figure 1 (reviewed by Mangelsdorf et al 1995). Most receptors function as homo- or heterodimers, and bind selectively to gene-regulatory sequences called hormone response elements (HREs) through which they regulate the expression of their target genes. These target genes, in turn, are responsible for controlling the growth, development, metabolism and homeostasis of higher eukaryotes. Nuclear-receptor-mediated transcriptional control is dependent on the paired assembly of the receptors, and on the sequence, orientation, and spacing of the half-site repeats in their HREs. These recognition events are the focus of this review. The functions of the receptors also involve their ligand occupancy (for the hormone/steroid receptors), and complex interaction with their mediators and the basal transcription machinery, which will not be discussed here.

A central and common feature of all nuclear receptors is their DNA binding domain (DBD). Like the nuclear receptors, which are typically dimeric in their functional forms, the HREs are arranged as 2-fold repeats. Efforts to understand the DNA targeting of these receptors at the molecular level have revealed a set of rules that describe the preference of certain dimeric receptors for HREs. A number of three-dimensional structures, derived from nuclear magnetic resonance (NMR) imaging and x-ray crystallography (see Table 1),

Molecular Biology of Steroid and Nuclear Hormone Receptors
Leonard P. Freedman, Editor
© 1998 Birkhäuser Boston

Table 1.—Summary of Nuclear Receptor DBD Structures

Receptor	HRE	Method	References
GR-DBD	—	NMR	Hard et al 1990, Baumann et al 1993
ER-DBD	—	NMR	Schwabe et al 1990
(GR-DBD)$_2$	pal4/pal3-GRE	x-ray	Luisi et al 1991
RAR-DBD	—	NMR	Katahira et al 1992
(ER-DBD)$_2$	pal3-ERE	x-ray	Schwabe et al 1993
RXR-DBD	—	NMR	Lee et al 1993
(ER-DBD)$_2$	ERE variant	x-ray	Schwabe et al 1955
RXR + TR DBD	TRE-dr4	x-ray	Rastinejad et al 1955
E/TR$_{GR}$	pal0-GRE	x-ray	Gewirth et al 1995

have allowed the visualization of the DBDs alone and in their functional assemblies on HREs, and have enhanced our knowledge of these rules.

In this review, a detailed look at the available DBD structure data base is used as the basis for understanding the general structural principles underlying DBD function. The stereochemical themes are then considered vis-à-vis the empirical rules derived from the wealth of molecular studies, in order to better understand the basis for HRE selection and nuclear receptor dimerization. The analysis explains how it is possible for relatively few HREs to be used in conjunction with highly conserved DBD sequences to generate the necessary specificity in this large superfamily.

The Common Receptor Architecture and the "core" DBD

Figure 1 shows that the receptors are composed of separate domains with distinct functions. The N-terminal region is a variable-sized, poorly conserved region that in some cases has a transcriptional activation function (Hadzic et al 1995, Pakdel et al 1993, Thompson et al 1989). The adjacent DBD region has dual functions: HRE binding and dimerization. The remarkable sequence conservation in the DBD has proven useful for cloning the complementary DNA sequences encoding the various nuclear receptors in this superfamily. The region immediately C-terminal to the DBD is referred to as the hinge region, and is variable in both size and sequence. As discussed below, the hinge region can also contribute significantly to the DNA binding and target specificity of the receptor. The carboxy termini of the nuclear receptors contain their ligand binding domain (LBD), which also elicits the transcrip-

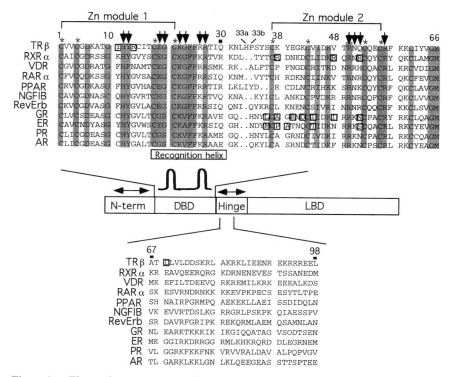

Figure 1.—The nuclear receptor architecture (center) and sequence comparisons at the DBD core region (top) and hinge region (bottom).

A common numbering scheme has been adopted, starting with the first conserved cysteine. Shaded aminoacid sequences are identical in more than 95% of the nuclear receptors. The residues belonging to the first and second zinc modules are indicated along with the zinc coordinating cysteines (asterisks). Arrows indicate the residues that make up the conserved "hooks" in DNA binding. Boxes indicate residues seen crystallographically at the subunit interfaces of $(GR)_2$, $(ER)_2$, and RXR-TR DBDs. TR: thyroid hormone receptor; RXR: 9-cis retinoic acid receptor; VDR: vitamin D receptor; RAR: all-trans retinoic acid receptor; PPAR: peroxisome proliferator activated receptor; NGFI-B: nerve growth factor induced receptor B; GR: glucocorticoid receptor; ER: estrogen receptor; PR: progesterone receptor; AR: androgen receptor.

tional response and maintains a secondary dimerization surface (Tsai and O'Malley 1994, Blanco et al 1995, Gloss et al 1995, Baniahmad et al 1995, Chen and Evans 1995, Burris et al 1995, Thompson et al 1989, Bourguet et al 1995, Renaud et al 1995, Wagner et al 1995, Rosen et al 1993, Barettino et al 1994).

By far the most highly conserved stretch of amino acid residues in the nuclear receptors consists of 66 residues within the DNA-binding region, referred to as the "core' region of the DBD (see Figure 1). To provide a simplified numbering scheme for the many receptor DBDs, a common frame of reference is adopted in this review whereby the first conserved cysteine residue in the core sequence is number 1, and the conserved methionine terminating the core is numbere 66 (shown in Figure 1). Domain-swapping experiments have shown that this region imparts the receptors" DNA specificity (Green and Chambon 1987, Green et al 1988). In addition, the DBDs form DNA-dependent intersubunit contacts (Perlmann et al 1993, Zechel et al 1994, Mader et al 1993). The core DBD contains two zinc-nucleated subdomains (Luisi et al 1991). Each zinc is coordinated in a tetrahedral geometry by the sulfurs from conserved cysteines (Freedman et al 1988, Figures 1 and 6). The DBDs of the 9-cis retinoic acid receptor (RXR), thyroid hormone receptor (TR), vitamin D3 receptor (VDR), and all-trans retinoic acid receptor (RAR) share some 40–50% sequence identity over this region and diverge to a large degree in the preceding and in the C-terminal flanking regions (see Figure 1).

The affinity of the DBD core regions for their half-sites is somewhat weaker than that of the full-length receptors (Zilliacus et al 1991, Lundback et al 1993, Hard et al 1990). This loss in DNA affinity is probably due to the absence of residues N- and C-terminal from the core DBD, which are likely to make significant DNA interactions. An example of this additional recognition comes from the crystal structure of the RXR-TR DBDs on the TRE element, where the hinge residues of TR make a sizable contribution to DNA binding (see below).

The DNA response elements dictate the dimeric assembly of the nuclear receptors into their functional forms. Interestingly, there are two distinct regions that maintain dimerization function. First, certain receptors can dimerize via their LBDs in the absence of DNA. Second, more functionally relevant dimerization interface is formed through a separate interface between the receptors' DBDs, and only in the presence of the appropriate HRE (Perlmann et al 1993, Zechel et al 1994, Mader et al 1993, Hard et al 1990). In fact, it has been shown that the DBDs alone can direct the binding of their hetero- or homodimers to the HRE half-sites (Perlmann et al 1993, Zechel et al 1994, Mader et al 1993). Thus, interactions between the LBDs may be the initial step towards the delivery of certain receptors into close proximity, from which the DBDs further coassemble in a response-element-dependent fashion.

THE STRUCTURE OF THE DBD

The successful overexpression and purification of nuclear receptor DBDs have made structural studies of these polypeptides feasible. Table 1 summarizes the various structures determined by two-dimensional NMR and x-ray diffraction studies reported to date. These include NMR structures of the glucocorticoid receptor (GR), estrogen receptor (ER), RXR, and RAR DBDs in the absence of DNA. The crystal structures of DBDs include those of $(GR)_2$, $(ER)_2$, and RXR+TR, along with an $(ER)_2$ variant, in their functionally revealing forms bound to appropriate DNA response elements. Because of the significantly larger molecular weight associated with the DBD complexes bound to HREs, structures of the dimeric DBD assemblies on DNA have so far been derived only from x-ray diffraction.

Figure 2 shows the overall folds of GR, RXR, and TR DBDs. The overall packing of the core DBD shows a compact architecture whose global features are determined by two packing events. First, rigid coordination of two zinc ions establishes two distinct substructures (referred to as module 1 and module 2 in Figure 1). Interestingly, these substructures are not derived from a duplication event in evolution (unlike the so-called "zinc fingers" of TFIIIA), and indeed represent distinct and complementary surfaces. Module 1 adopts the S-configuration in its chirality with respect to the zinc coordination, whereas module 2 adopts the R-configuration. This handedness of the zinc coordinations has been consistently observed in all of the reported structures, except for the NMR structure of RXR DBD (Lee et al 1993).

The second feature determining the overall fold is the perpendicular packing of two amphipathic alpha-helices, the first encompassing residues 18–30, and the second encompassing residues 53–64. The interhelical contacts effectively bury a cluster of hydrophobic side-chains, particularly those from a pair of conserved phenylalanines (residues 24 and 25, Figure 1), thus establishing the central hydrophobic core of the polypeptide. While the hydrophobic face of the first helix (residues 18–30) participates in packing, its other face, containing charged residues, acts as an ideal contact surface for major groove recognition at the DNA half-site, and is therefore referred to as the DNA recognition helix (see Figure 1).

The high resolution structures of the five nuclear receptor DBDs determined to date show a remarkably similar three-dimensional fold when their backbone atoms are compared (Rastinejad et al 1995). The structural comparison of the crystal structures with their counterparts derived from NMR has revealed reasonable agreement, particularly for the ER and GR subunits (Hard et al 1990, Baumann et al 1993, Schwabe et al 1990, Schwabe et al 1993, Luisi et al 1991). There are, however, some noteworthy differences between the DBD structures on- versus off-DNA. First, there is a distorted alpha-helix (residues 42–55) in module 2, when bound to DNA as seen in the crystal structures, but absent in the free form as determined by NMR. This distorted helix appears to facilitate a number of conserved contacts to the DNA backbone. Second, there is a limited structure deviation between the second zinc module seen in the co-crystal structures and the NMR structures off-DNA. This difference is particularly striking for RXR, but has also been

Fraydoon Rastinejad

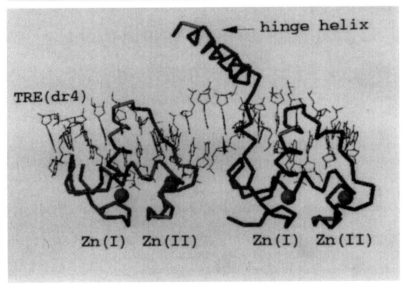

Figure 2.—A: The crystal structure of the GR DBD on a glucocorticoid response element (GRE) showing the symmetric head-to-head dimer interface (Luisi et al 1991). B: The crystal structure of the RXR-TR DBD heterodimer bound to a thyroid response element (TRE) showing an asymmetric head-to-tail dimer interface (Rastinejad et al 1995). RXR is the subunit on the left shown in lighter shade. The zinc atoms are shown as spheres. The RXR DBD subunit occupies the 5' half-site.

noted for GR and ER. These structural differences may arise from the fact that on-DNA, the second zinc module is able to provide a number of phosphate contacts and nearly the entire dimer interface for RXR, ER, and GR. Interestingly, the TR core DBD contains a two-residue insertion relative to all other DBDs (labeled as 33a and 33b in Figures 1 and 6; Rastinejad et al 1995). However, this additional sequence does not alter the overall three-dimensional structure, since together with residue 34, a short 3_{10} helix is formed to compensate for the increased contour length of TR-DBD.

DISCRIMINATION OF THE RESPONSE ELEMENT HALF-SITE SEQUENCE

The nuclear receptor dimers bind to bipartite DNA targets (Umesono et al 1991). A limited number of features distinguish these HREs. One is the sequence of the half-sites (see Figure 3). Despite the large number of nuclear receptors, only two types of consensus hexameric half-sites appear to be used, and these two types differ only in their central two base pairs (Klock et al 1987). The steroid receptors, notably the receptors for glucocorticoids, mineralocorticoids, androgens, and progesterone, bind to 5'-AGAACA-3' half-sites—and this is the "GR" element in Figure 3. The estrogen receptor (ER), RXR, TR, VDR, RAR, PPAR, COUP-TF, and NGFI-B preferentially bind to consensus 5'AGGTCA-3' half-sites (Klock et al 1987, Umenoso et al 1991), in what is referred to as the "ER" element in Figure 3.

The knowledge gained from the co-crystal structures of three cognate DBD complexes bound to consensus HREs (Table 1, Figure 2) together with the sequence alignment of the DBDs (Figure 1) now allows us to derive the basis for half-site discrimination, and to predict the half-site preference for nuclear receptor. Three residues, called the P-box, in the recognition helices of these receptors (numbered 19, 20, and 23 in Figure 1) predispose receptors to preferentially bind the GR versus the ER elements (Mader et al 1989). Receptors recognizing the AGAACA half-sites contain a GS..V in their P-box, while those that recognize AGGTCA half-sites typically contain EG..G residues at the corresponding positions.

Figures 4a and 4b show a comparison of the GR DBD/GRE versus ER DBD/ERE half-site interactions, and the stereochemical basis for cognate HRE discrimination. An important discerning interaction comes from Val23 of GR, which makes a favorable hydrophobic interaction with the 5-methyl major-groove functional group of a central thymine in GRE (see Figure 4A). The other two residues in the GR P-box, namely Gly19 and Ser20, do not contain sufficiently long side-chains for significant interaction with the GRE. However, for the ER, RXR, and TR DBDs, the P-box residue Glu19 plays a

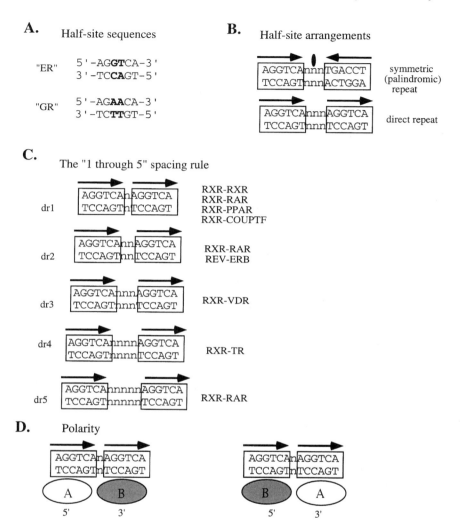

Figure 3.—A: Comparison of the two consensus HRE half-site sequences. The "ER" half-site is selected by ER, RXR, TR, VDR, RAR, PPAR, Rev-Erb, NGFI-B, and many other nonsteroid nuclear receptors. The "GR" half-site is selected also by GR, PR, AR, and MR. B: The two most common half-site arrangements are the symmetric (pal) repeat and the direct repeat (dr). The symmetry axis is shown by the oval for the palindromic HRE. C: The correspondence of various RXR dimers with their cognate direct repeats differing in the size of their inter-half-site spacer. D: The two distinct polarities of heterodimeric receptor assembly on direct repeat HREs.

remarkable role in networking direct and water-mediated hydrogen bonds to the central base-pairs of the half-site. Moreover, the side-chains of Glu19 and Lys22 form a hydrogen bond between them, which simultaneously positions both side-chains for optimal contacts to the major-groove surface of the central two base-pairs. The favorable hydrophobic contact made by GR's Val22 is impossible in the ER/ERE, because both the side-chain (Gly22) and the major-groove hydrophobic group (the thymine now switched to adenine) are absent.

There are two crystal structures that speak to the issues of nonconsensus (Shwabe et al 1995) and noncognate (Gewirth et al 1995) interactions. In the former study, the DNA construct contains a single base substitution in one-half of the palindromic estrogen response element, which results in a 10-fold loss of affinity. Comparison of this structure with the same structure on the consensus element suggests that there is a rearrangement of Lys22 to form an alternative type of DNA contact. Gewirth and colleagues (1995) investigated the basis of half-site specificity by crystallographically visualizing the non-cognate interactions of a DBD construct designed to target EREs but bound to a GRE consensus sequence. The study revealed a subtle distortion of the helical geometry of the half-site, which opens a gap at the interface filled by at least five additional water molecules.

THE CONSERVED "HOOKS": THE SHARED SET OF DBD-DNA CONTACTS

There is a set of highly conserved amino acid positions in the core DBD that have been visualized independently in the ER, RXR, and TR DBD complexes with their cognate half-sites. These hooks take the form of three distinct clusters that are expected to be invariant in all nuclear receptors, as shown in Figure 1. The first hook cluster is derived from certain side-chains along one face of the recognition helix (residues 18–27). The four residues here that make the greatest contribution to DNA binding (shown in black ovals in Figures 4A–4D) are residues 19, 22, 26, and 27. Importantly, these residues interconnect with several well-positioned and structurally conserved water molecules to form an intricate network of concerted base-specific and phosphate contacts with the DNA.

Two additional clusters of conserved hooks are highlighted in Figure 1. First, a cluster of residues (including residues 50–52 and 57) in the second zinc module is seen, in three different structures, to make contacts with the phosphate backbone of the DNA at the last two base pairs of the hexameric half-site. Second, two well-conserved residues in the first zinc module, His12 and Tyr13, along with the backbone atoms of a nonconserved residue 11, make a clustered set of contacts for GR, ER, RXR, and TR at the phosphate

A: GR - GRE

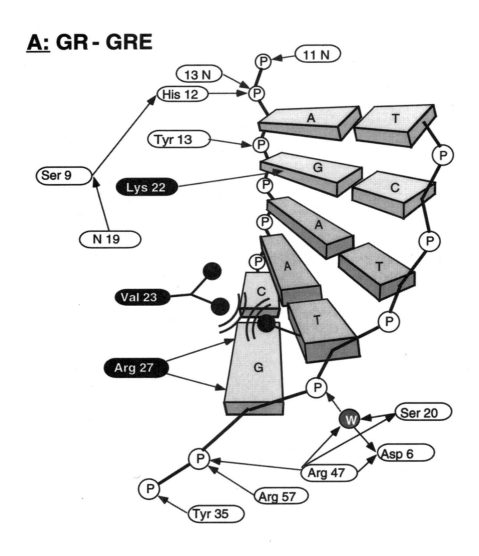

A

Figure 4A.—Schematic description of the protein-DNA contacts in the A: GR-DBD/ GRE$_{1/2}$, B: ER-DBD/ERE$_{1/2}$, C: RXR-DBD/TRE$_{1/2}$, and D: TR-DBD/TRE$_{1/2}$. The DNA is shown underwound for clarity. Water molecules are shown as circles. The hydrogen bond acceptor is shown at the head of the arrows. The van der Waals contact in the GR-DBD/GRE1/2 interaction is shown by semicircles.

<u>B:</u> ER - ERE

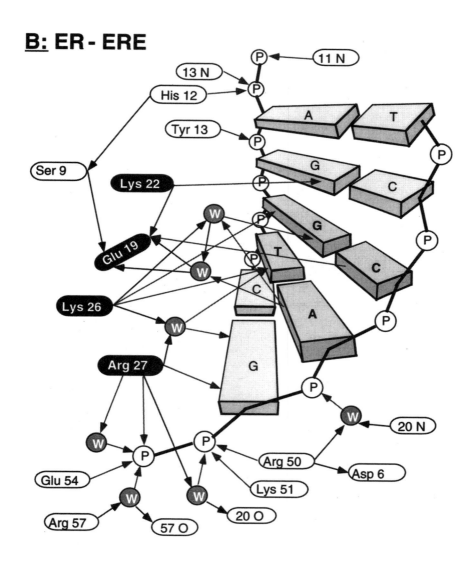

B

Figure 4B.—ER DBD/ERE$_{\frac{1}{2}}$

<u>C:</u> RXR - RXRE

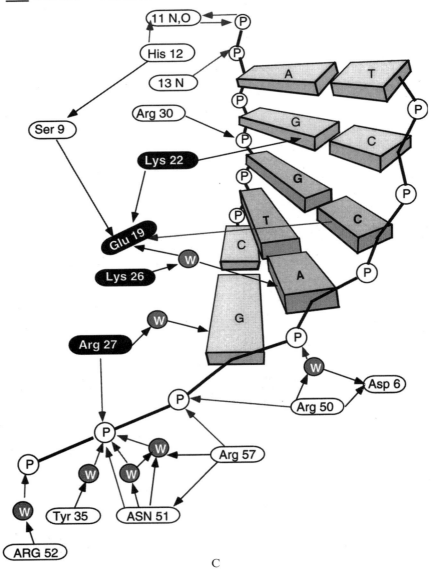

Figure 4C.—RXR DBD/TRE$_{\frac{1}{2}}$

D: TR - TRE

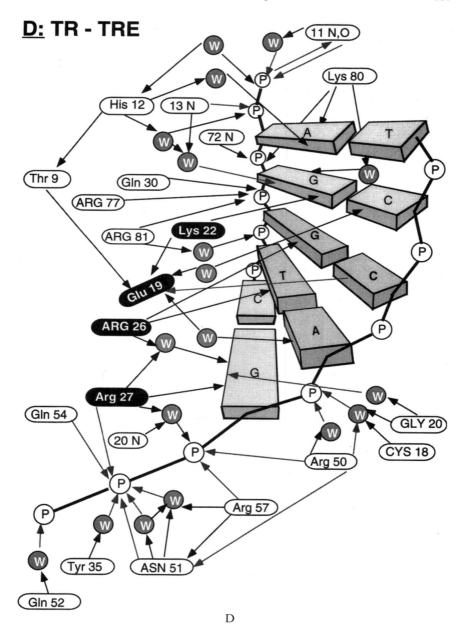

D

Figure 4D. — TR DBD/TRE$_{\frac{1}{2}}$

oxygens of the first adenine in 5'-AGXXCA-3' (see Figures 4a–4d). Interestingly, a side-chain oxygen atom from a conserved Ser/Thr at position 9 helps to align two of the three clusters by buttressing residues 12 and 19.

HALF-SITE ORIENTATION AND THE DIMER INTERFACE

Figure 3 shows that the second distinguishing feature of an HRE is the relative orientation of the DNA half-sites. Whereas steroid receptors such as GR, ER, MR, PR, and AR can bind to symmetrical/palindromic repeats of their respective half-sites (the palindromic HREs are referred to as pal3, where the number designates the base-pair separation of the half-sites), other nuclear receptors such as RXR, TR, VDR, RAR, PPAR, and Rev-Erb bind to a tandem of direct repeats (these are refereed to as dr$n$ sequences, where the number n designates the base-pair size of the separation between the repeats; Umesono et al 1991, Kliewer et al 1992, Marks et al 1992, Perlmann et al 1993, Mader et al 1993, Zechel et al 1994, Harding and Lazar 1995). The symmetric (palindromic) response element demands a symmetric assembly of the receptors' cognate DBDs, and this can only be derived from a homodimeric assembly. A tandem repeat, however, requires an asymmetric interface, which may come from a homodimer where the subunits are in a head-to-tail arrangement (e.g., RXR homodimer and Rev-Erb homodimer), or from the many possible heterodimers where RXR is a common partner like RXR-TR, RXR-VDR, and RXR-RAR (Yu et al 1991, Kliewer et al 1992, Marks et al 1992).

Figure 5 illustrates the crystallographically observed subunit interactions: one for the symmetric interface of GR, and one for the tandem head-to-tail interface of RXR-TR. Each GR DBD subunit makes its homodimeric contacts via residues from the second zinc module to form a head-to-head symmetric dimer (Figure 2A). The interactions involve side-chain interactions (between Leu36 and Ile48, Arg40 and Asp42, Asn52 and Asn52) as well as main-chain interactions between Ala38 and Ile44. Because the latter interaction does not involve any side-chains, it is possible that other residues could substitute in this interaction.

Figure 5B shows the nature of the head-to-tail interactions of RXR-TR DBD heterodimer on a dr4. Residues from the second zinc module of RXR DBD are positioned at the upstream side of the subunit interface, and form a number of hydrogen bonds with residues from the first zinc module and hinge region of the TR. Interestingly, RXR uses the long side-chains of three arginines. Arg38 and Arg48 are salt-bridged to aspartates from the TR, and Arg52 donates a hydrogen bond to the main-chain carbonyl of TR's Arg14, which in turn reciprocates this interaction. A tyrosine from TR (Tyr11) makes a stacking interaction to support RXR's Arg52 at the dimer interface.

A

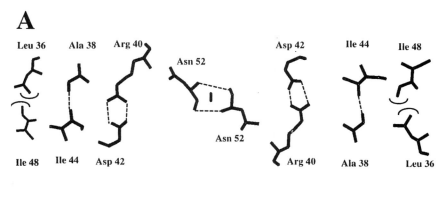

B

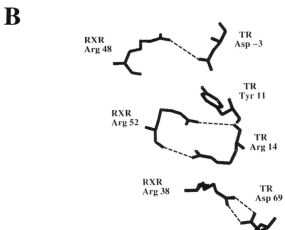

Figure 5.—The dimerization interface in A: (GR-DBD)$_2$/GRE complex, and B: RXR-TR DBD heterodimer/TRE complex. The symmetry axis is shown in A.

DISCRIMINATION OF THE INTER-HALF-SITE SPACE

Another feature that distinguishes a DNA response element is the size of the intervening sequence between the tandem half-sites (Umesono et al 1991, Kliewer et al 1992). The HREs of the RXR homo- and heterodimers are composed of direct repeats of their consensus half-site 5′-AGGTCA-3′ separated by one to five base pairs of "space." Figure 3 shows the "1–5 rule," which relates the targeting of various RXR heterodimers to the same direct repeats, differing only in the size of their inter-half-site spacing (Umesono et al 1991). Moreover, the intervening element is a true "spacer" in that it acts independently of its sequence (Umesono et al 1991). The implications of the 1–5 rule are extraordinary, suggesting that addition or subtraction of even a single neutral base pair in the spacer may be sufficient to interconvert the hormonal signaling pathways of vitamin D_3, triiodothyronine, retinoids, and prostanoids.

Because of the inherent pitch in the DNA structure, removing or deleting a single base pair from the inter-half-site element means changing both the displacement (3.4 Å) and the rotation angle (36°) between the two half-complexes rather significantly. This readjustment forces the subunits to adopt a new relative orientation at their interface. If this interface did not change in concert with the increased spacing, one would expect that at least one subunit would depart from its complementary register with respect to its half-site. For example, adding a fourth base pair in the intervening sequence of the GRE produced a complex in which one subunit was displaced out of the proper register with an AGAACA half-site but which maintained the symmetric dimer interface (Luisi et al 1991). It is worth noting that when compared to the GR/GRE complex, the RXR-TR DBD heterodimer maintains fewer inter-subunit contacts (Figure 5) and significantly more DNA contacts at the half-sites (Figure 4). These observations suggest that, unlike the $(GR)_2$/GRE complex, favorable interactions of RXR and TR DBDs on a misspaced response element may not be maintained at the expense of DNA misregistration.

Although all direct repeat elements require head-to-tail interactions, each of these differs by the degree of rotation and displacement by which the complementary subunits are aligned. We are far from a definitive structural understanding of how a limited subset of all possible RXR heterodimers is selected by the size of the spacing alone. As a start, the RXR-TR/dr4 structure provides some initial insights about the way in which a unique subunit interface is able to indirectly discern the spacing in a thyroid response element. The sequence alignment of the observed subunit contacts in these two receptors and other nuclear receptors (Figure 1) shows that RXR DBDs Arg38, 48, and 52 are not conserved in tandem in any other nuclear receptor. A wider comparison shows that the only exception is USP, the RXR homologue in Drosophila (Yao et al 1992). Similarly, the residues used by TR DBD in this complex are also not seen conserved in unison by any other receptor. This nonconservation ensures their unique mode of interaction on dr4. By extension, it may be assumed that the RXR-RAR contacts used on dr5 and the RXR-VDR contacts used on dr3 must be uniquely encoded in the amino acids of their respective DBDs. In the case of RXR-TR DBDs, although a number of phosphate contacts made to the spacer of dr4 are seen, no direct base contacts were seen in the crystal structure (Rastinejad et al 1995). This is consistent with the notion that the spacer functions as a neutral element, and in a sequence-independent manner.

Figure 3 illustrates that the 1–5 spacing rule contains two types of degeneracies. First, the consensus dr1 sequence is the high-affinity response element for the RXR homodimer, RXR-RAR, RXR-PPAR, and others—all of which can regulate transcription from this element (Mangelsdorf and Evans 1995, Leblank and Stunnenberg 1995, Kurokowa et al 1995). The biological significance of a shared HRE remains unclear. It is not known if a common set of inter subunit contacts will form for each dimer, or if entirely different dimer interfaces form. A second type of degeneracy is also apparent in Figure 3, where RXR-RAR targets and regulates transcription from three different response elements (dr1, dr2, and dr5) (Heery et al 1994, Kurokowa et al 1995, Zechel et al 1994, Mader et al 1993). This suggests that the DBDs of RAR and RXR form three alternate dimerization interfaces, one for each response element.

To fully break the chemical code underlying tandem assembly/space discrimination, there is need for a larger sampling of tandem DBD structures assembled on their HREs. In the absence of such structures, it has been possible to take advantage of the closely conserved three-dimensional fold of the DBDs and their known half-site registrations to model the presumptive interfaces on dr3 and dr5. One modeling study (Rastinejad et al 1995), which assumed no large overall differences in the backbone positions of RXR/RAR/TR/VDR DBDs, and which forced the correct half-site registration for the heterodimers, has provided a set of plausible interacting surfaces for RXR-VDR and for RXR-RAR. Importantly, nearly all the residues implicated in the dimerization of VDR and RAR DBDs were also unique to these respective DBDs and their isoforms.

Figure 1 illustrates the fact that, unlike the P-box residues involved in half-site discrimination, which are all highly conserved, the dimerization contacts for RXR heterodimers are likely to be distributed along the many non-conserved positions of the DBD sequences. The underlying stereochemical basis for the "1–5" rule is encoded in the pattern of unique and nonconserved amino acid positions, which in the folded three-dimensional structures present themselves to interact upon the correct racheting of the half-complexes. Thus, the propensity of certain RXR heterodimers to make favorable side-chain contacts acts to *indirectly* read the size of the inter-half-site spacing. On the other hand, the RXR-TR DBD structure also points to a possible direct readout of the inter-half-site space by which the hinge region helix of TR acts as a ruler (see below).

THE ROLE OF THE RECEPTOR HINGE REGION IN DNA BINDING AND SPACE DISCRIMINATION

The hinge region is suitably named for two reasons. First, it tethers together two globular domains of the receptors, namely the DBD and the LBD. Second, it is extremely sensitive to proteolysis (Bhat et al 1993, Leid et al

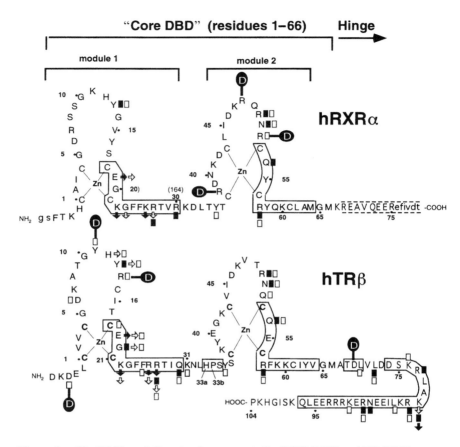

Figure 6.—The DNA and dimerization contacts for RXR DBD and TR DBD as seen in the crystal structure.

Dark circles indicate residues involved in dimerization of these DBDs. Filled and open arrows indicate residues making direct and water-mediated contacts to the DNA functional groups, respectively. Filled and open boxes indicate residues making direct and water-mediated contacts to the phosphate atoms of the DNA, respectively. Note that many residues making intersubunit contacts are simultaneously making contacts to the DNA, and others are clustered with adjacent contact points to the DNA.

1994). The hinge regions of VDR and GR have been shown to contain their respective nuclear localization signals (Luo et al 1994, Madan and DeFranco 1993), and thus nuclear localization may be this region's function in the TR and other nuclear receptors. The crystal structure shows that the TR hinge region contains an important structure that forms a considerable, yet totally unexpected, extension of the DNA-binding surface (Rastinejad et al 1995). In particular, residues 74–104 form an extended alpha-helix that forms along the minor groove of the four base-pair spacer, making a total of 15 contacts with the DNA, mostly to the phosphate oxygens and minor-groove edges of this sequence (see Figure 6). This region also contributes an important inter-subunit contact with RXR via Asp69 (Figure 5B).

Figure 7 shows that perhaps most importantly, the hinge region of TR may act as a molecular ruler that directly measures the half-site spacing of the thyroid response element by steric exclusion. Model building, using the crystallographic coordinates as a starting point, suggests that if the spacer size in this structure were reduced from four to fewer base pairs, the hinge helix could sterically block the upstream binding of RXR (see Figure 7). On the other hand, RXR-RAR and RXR-VDR can accept these lesser spacings while adopting the same polarity. Therefore, we expect the hinge regions of these receptors to form structures that are different from the TR's hinge helix, in order to permit their coassembly with RXR.

Interestingly, the RXR-TR DBD/TRE structure strongly suggests that only the receptor occupying the downstream (3′) half-site is able to use its hinge region effectively to "read" the size of the inter-half-space. This is because the hinge region of the 5′ partner points at the 5′ end of the HRE, and not towards the spacing (see Figure 2B). Since biochemical studies suggest that RXR mostly resides in the upstream repeat, this would mean that the space discrimination is largely the function of the downstream partner (see section on polarity). A comparison of the first 20 residues of the hinge region shows little sequence conservation (Figure 1). This lack of conservation would suggest that the corresponding spacing rules are differentially encoded into each receptor. Moreover, the hinge structures will differentially contribute to DNA binding and intersubunit association. In the case of NGFI-B, a study has already shown that the NGFI-B hinge region also acts as a contact surface for DNA. This allows stable monomeric binding on the extended NGFI-B half-site (Wilson et al 1992).

Although the large number of DNA interactions made by the TR hinge helix clearly stabilizes its structure, two observations suggest that the hinge region must afford a great deal of flexibility in the absence of DNA. First, appropriate LBDs form dimers in solution, which in the case of the RXR homodimer is achieved through a symmetric interface. The hinge region of one of the subunits must, therefore, afford enough flexibility to allow the rotation of a DBD, in order to make nonsymmetric head-to-tail interactions at the HRE while maintaining a symmetrically oriented LBD interface. This expectation is supported by a number of crystallographic observations. First, parts of the hinge regions of TR and RXR are contained in the "LBD" constructs whose structures were recently determined, but the electron density in this region was disordered and not traceable, suggesting flexibility (Bour-guet et al 1995, Wagner et al 1995). Second, while the structure clearly shows the long hinge-helix of TR making an extensive set of DNA contacts, it points completely away from the "core" DBD and thus completely lacks tertiary contacts (Rastinejad et al 1995).

This observation brings into question the conformation of the hinge region sequence off-DNA. A DNA induced conformation may be of importance for allowing free rotation of the DBDs relative to LBDs for slight adjustments in head-to-tail assembly on the direct repeats. The proposed DNA-induced secondary structure is not without precedent. A similar hinge region connecting the DNA-binding region of the lac repressor to its tetramerization domain has been shown to be induced into an alpha-helical configuration upon binding to the lac operator DNA (Spronk et al 1996).

POLARITY OF SUBUNIT INTERACTION

With the same consensus half-sites flanking a neutral spacer element in a direct-repeat fashion, the subunits can in theory adopt two alternate polarities, depending on which subunit occupies the 5′ half-site (see Figure 3). Biochemical studies have revealed that in the RXR-VDR, RXR-TR, and RXR-RAR complexes on dr3, dr4, and dr5 response elements, respectively, RXR is always positioned along the 5′ half-site (Perlamann et al 1993, Zechel et al 1994, Mader et al 1994). The crystal structure of the RXR-TR DBD on dr4 has indeed confirmed this information for the RXR-TR heterodimer. While the structure cannot directly explain why the alternate polarity is unfavored, it would seem that an equally favorable subunit interface is not possible when the subunits interchange their relative positions.

Some biochemical data suggest that the polarity of RXR-RAR may be response-element-dependent (Kurokawa et al 1995). On dr5, RXR binds upstream and RAR downstream. Moreover, in this conformation, the heterodimer may bind only to the RAR-specific ligand(s) to *activate* transcription (Kurokawa et al 1995). On a dr1 element, however, the subunits may occupy the reversed positions, and the heterodimer, which is then allosterically blocked from binding to RAR-specific ligand(s), *represses* transcription (Kurokawa et al 1995). Thus, ligand-dependent control of gene expression depends on the polarity of the receptor. Although the determinants of the polarity in the RXR-TR heterodimer have been visualized, it is not known what protein-protein contacts set the unique polarity in the other RXR heterodimers.

Figure 7.—The possible role of the TR hinge region in spacer discrimination. The crystal structure of the RXR-TR DBD complex is modeled on dr3, dr2, and dr1 elements. The size of the inter-half-site spacing may be directly read by the steric clashes of the TR hinge region with the upstream RXR subunit.

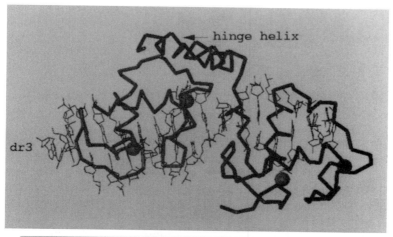

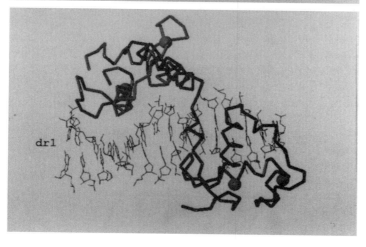

THE MECHANISM OF COOPERATIVE DBD DIMERIZATION ON THE RESPONSE ELEMENT

The DBDs in their isolated forms are monomeric in solution (Hard et al 1990, Schwabe et al 1990, Lee et al 1993). Upon being presented to the correct response element, ER and GR each associate to form their respective homodimers with high cooperativity, and RXR forms homodimers (dr1) or heterodimers, with VDR (dr3), RAR (dr1, dr2, dr5), or TR (dr4), for example. This suggests that the contacts of each of these receptors with their respective half-sites are too weak to stabilize monomeric binding. Moreover, the intersubunit contacts are too weak to allow stable dimer formation off-DNA. The presentation of the correct half-site sequences in combination with the proper orientation and spacing allows the subunits to form their subunit contacts, and these contacts in turn reinforce the registration of the DBDs at the half-site.

Because the subunit interface in the receptor DBD/DNA complexes will always form directly on top of the DNA spacer element, many of the residues forming the subunit interface can simultaneously interact with DNA phosphates. For example, Arg52 of RXR and TR residues Asp3, Tyr12, and Arg14 are all seen in the crystal structure to form water-mediated phosphate contacts to the DNA while making the asymmetric protein contacts between the two receptors. This ensures that the protein-protein interactions and protein-DNA interactions are linked and mutually supportive. Even residues involved only in subunit interactions (Arg48 of RXR) sit directly adjacent to clusters forming phosphate contacts (from residues 50–52 of RXR, see Figure 6) so that their respective interactions are mutually supportive.

Interestingly, some receptor DBDs such as TR, NGFIB, and Rev-Erb are also able to bind with efficiency as monomers (Schrader et al 1994, Wilson et al 1992, Perlamann and Jansson 1995, Harding and Lazar 1995). Figures 4A–D show that TR DBD makes a significantly more intricate network of interactions at its half-site than do RXR, ER, or GR DBDs. A great many of the extra contacts are derived from the hinge region just beyond the C-terminal DBD region (see below), and these largely account for the stable monomeric interactions of TR DBD seen in gel-shift experiments. The TR hinge region adds some 15 additional contacts with the DNA (see Figures 4D and 6). One of these contacts is particularly striking: Lys80 slides between the DNA bases along with an ordered water molecule to make two minor-groove contacts and a hydrogen bond with the DNA backbone. Moreover, a comparison of Arg26 or TR with a corresponding lysine in RXR indicates that the TR arginine makes two direct base-contacts not made by its counterpart in the RXR DBD. Other TR residues within the core DBD, such as Gln30 and Gln54, make additional contacts not mediated by their counterparts in RXR. In general, the extra interactions made by these core residues and those in the TR hinge act to orient and align the conserved residues for optimal interactions.

FUTURE PERSPECTIVES

The structural and functional studies to date have revealed much of the basis for HRE discrimination. Nevertheless, we are far from a complete understanding of how DNA targeting is achieved by the nuclear receptors. Several important questions remain. In the area of direct repeats, it remains to be fully uncovered how all the multitudes of distinct dimerization surfaces are generated within the conserved DBDs, and how these interactions specify the inter-half-site space of the response element. The crystallographic results clearly point to the possibility that the hinge region is being overlooked in terms of its contributions to DNA binding and space discrimination in receptors other than TR. Therefore, a more thorough analysis of this receptor region is in order.

Concerning the symmetric repeat DNA targets of the steroid receptors, it is still unclear how a number of different homodimers can derive specificity and regulate transcription from a single consensus HRE (a symmetric repeat of "GR" half-sites with three-base-pair spacing). The discrimination is likely to reside in the subtle departures from the consensus sequence seen in naturally occurring HREs. Therefore, it is important to pursue the receptor structures involving naturally occurring HREs to understand how sequence divergence can modulate the binding of the nuclear receptors.

ACKNOWLEDGMENTS

The author is grateful to Dan Gewirth (Yale University) for his assistance with figures, and to Sepideh Khorasanizadeh (University of Maryland) for her critical reading of the manuscript.

REFERENCES

Baniahmad A, Leng X, Burris TP, Tsai SY, Tsai MJ, and O'Malley BW (1995): The tau 4 activation domain of the thyroid hormone receptor is required for release of a putative corepressor(s) necessary for transcriptional silencing. *Molecular & Cellular Biology* 15: 76–86.

Barettino D, Vivanco-Ruiz MM, and Stunnenberg HG (1994): Characterization of the ligand-dependendent transactivation domain of thyroid hormone receptor. *EMBO J,* 13: 3039–49.

Baumann H, Paulsen K, Kovacs H, Berglund H, Wright AP, Gustafsson JA, and Hard T (1993): Refined solution structure of the glucocorticoid receptor DNA-binding domain. *Biochemistry* 32: 13463–71.

Beitel LK, Prior L, Vasiliou DM, Gottlieb B, Kaufman M, Lumbroso R, Alvarado C, McGillivray B, Trifiro M, and Pinsky L (1994): Complete androgen insensitivity due to mutations in the probable alpha-helical segments of the DNA-binding domain in the human androgen receptor. *Human Molecular Genetics* 3: 21–7.

Blanco JC, Wang IM, Tsai SY, Tsai MJ, O'Malley BW, Jurutka PW, Haussler MR, and Ozato K (1995): Transcription factor TFIIB and the vitamin D receptor cooperatively activate ligand-dependent transcription. *Proceedings of the National Academy of Sciences of the United States of America* 92: 1535–9.

Bourguet W, Ruff M, Chambon P, Gronemeyer H, and Moras D (1995): Crystal structure of the ligand binding domain of the human nuclear receptor RXR-a. *Nature* 375: 377–382.

Bugge TH, Pohl J, Lonnoy O, and Stunnenberg HG (1992): RXR alpha, a promiscuous partner of retinoic acid and thyroid hormone receptors. *EMBO J*, 11: 1409–18.

Burris TP, Nawaz Z, Tsai MJ, and O'Malley BW (1995): A nuclear hormone receptor-associated protein that inhibits transactivation by the thyroid hormone and retinoic acid receptors. *Proceedings of the National Academy of Sciences of the United States of America* 92: 9525–9.

Chen JD, and Evans RM (1995): A transcriptional co-repressor that interacts with nuclear hormone receptors. *Nature* 377: 454–457.

Freedman LP, Luisi BF, Korszun LR, Basavappa R, and Sigler PB (1988): The function and structure of the metal coordination sites within the glucocorticoid receptor DNA binding domain. *Nature* 334: 543–546.

Gewirth DT, and Sigler PB (1995): The basis for half-site specificity explored through a non-cognate steroid receptor-DNA complex. *Nature Struct. Biol.* 2: 386–394.

Gloss B, Kurokawa R, Ryan A, Kamei Y, Soderstrom M, Glass CK, and Rosenberg MG (1995): Ligand-independent repression by the thyroid hormone receptor mediated by a nuclear receptor co-repressor. *Nature* 377: 397–404.

Green S, and Chambon P (1987): Oestradiol induction of a glucocorticoid responsive gene by a chimaeric receptor. *Nature* 325: 75–78.

Green S, Kuman V, Theulaz I, Wahli W, and Chambon P (1988): The N-terminal DNA binding zinc finger of oestrogen and glucocorticoid receptors determines target gene specificity. *EMBO J.* 7: 3037–3044.

Hadzic E, Desai-Yajnik V, Helmer E, Guo S, Wu S, Koudinova N, Casanova J, Raaka BM, and Samuels HH (1995): A 10-amino-acid sequence in the N-terminal A/B domain of thyroid hormone receptor alpha is essential for transcriptional activation and interaction with the general transcription factor TFIIB. *Molecular & Cellular Biology* 15: 4507–17.

Hard T, Kellenbach E, Boelens R, Maler BA, Dahlman K, Freedman LP, Calrstedt-Duke J, Yamamoto KR, Gustafsson JA, and Kaptein R (1990): Solution structure of the glucocorticoid receptor DNA-binding domain. *Science* 249: 157–160.

Harding HP, and Lazar MA (1995): The monomer-binding orphan receptor Rev-Erb represses transcription as a dimer on a novel direct repeat. *Molec. Cell. Biol.* 15: 4691–4802.

Heery DM, Pierrat B, Gronemeyer H, Chambon P, and Losson R (1994): Homo- and heterodimers of the retinoid X receptor (RXR) activated in yeast. *Nucleic Acids Research* 22: 726–31.

Katahira M, Knegtel R, Schilthius J, Boelens R, Eib D, van der Saag P, and Kaptein R (1992): The structure of the human retinoic acid receptor-beta DNA-binding domain determined by NMR. *Nucleic Acids Symposium Series* 27: 65–6.

Kliewer SA, Umesono K, Mangelsdorf DJ, and Evans RM (1992): Retinoid X receptor interacts with nuclear receptors in retinoic acid, thyroid hormone and vitamin D3 signaling. *Nature* 355: 446–9.

Kliewer SA, Umesono K, Noonan DJ, Heyman RA, and Evans RM (1992): Convergence of 9-cis retinoic acid and peroxisome proliferator signalling pathways through heterodimer formation of their receptors. *Nature* 358:771–4.

Klock G, Strahle U, and Schutz G (1987): Oestrogen and glucocorticoid responsive elements are closely related but distinct. *Nature* 329: 734–736.

Kurokawa R, Soderstrom M, Horlein A, Halachmi S, Brown M, Rosenfel MG, and Glass CK (1995): Polarity-specific activities of retinoic acid receptors determined by a co-repressor. *Nature* 375: 451–454.

Leblanc BP, and Stunnenberg HG (1995): 9-cis retinoic acid signaling: changing partners causes some excitement. *Genes & Development* 9: 1811–6.

Lee MS, Kliewer SA, Provencal J, Wright PE, and Evans RM (1993): Structure of the retinoid X receptor alpha DNA binding domain: a helix required for homodimeric DNA binding. *Science* 260: 1117–112.

Leid M, Kastner P, and Chambon P (1992): Multiplicity generates diversity in the retinoic acid signalling pathways. *Trends in Biochemical Sciences* 17: 427–33.

Luisi BF, Xu WX, Otwinowski Z, Freedman LP, Yamamoto KR, and Sigler PB (1991): Crystallographic analysis of the interaction of the glucocorticoid receptor with DNA. *Nature* 352: 497–505.

Lundback T, Cairns C, Gustafsson J.A, Carlstedt-Duke J, and Hard T (1993): Thermodynamics of the glucocorticoid receptor-DNA interaction: binding of wild-type GR DBD to different response elements. *Biochemistry* 32: 5074–5082.

Luo Z, Rouvinen J, and Maenpaa PH (1994): A peptide C-terminal to the second Zn finger of human vitamin D receptor is able to specify nuclear localization. *European Journal of Biochemistry* 223: 381–7.

Madan AP, and DeFranco DB (1993): Bidirectional transport of glucocorticoid receptors across the nuclear envelope. *Proceedings of the National Academy of Sciences of the United States of America* 90: 3588–92.

Mader S, Chen JY, Chen Z, White J, Chambon P, and Gronemeyer H (1993): The patterns of binding of RAR, RXR and TR homo- and heterodimers to direct repeats are dictated by the binding specificites of the DNA binding domains. *EMBO J*, 12: 5029–41.

Mader S, Kumar V, de Verneui H, and Chambon P (1989): Three amino acids of the oesrogen receptor are essential to its ability to distinguish an oestrogen from a glucocorticoid-responsive element. *Nature* 338: 271–274.

Mangelsdorf DJ, Evans RM (1995): The RXR heterodimers and orphan receptors. *Cell* 83: 841–850.

Mangelsdorf DJ, Thummel C, Beato M, Herrrlich P, Schutz G, Umesono K, Blumberg B, Kastner P, Mark M, Chambon P, and Evans RM (1995): The nuclear receptor superfamily: the second decade. *Cell* 83: 835–839.

Marks MS, Hallenbeck PL, Nagata T, Segars JH, Appella E, Nikodem VM, and Ozato K (1992): H-2RIIBP (RXR beta) heterodimerization provides a mechanism for combinatorial diversity in the regulation of retinoic acid and thyroid hormone responsive genes. *EMBO J*, 11: 1419–35.

Pakdel F, Reese JC, and Katzenellenbogen BS (1993): Identification of charged residues in an N-terminal portion of the hormone-binding domain of the human estrogen receptor important in transcriptional activity of the receptor. *Molecular Endocrinology* 7: 1408–17.

Perlmann T, Rangarajan PN, Umesono K, and Evans RM (1993): Determinants for selective RAR and TR recognition of direct repeat HREs. *Genes & Development.* 7: 1411–22.

Perlmann T, and Jansson L (1995): A novel pathway for vitamin A signalling mediated by RXR heterodimerization with NGFI-B and NURR1. *Genes and Develop.* 9: 769–782.

Rastinejad F, Perlmann T, Evans RM, and Sigler PB (1995): Structural determinants of nuclear receptor assembly on DNA direct repeats. *Nature* 375: 203–11.

Renaud JP, Rochel N, Ruff M, Vivat V, Chambon P, and Moras D (1995): Crystal structure of the RARg ligand-binding domain bound to all-trans retinoic acid. *Nature* 378: 681–689.

Rosen ED, Beninghof EG, and Koenig RJ (1993): Dimerization interfaces of thyroid hormone, retinoic acid, vitamin D, and retinoid X receptors. *Journal of Biological Chemistry* 268: 11534–41.

Schrader M, Becker-Andre M, and Carlberg C (1994): Thyroid hormone receptor functions as monomeric ligand-induced transcription factor on octameric half-sites. Consequences also for dimerization. *Journal of Biological Chemistry* 269: 6444–9.

Schwabe JW, Chapman L, Finch JT, and Rhodes D (1993): The crystal structure of the estrogen receptor DNA-binding domain bound to DNA: how receptors discriminate between their response elements. *Cell* 75: 567–78.

Schwabe JW, Chapman L, and Rhodes D (1995): The oestrogen receptor recognizes an imperfectly palindromic response element through an alternative side-chain conformation. *Structure* 3: 201–213.

Schwabe JW, Neuhaus D, and Rhodes D (1990): Solution structure of the DNA-binding domain of the oestrogen reeptor. *Nature* 348: 458–46.

Spronk CAEM, Slijper M, van Boom JH, Kaptein R, and Boelens R (1996) Formation of the hinge helix in the lac repressor is induced upon binding to the lac operator. *Nature Struc. Biol.* 3: 916–919.

Thompson CC, and Evans RM (1989): Trans-activation by thyroid hormone receptors: functional parallels with steroid hormone receptors. *Proceedings of the National Academy of Sciences of the United States of America* 86: 3494–3498.

Tsai MJ, and O'Malley BW (1994): Molecular mechanisms of action of steroid/thyroid receptor superfamily members. *Annual Review of Biochemistry* 63: 451–86.

Umesono K, Murakami KK, Thompson CC, and Evans RM (1991): Direct repeats as selective response elements for the thyroid hormone, retinoic acid, and vitamin D3 receptors. *Cell* 65: 1255–66.

Wagner RL, Apriletti JW, McGrath ME, West BL, Baxter JD, and Fletterick RJ (1995): A structural role for hormone in the thyroid hormone receptor. *Nature* 378: 690–697.

Wilson TE, Paulsen RE, Padgett KA, and Milbrandt J (1992): Participation of non-zinc finger residues in DNA binding by two orphan receptors. *Science* 256: 106–110.

Yao TP, Segraves WA, Oro AE, McKeown M, and Evans RM (1992) Drosophila ultraspiracle modulates ecdysone receptor function via heterodimer formation. *Cell* 71(1):63–72.

Yu VC, Delsert C, Andersen B, Holloway JM, Devary OV, Naar AM, Kim SY, Boutin JM, Glass CK, and Rosenfeld MG (1991): RXR beta: a coregulator that enhances binding of retinoic acid, thyroid hormone, and vitamin D receptors to their cognate response elements. *Cell* 67: 1251–66.

Zechel C, Shen XQ, Chen JY, Chen ZP, Chambon P, and Gronemeyer H (1994): The dimerization interfaces formed between the DNA binding domains of RXR, RAR and TR determine the binding specificity and polarity of the full-length receptors to direct repeats. *EMBO J.* 13: 1425–33.

Zechel C, Shen XQ, Chambon P, and Gronemeyer H (1994): Dimerization interfaces formed between the DNA binding domains determine the cooperative binding of RXR/RAR and RXR/TR heterodimers to DR5 and DR4 elements. *EMBO J.* 13: 1414–24.

Zilliacus J, Dahlman-Wright K, Gustafsson JA, and Carlstedt-Duke J (1991): DNA binding specificity of mutant glucocorticoid receptor DNA-binding domains. *J. Biol. Chem* 266: 3101–3106.

5

Modulation of Steroid/Nuclear Receptor Dimerization and DNA Binding by Ligands

BORIS CHESKIS AND LEONARD FREEDMAN

INTRODUCTION

Transcription is a primary regulatory step of gene expression. Differential gene expression is controlled by a complex regulatory network of highly specialized transcription factors. Sequence-specific transcription factors modulate the formation of the preinitiation complex and thus control the rate of gene transcription. The concentration of the various transcription factors and their activities determines whether their target genes are transcribed, and to what extent. Therefore, the regulation of these regulators is of crucial importance for differential gene expression during development and in terminally differentiated cells.

Nuclear receptor concentrations, and the concentrations of intracellular proteins in general, may be regulated at any of the steps leading from DNA to protein, including transcription, RNA processing, mRNA degradation, and translation. Nuclear receptors bind DNA generally as homo- or heterodimers. Dimerization is a general mechanism among transcription factors to increase affinity and specificity of DNA binding. Interestingly, homo- and heterodimers play distinct and sometimes even opposite roles in transcriptional regulation. Ligand binding appears to modulate the dimerization status of nuclear receptors, affecting monomer-dimer and/or homo- versus heterodimer equilibrium and, as a consequence, concentration of the effector complex.

The activity of nuclear receptors is controlled by posttranslational modification (e.g., phosphorylation, see Chapter 9) and of course by the binding of

Molecular Biology of Steroid and Nuclear Hormone Receptors
Leonard P. Freedman, Editor
© 1998 Birkhäuser Boston

the corresponding ligand. Ligand binding induces conformational changes (see chapter 3), remodeling the ligand binding domain (LBD). Different ligand binding domain conformations reflect differences in the bound ligand structure and, as a consequence, could result in the different biological activities elicited.

In this chapter, we will discuss details of the molecular mechanism of ligand-dependent activation: how binding of the corresponding hormones modulates the activity of the target receptor, and how it affects the target receptor's dimerization status and DNA binding, all of which inevitably change the rate of transcription of specific subsets of target genes.

STEROID HORMONE RECEPTORS

The classical model of gene activation by steroid hormone receptors is deceptively straightforward. In the absence of hormone, the steroid receptors exist as large oligomeric complexes with heat shock proteins, including Hsp90, Hsp56, Hsp70 and p23. (The role of heat shock proteins is discussed in Chapter 1.) Binding of the hormone induces a conformational change in the ligand binding domain (LBD), reducing the affinity of the interaction with heat shock proteins and initiating dissociation of the multiprotein complex and exposure of the LBD dimerization interface. The homodimeric receptor then binds specific DNA, characteristically inverted palindromic repeats separated by three nucleotides consisting of AGGTCA half-sites for the estrogen receptor (ER), or AGAACA half-sites for the glucocorticoid (GR), mineralocorticoid (MR), progesterone (PR), and androgen receptors (AR). (Martinez et al 1987, Klock et al 1987, Picard et al 1990, Bohenet et al 1995).

Recent advances have clarified some of the molecular details of the modulation of steroid receptor activity, as well as their dimerization status by ligand binding. These advances have, however, added new levels of complexity to the model described above, which we will attempt to describe.

Estrogen Receptor

It has traditionally been believed that estrogen induces dimerization of ER and hence its DNA binding. The colocalization of the C-terminal dimerization domain with the ligand binding domain is consistent with the role of ligand in the regulation of dimerization (Fawell et al 1990a). ER, however, has been shown to form stable homodimers in solution ($K_D \approx 0.3$ nM) in the absence of estradiol (Notides et al 1985, Gordon and Notides 1986, Redeuilh et al 1987). It has also been reported that the human ER LBD dimerizes independently of ligand binding (Salomonsson et al 1994). However, steroid molecules with the large 7α-alkyl-amide extension, such as ICI 164,384, have been found to interfere with ER dimerization and DNA-binding, possibly by steric hindrance (Fawell et al 1990b). This result is consistent with the finding that the length of the side chain at position 7 is critical in the activity of this family of steroids (Bowler et al 1989). Using a two-hybrid assay, Wang et al. (1995) have recently demonstrated that ER-dimerization *in vivo* is hormone-dependent. Furthermore, they have shown that dimerization can be induced by both the "partial" antagonist tamoxifen and "pure" antagonist ICI 182,780.

The idea that ER binds DNA only as a homodimer is being reexamined since it has been shown that ER is capable of binding to a thyroid hormone response element (TRE) consisting of an inverted palindrome without spacing, as a monomer (Hirst et al 1992), and that ERE binding requires neither homodimerization nor estrogen (Furlow et al 1993). Affinity of this binding is, however, unknown. In addition, it was shown that ER also binds as a monomer to a half-palindromic ERE (Medici et al 1991), and that at the ERE half-site in the c-jun gene promoter is a strong regulatory element in response to estrogen induction (Hyder et al 1995).

It is also very controversial whether or not estrogen affects the receptor's ability to bind specific DNA (ERE). Initially, *in vitro* analysis performed with hER expressed in HeLa cells, *Xenopus oocytes*, or yeast, or produced by *in vitro* transcription/translation, indicated that binding of ER to an ERE was hormone-dependent, and that estradiol induced the formation of receptor homodimers (Kumar and Chambon 1988). It was subsequently found that this protein (HEO) has an artifactual mutation (Gly 400 → Val) that decreases hormone binding at 25°C, but not at 4°C (Tora et al 1989). Later, it was reported that wild-type hER *in vitro* binds DNA in absence of ligand (Walter et al 1985, Reese and Katzenellenbogen 1991). Hormone-independent formation of an ER-ERE complex was also reported with crude extracts (Klein-Hitpass et al 1989) and with purified ER from calf uterus (Sabbah et al 1991), rat uterine extracts (Furlow et al 1993), mouse uterine extracts (Curtis and Korach 1991), transfected COS-1 cells (Arbuckle et al 1992, Dauvois et al 1992), and SF9 cells infected with recombinant baculovirus (Arbuckle et al 1992). In contrast, ligand-induced DNA binding was reported with hER, produced by *in vitro* transcription/translation or in SF9 cells (Beekman et al 1993). *In vivo,* ligand-dependent and ligand-independent ERE association was reported for hER and *Xenopus* ER, using a promoter interference assay (Reese and Katzenellenbogen 1992, Xing and Shapiro 1993). Conversely, it was also shown that hormone may be required to promote DNA binding at low, but not at high, concentrations of ER (McDonnell et al 1991). Genomic footprinting has indicated that occupation of the ERE present in the apoVLDLII promoter region is hormone-dependent (Wijnholds et al 1988), which indeed suggests that the hormone is affecting ER interaction with ERE in the nucleus.

Recently, ER-ERE interactions were examined using surface plasmon resonance (SPR) methodology (Cheskis et al 1997). Using this approach, it was found that ligand binding dramatically affects the kinetics of hER interaction with specific DNA. It was demonstrated that binding of estradiol induces rapid formation of an unstable ER-ERE complex, and furthermore, as shown in Figure 1, that binding of a "pure" antagonist such as ICI-182,780 results in a slow formation (k_a is approximately 100 times lower) and a very stable receptor-DNA complex (k_d is almost 100 times lower). Most importantly, it was shown that there is a good correlation between the kinetics of hER-ERE interaction induced by a hormone and its transactivation function.

The precise mechanism whereby differences in the kinetics of receptor-DNA interaction induced by the binding of the ligand occur and how that relates to the observed behavior of the estrogen receptor *in vivo* is not yet known. It is clear, however, that binding of estradiol magnifies the frequency of ER-DNA complex formation more than 50-fold, compared with unliganded ER, and more than 1000-fold compared to ER liganded with ICI-182,780. It is therefore feasible that a correlation exists between the rate of gene transcription and the frequency of receptor-DNA complex formation. In this work (Cheskis et al 1997), the role of ER ligands was investigated using a consensus ERE. The pattern found may, however, be different for other EREs, or it may be modified by other transcription factors interacting with ER, which could explain tissue-specific effects of different analogs of estradiol.

Progesterone and Glucocorticoid Receptors

Human progesterone receptor (PR) is expressed from a single gene as two distinct molecular forms, PR-A (94 kDa) and PR-B (120 kDa). The B form contains an additional 164-amino-acid N-terminal sequence; otherwise, the two proteins are identical (Kastner et al 1990), and may (Horwitz et al 1988) or may not (Ilenchuk et al 1987) be present in equimolar concentrations in tissues. There is evidence suggesting that the cellular pathways used by hPR-A and hPR-B are different (Vegeto et al 1993)

DeMarzo and co-workers (1991), using a coimmunoprecipitation assay, have demonstrated that PR-A is capable of dimerization in solution with PR-B, and that solution dimerization correlates with the dissociation of the Hsp90 and the ability of PR to bind to specific DNA. This suggests that dimerization in solution is required for PR binding to specific DNA. Dimerization of PR monomers *in vivo* does not occur after deletion of the LBD (Guiochon-Mantel et al 1989). The steroid analog RU 486 is a progestin antagonist that is an important clinical compound as well as a useful tool for probing the molecular mechanism of PR action (Philibert et al 1984, Baulieu 1989). Several reports have suggested that RU 486 both *in vivo*

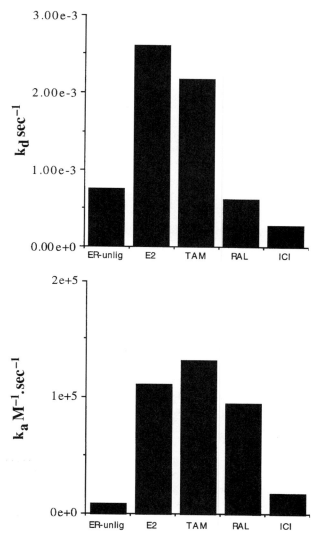

Figure 1.—Effects of different estrogen agonists and antogonists on the kinetics of ERE-ER interactions.

Serial injections of unliganded ER and ER incubated overnight with 10–6 M estradiol (E2), 4(OH)-tamoxifen (TAM), raloxifene (RAL), and ICI-182,780 (ICI) at protein concentrations ranging from 35 nM to 270 nM were run over a sensor chip with an immobilized surface gradient of the vitellogenin A_2 ERE (Cheskis et al 1997). Sensograms of the injections were analyzed using the BIA evaluation 2.1 program. Values of apparent dissociation (k_d) and association (k_a) rate constants determined for the ER-ERE interactions in the absence and presence of different ligands are presented.

and *in vitro* does not impair PR dissociation from Hsp proteins, dimerization, or DNA binding. Accordingly, interaction with other transcription factors has been suggested to be the primary cause for the failure of RU 486-bound PR to activate transcription (Batch et al 1988, El-Ashry et al 1989, Meyer et al 1991, Hurd et al 1991, Moudgil et al 1989, Klein-Hitpass et al 1991, Skafar, D. F., 1991a).

Using cooperativity of ligand binding to evaluate affinity of PR dimerization, Skafar has shown that calf uterine PR bound to RU 486 has a higher dimerization affinity ($K_D = 2.6$ nM) (Skafar 1993) than PR bound to progesterone ($K_D = 7.0$ nM) (Skafar 1991b). DeMarzo and co-workers (1992) have demonstrated that RU 486 promotes a 2-fold increase over R5020 (agonist) in solution concentration of PR dimers and also increases homodimer stability to salt dissociation. Evidence suggests that RU 486 binding alters PR conformation distinct from that induced by progesteron agonists (Mullick and Katzenellenbogen 1986, DeMarzo et al 1992, El-Ashry et al 1989) and that it most likely involves the PR C-terminus (Weigel et al 1992), which harbors the dimerization interface (Fawell et al 1990, Kumar and Chambon 1988). The binding of agonists, or RU 486, is required for efficient DNA binding of PR in gel retardation assay (Meyer et al 1990, Bagchi et al 1990, El-Ashry et al 1989).

Direct evidence has been presented by Tsai and co-workers that GR binds to a GRE as a homodimer. (Tsai et al 1988). It was subsequently demonstrated that dimers can form in the absence of DNA (Wrange et al 1989). Furthermore, homodimer formation was shown to be rate-limiting for high-affinity GRE binding and hormone-dependent activation of transcription (Drouin et al 1992). *In vivo*, hormone is a prerequisite for occupation of the HRE by the GR (Becker et al 1986). However, it was demonstrated that hormone-free receptor binds specifically to the GRE from the MMTV LTR *in vitro* (Willman and Beato 1986).

Beato and co-workers (Schauer et al 1988) have demonstrated that binding of a functional steroid to either glucocorticoid or the progesterone receptors influences the kinetics of protein-DNA interaction *in vitro*. In the presence of hormone, the on-rate of the receptor binding to DNA is accelerated 2- to 5-fold, and the off-rate is accelerated 10- to 20-fold. It was also shown that receptors complexed to an antihormone bind DNA with kinetics intermediate between those of steroid-free and hormone-bound receptor. Binding of the antihormone RU 486 had little effect on the dissociation rate, which was similar to that of apo receptors. Similar results with PR were also obtained by El-Ashry et al (1989). They found that PR bound to RU 486 displayed an off-rate that was 2-fold slower when compared to receptor bound to RU 5020.

Several years ago, von Hippel (Berg et al 1981) proposed that specific DNA binding of transcriptional regulatory proteins is accomplished by a scanning trial and error mechanism involving multiple association and dissociation steps. If such a mechanism applies to hormone receptors *in vivo*, binding of hormone would significantly accelerate the kinetics of DNA searching, which would facilitate the hormone-receptor complex's ability to localize a required sequence (Schauer et al 1988).

Along these lines, an interesting result has been described by Horwitz and co-workers. While PR occupied by the antagonist RU 486 and ZK 112993 is transcriptionally inactive, the antagonist-occupied receptor, in presence of 8-Br-cAMP, becomes a strong activator of transcription in a human breast cancer cell line (Sartorius et al 1993). It was suggested that cAMP, which is an activator of protein kinase A, regulates a transcription factor(s) that can modify the transcriptional effects of antagonist-occupied hPR.

Androgen Receptor

The androgen receptor (AR) mediates male sex development. AR binds with high affinity the two biologically active androgens, testosterone and dihydrotestosterone. Although these two hormones have almost identical affinity, the kinetics of their binding to AR are very different. Rates of association and dissociation of testosterone are about three times faster than those of dihydrotestosterone (Wilson and French 1976). In its unliganded form, AR resides in the cytoplasm, where it rapidly degrades. Androgen binding significantly increases receptor stability (Kemppainen et al 1992). Faster dissociating testosterone is less effective than dihydrotestosterone in stabilizing AR against degradation (Zhou et al 1995).

Hormone binding induces AR translocation to the nucleus and triggers AR dimerization and subsequent DNA-binding. Dimerization of human AR was shown to be mediated through direct intermolecular interactions between the androgen-bound LBD and the amino-terminal region. It has been proposed that androgen-dependent conformational changes can affect the affinity of AR antiparallel dimerization and may be required for high affinity DNA binding that results in transcriptional activation (Langley et al 1995).

NONSTEROID RECEPTORS

Unlike steroid receptors, the nonsteroid hormone receptors do not bind to heatshock proteins (Pratt 1993, Dalman et al 1990, Dalman et al 1991). Thyroid hormone receptor (TR), retinoic acid receptor (RAR), and vitamin D receptor (VDR) have all been found to bind DNA as homodimers, but high-affinity binding to cognate HREs requires heterodimerization with the retinoid X receptor (RXR) (Mangelsdorf et al 1990, Yu et al 1991, Leid et al 1992, Mangelsdorf et al 1992). Heterodimerization with RXR alters the specificity of receptor interactions with response elements (Glass 1994). The most potent of these HREs are direct repeats (DRs) of the AGGTCA half-site for which receptor specificity is determined by the so-called "1-to-5 rule" specifying the optimal half-site spacing of 1, 2, 3, 4, and 5 nucleotides for the PPAR, RAR, VDR, TR, and RAR response elements respectively (Näär et al 1991, Umesono et al 1991).

Following the asymmetry of the direct repeat binding sites, heterodimer receptors bind to direct repeats in an asymmetric fashion. It has been established that on DR3, DR4, and DR5, RXR occupies the 5′ half-site, and the partner (e.g., VDR, TR, and RAR respectively) occupies the 3′ half-site (Perlmann et al 1993, Kurokawa et al 1993, Zechel et al 1994, Lemon and Freedman 1996). On a DR1, RXR can bind both as a homodimer and as a heterodimer with RAR. Interestingly, the polarity of the RXR/RAR heterodimer on a DR1 element is reversed (Kurokawa et al 1994, Forman and Evans 1995).

It is therefore uncertain whether homo- and heterodimers play similar roles in transactivation. The presence of two receptor molecules with different ligand specificities within the same heterodimer species raises the questions of (1) what the role of each ligand is, (2) whether a heterodimeric complex can be activated by both ligands, and, (3) if so, whether they act synergistically.

The actual situation appears to be quite complex and dependent on the particular receptor combination.

Thyroid Hormone Receptor

The major forms of thyroid hormones are 3,5,3′,5′-tetraiodo-L-thyronine (L-T_4) and 3,3′,5-triiodo-L-thyronine (L-T_3), the latter being the most active form. The biological effects of these hormones are mediated by specific nuclear thyroid hormone receptors (TRs). Two genes encoding two different receptor subtypes, TRα and TRβ, have been characterized (Glass et al 1990, Chin 1991, Lazar 1993, Piedrafita and Pfahl 1994). TRs have dual regulatory roles and can function as transcriptional activators as well as transcriptional repressors (Pfahl 1994, Graupner et al 1989). The characterization of TREs revealed that several configurations of two half-sites of the sequence AGGTCA (or derivatives of this sequence) are possible, including a palindrome with no spacer (TREpal) (Glass et al 1988), direct repeats with 4 base pair (bp) spacer (DR4) (Desvergne, et al 1991, Näär et al 1991, Umesono et al 1991), and inverted palindromes/everted repeats (IP) spaced by 4–6 bp (Baniahmad, et al 1990, Saatcioglu et al 1993, Bendik and Pfahl 1995, Piedrafita et al 1995).

Considerable evidence has been provided demonstrating that TR binds native and synthetic DNA as a monomer, homodimer, and heterodimer with RXR (reviewed by Glass 1994). It was also reported that TRs can form heterodimers with other receptors (Schräder et al 1994, Bogazzi et al 1994), but this is controversial and not widely viewed as biologically relevant. Formation of a TR homodimer in solution has not been reported; therefore, it is conceivable that TREs enhance the affinity of homodimerization. Cooperative TR dimerization was demonstrated on the rat growth hormone (GH) TRE, and on rat α-myosin heavy-chain (rMHC) TRE (Williams et al 1991, Brent et al 1992, Miyamoto et al 1993), both consist of three repeats of the hexamer (AGGTCA). The first and second hexamers constitute a direct repeat, and the second and third form a palindrome. In cooperative binding, a receptor monomer binds to a half-site, and the second monomer then binds with greater affinity because of protein-protein interactions between the receptor molecules. Oligomer formation can also be cooperative, in which case a preformed dimer interacts with another monomer bound to the third hexamer. A strong cooperativity effect was also observed with inverted palindromes spaced by 6 bp (IP-6), but not with DR4 or TREpal elements (Piedrafita et al 1995). Monomer and dimer binding was reported on the rat malic enzyme TRE (ME-TRE). This TRE consists of the three direct repeats of the same core hexamer with three or four nucleotide spacing.

T_3 has diverse effects on homodimer formation. Hormone binding enhances monomeric but disrupts TR dimeric DNA binding to the DR- and IP-TREs with 4–6 bp spacers (Piedrafita et al 1995), but not on palindromic elements (Williams et al 1991, Miyamoto et al 1993, Yen et al 1992). It is therefore conceivable that conformational changes induced by hormone binding reduce the affinity of TR homodimerization on DR and IP elements, while having no detectable effect on palindromic elements. TR homodimers were found to bind with high affinity to IP types of TREs, while the TREpal and DR-4, type elements bind TR in the absence of RXR with low affinity (Wahlstrom et al 1992). The chicken lysozyme TRE (F2-TRE), an IP-6-type response element, initially characterized as a silencer (Baniahmad et al 1990), can be bound by TR/TR homodimers and TR/RXR heterodimers, and the TR/TR homodimers form more stable complexes with this TRE. TR/TR homodimer binding to the F2-TRE, however, is inhibited by T_3 (Yen et al 1992), allowing the homodimers to function as T_3-sensitive repressors (Piedrafita and Pfahl 1994, Burnside et al 1990). Therefore, ligand binding prevents homodimer-DNA interactions on IP-TREs, which precludes TR homodimers from functioning as hormone-dependent activators from these sites. Consistent with this, a natural IP from the TRβ promoter that only binds TR homodimers showed no transcriptional activation by T_3 (Piedrafita et al 1995). Furthermore, the TR homodimer can function as a repressor in the absence of ligand, blocking transcription. T_3 inhibits TR binding to these sites, allowing activation of transcription presumably by affecting formation of TR-RXR heterodimer. TR repressor activity was also observed on the TREpal (Brent et al 1989, Damm et al 1989, Grauper et al 1989, Zhang et al 1991), which has higher affinity for heterodimer binding (Zhang and Pfahl

1993). On such a response element, a nonactivated heterodimer can repress a ligand-activated heterodimer by competing for the binding site (Hermann et al 1993). The v-erbA oncogene product, a mutated TRα that no longer binds T$_3$, can form transcriptionally inactive heterodimers with RXR that act in a dominant negative fashion to TR/RXR in the presence of T$_3$ (Damm et al 1989) by the same mechanism.

TR/RXR heterodimers were shown to form in solution in the absence of ligand (Kliewer et al 1992, Leid, et al 1992, Bugge et al 1992), while the presence of specific DNA binding sites like the MHC-TRE or the malic enzyme TRE (ME-TRE) were shown to strongly enhance this heterodimerization (Hermann et al 1992).

Thus TREs appear to allow homo- as well as heterodimer action, presumably depending on the availability of RXR and ligands. Since RXR interacts with many different receptors, its concentration can limit TR/RXR complex formation. The presence of 9-cis retinoic acid (9-cis RA) can further reduce the concentration of available RXR, since it induces formation of RXR homodimer (Zhang et al 1992, Lehmann et al 1993, Kersten et al 1995). Consistent with this model, RXR enhancement along with 9-cis RA-dependent attenuation of the thyroid-hormone-dependent transactivation was reported from TRE-containing reporters (Lehmann et al 1993, Leng et al 1994). In contrast, costimulation by 9-cis RA and the RXR-specific ligand SR 11237 upon overexpression of RXR was demonstrated on the rGH TRE, which was composed of both, in combination with an inverted palindrome and direct repeat (Rosen et al 1992, Davis et al 1994).

Vitamin D$_3$ Receptor

1,25-Dihydroxyvitamin D$_3$ (1,25(OH)$_2$ D$_3$), the most active metabolite of vitamin D$_3$, plays a critical role in regulating multiple physiological processes, including calcium and phosphorus absorption in intestine, bone remodeling, and differentiation of hematopoetic cells. The diverse genomic effects of the 1,25(OH)$_2$ D$_3$ are mediated by a single nuclear protein the vitamin D$_3$ receptor (VDR).

Although a variety of spacing and orientations has been reported (Schrader et al 1995), the consensus high-affinity VDR recognition element (VDRE) consists of two hexameric half-sites arranged as a direct repeat with a spacing of three base pairs (DR3) (Umesono and Evans 1989). Different VDREs, however, have been identified in the promotors from the human and rat osteocalcin gene (Kerner et al 1991, Ozono et al 1990, Demay et al 1990, Markose et al 1990), mouse secreted phosphoprotein 1 gene (Spp-1, also known as osteopontin) (Noda et al 1990), rat calbindin D-9K (Darwish and Deluca 1992), avian integrin β_3 subunit (Cao et al 1993), rat 25-hydroxyvitamin D$_3$ 24-hydroxylase gene (Zierold et al 1994), and the human p21[WAF1/]

[CIP1] gene (Liu et al 1996). Except for the mSpp-1, these VDREs are degenerate and do not have perfect direct repeats. It has been reported that VDR homodimers can bind to the Spp-1 VDRE with high affinity (Freedman et al 1994, Nishikawa et al 1994, Cheskis and Freedman 1994, Koszewski et al 1996) but that heterodimerization with RXR is a necessary prerequisite for VDR binding to the other VDREs, which suggests that the specificity and affinity of VDR binding to DNA can be altered by heterodimerization with RXR.

VDR directly derived from mammalian cells has been shown to bind to a VDRE (DR3) with high affinity. An equilibrium dissociation constant of approximately 0.2 nM was calculated from gel-shift analysis and affinity chromatography (Sone et al 1991). Interestingly, VDR-DNA binding was shown to be enhanced in the presence of hormone (Liao et al 1990, Sone et al 1990, Sone et al 1991). It was subsequently demonstrated that high-affinity DNA binding by purified VDR required heterodimerization with a protein present in mammalian cell extract. The affinity of this interaction was demonstrated to be increased 10-fold in the presence of $1,25(OH)_2D_3$ (Sone et al 1991). This protein was later identified as RXR.

Using overexpressed and purified human vitamin D_3 receptor, we (Cheskis and Freedman 1994) have demonstrated that VDR is a monomer at low protein concentration. This protein, however, is able to homodimerize ($K_D \approx 0.5\,\mu M$) in the absence of DNA. Specific DNA binding induces conformational changes, and significantly increases the affinity of VDR homodimerization. Formation of VDR homodimer complex with Spp-1 VDRE can be detected at low nonamolar concentrations of VDR. Interestingly, $1,25(OH)_2D_3$ binding reduces the affinity of VDR homodimerization (Cheskis and Freedman 1996), making the homodimer less stable and reducing the rate of monomer-monomer association. Reduced homodimer binding to DNA, in the presence of $1,25(OH)_2D_3$, is probably due to its lower concentration in solution. In contrast, VDR/RXR heterodimerization can be enhanced by the $1,25(OH)_2D_3$ (Sone et al 1991, Cheskis and Freedman, 1994, Cheskis and Freedman 1996). Formation of the heterodimeric complex can be detected in solution with no DNA added. Using real-time interaction analysis, we demonstrated that ligand binding does not alter stability of the heterodimer (dissociation rates are very similar); instead, the rate of VDR-RXR association in the presence of $1,25(OH)_2D_3$ is increased more than seven times, which significantly increases the overall affinity of this interaction. Therefore, conformational changes induced by ligand binding prompt VDR-RXR intermolecular recognition. In addition, DNA binding further increases the affinity of VDR-RXR heterodimerization.

In the presence of 9-cis RA, much less of the VDR-RXR-DNA complex can be detected, yet at the same time the affinity of this interaction is only 2-fold lower. This may imply that RXR-homodimerization induced by 9-cis RA may limit VDR-RXR complex formation, reducing the concentration of monomeric form of RXR (Zhang et al 1992, Lehmann et al 1993, Cheskis and Freedman 1996). Interestingly, when both receptors are liganded, further reduction in the amount of heterodimer formed is detected (Cheskis and Freedman 1996). The possible reasons for this are that the conformational changes induced by both ligands binding reduce the rate of heterodimer binding to DNA, and/or VDR-RXR heterodimerization on the DNA.

Several analogs of vitamin D_3 that more potently induce differentiation of the myeloid-leukemic cell without inducing hypercalcemia were tested for their effects on VDR dimerization using real-time interaction analysis with Biacore (Cheskis et al 1996). We found that the affinity and kinetics of VDR-RXR heterodimerization, as well as DNA binding, are differentially affected by the binding of these analogs. One of the compounds tested was shown to confer a higher affinity of heterodimer-DNA interaction that correlates with the analogs higher potency on a VDRE-regulated reporter gene in a transient transfection experiment. Hence, ligand structure through conformational changes influences receptor functions.

To test the model derived from *in vitro* studies, a VDRE was linked to one reporter, and an RXRE to another reporter, and they were singly or simultaneously introduced into cells in response to various combinations of overexpressed and endogenous receptors and specific ligands (Lemon and Freedman 1996). In single-reporter transfections, RXR enhanced $1,25(OH)_2D_3$ stimulation from a VDRE-regulated reporter by both endo-

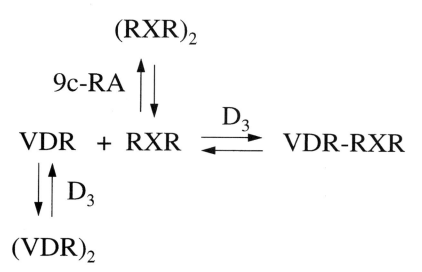

Figure 2.—Ligands modulate the dimerization state of vitamin D receptor. See text for details.

genous and overexpressed VDR, and 9-cis RA strongly inhibited VDRE-driven transactivation. When both the VDRE and RXRE-containing reporter were simultaneously transfected, 9-cis RA again attenuated transactivation from the VDRE, and $1,25(OH)_2D_3$ attenuated the response to 9-cis RA from the RXRE-regulated reporter. The attenuation of $1,25(OH)_2D_3$ responsiveness by 9-cis RA from the VDRE-regulated reporter and of 9-cis RA responsiveness by $1,25(OH)_2D_3$ from the RXRE-regulated reporter was apparent only when VDR was overexpressed relative to RXR. Increased expression of RXR abolished the attenuation, and in some cells conferred a costimulation that was interpreted to be from RXR homodimers binding to and transactivating from the VDRE (Lemon and Freedman 1996). A similar RXR enhancement and 9-cis RA-dependent attenuation of $1,25(OH)_2D_3$-dependent transactivation was also reported for the osteocalcin VDRE (MacDonald et al 1993). Taken together, these results are consistent with those obtained *in vitro* and suggest that the VDR-RXR heterodimer is the transcriptionally active receptor complex *in vivo*, and that ligands of VDR and RXR modulate the intracellular concentration of this complex (Figure 2). Most importantly, the intracellular concentration of RXR may be a limiting factor for the expression of VDR-regulated genes.

Retinoic Acid Receptors

The retinoid response pathway is mediated by two families of nuclear receptors, the RXRs and RARs. Each consists of three isotypes (α, β, γ) encoded by separate genes. This may emphasize the exceptional role played by retinoids in vertebrate development and homeostasis (for reviews, see Chambon 1996, Kastner et al 1994, Keaveny and Stunenberg 1995, Giguere 1994, Mangelsdorf et al 1995, Gronemeyer and Laudet 1996). The RAR family can be activated both by all-trans (ATRA) and 9-cis RA. The RXR family is activated by 9-cis RA.

Early studies with partially purified recombinant receptors showed that RAR and RXR can bind RAREs and RXREs as homodimers when used at high protein concentration (Mangelsdorf et al 1991, Yang et al 1991, Mader et al 1993). It was subsequently reported that RXR can homodimerize in solution, and that 9-cis RA promotes the formation and binding of RXR homodimers to several HREs (Zhang et al 1992). However this 9-cis RA effect could not be reproduced with RXR prepared from a variety of sources (Chen et al 1994). It was also shown that RXR in solution forms dimers and tetramers (Kersten et al 1995a), and that binding of 9-cis RA results in the dissociation of tetramers to dimers. Strong positive cooperativity was detected between tetramer dissociation and 9-cis RA binding (Kersten et al 1995b).

RAR and RXR can also heterodimerize in solution; the affinity of this interaction can be significantly enhanced upon binding to DNA. The presence of ligands, however, does not appear to affect heterodimerization in solution (Nagpal et al 1993, Leid et al 1992), as detected by gel-shift analysis.

A comparison of the *in vitro* DNA binding of RAR and RXR homodimers and of RXR-RAR heterodimers indicates not only that heterodimers bind more efficiently to their response elements, but also that their binding is more selective than that of homodimers. The efficiency and selectivity of DNA binding are determined by the orientation, spacing, and actual sequence of the hexameric motifs, and also to some extent by the sequence of the spacer and flanking sequences. The affinity of RAR-RXR binding to DNA *in vitro* increases in the following order: DR1, DR2, and DR5 (Giguere 1994, Mangelsdorf et al 1994, Glass 1994, Gronemeyer and Laudet 1996).

In the presence of 9-cis RA, RXR homodimers bind strongly to the DR1 RXRE located in the promoter of the CRBP-II and ApoA1 genes and to synthetic TREpal, they have also been shown to bind weakly to DR-5 RARE. The binding of RXR homodimers to the CRBP-II RXRE leads to transcriptional activation, whereas the binding of RXR/RAR heterodimers to the CRBP-II RXRE results in transcriptional repression (Mangelsdorf et al 1991). The expression of this gene has been shown to be induced by an RXR-specific ligand in the absence of an RAR-specific ligand, which indicates the possible involvement of ligand-activated RXR homodimers (Nakshatri and Chambon 1994, Allegretto et al 1995). This repression mechanism by RXR/RAR heterodimers may be unique to this RXRE, since RXR/RAR can apparently activate transcription by binding to the DR1 RXRE in the mouse CRABP-II promoter (Durand et al 1992). Therefore, it seems feasible that the relative concentration of 9-cis RA and ATRA could control the affinity of homo- versus heterodimerization and could determine hence which of the two retinoid pathways is activated.

Because the two heterodimerization partners, RAR and RXR, asymmetrically bind to DR response elements, and interacting in part through dimerization interfaces located in their LBD's, the question arose as to whether this interaction, through allosteric effects, could influence each receptor's functions. Conflicting results have been reported relating to the ligand binding and transactivation properties of RXR. It has been proposed that RXR, within an RXR-RAR heterodimer bound to a DR5, (1) cannot be liganded *in vitro*, (2) is inactive transcriptionally, and (3) essentially serves as a silent partner, whose main function is to enhance the affinity of the other partner's interaction with DNA (Forman et al 1995, Kurokawa et al 1994, Leblane and Stunenberg 1995). In contrast, it was reported that both receptors within a DR5-RXR-RAR complex are liganded *in vitro* and are transcriptionally active in transfected cells (Apfel et al 1995). Similarly, it has been reproted that RXR is silent in DR3- and DR4-bound VDR-RXR and TR-RXR heterodimers, respectively (Forman et al 1995). However, numerous observations, discussed previously in this chapter, suggest that RXR is a ligand-regulated partner for TR and VDR.

Using a direct analysis of ligand binding, Kersten et al. (1995) demonstrated that heterodimerization in fact does not alter the affinity of ligand binding by RXR. Rather, they found that the affinity of RAR for 9-cis RA is considerably higher than that of RXR, and that this difference is retained within the RAR-RXR heterodimer and is not affected by association with cognate DNA.

These results are in agreement with those reported by Chen et al (Chen et al 1996). These authors showed that addition of RAR-specific and RXR-specific ligands leads to transcriptional synergy, which is dependent on the AF-2 domain of RXR. In their experiments, cotreatment of NB4 cells with RXR- and RAR-specific ligands stimulated occupancy of the RARβ2 promoter *in vivo*. Similar results were reported by Ozato and co-workers (Minucci et al 1997). They demonstrated that both RAR and RXR in a heterodimer bind their specific ligands in the presence of specific DNA and undergo conformational changes detectable by limited proteolysis analysis. Furthermore, using genomic footprinting analysis, they found that ligands of both receptors affect response element occupancy. In transfection experiments in cultured cells, the RAR-specific ligand alone activated transcription, while an RXR-specific ligand did not. Co-addition of both RAR and RXR ligands led to a synergistic activation of transcription.

It is feasible, therefore, that ligand binding by RAR, modulating affinity of its interaction with corepressors (Chen and Evans 1995, Horlein et al 1995, Kurokawa et al 1995), is an absolute prerequisite for transactivation. It probably results in the release of the corepressor and promotion of RAR interactions with coactivators and/or basal transcription factors. Corepressors have been reported to associate with unliganded RAR and TR, but not with RXR. Ligands of RXR may influence receptor interactions with other transcription factors and therefore affect (enhance or repress) transactivation.

There are a number of reports suggesting that RAR and RXR can act additively and sometimes synergistically through DR1, DR2, and DR5 elements, and that RXR-selective ligands cooperate with RAR-selective ligands (Angulo et al 1996, Apfel et al 1996, Chen et al 1996, Clifford et al 1996, Horn et al 1996, Nagy et al 1995, Roy et al 1995, Nakshatri and Chambon 1994, Durant et al 1992, Durant et al 1994).

Recently a new ligand of RXR was found that antagonizes the transcriptional effect of RXR homodimers (Lala et al 1996) and at the same time acts as an agonist for RXR-PPAR and RXR-RAR heterodimers. As a result, this dimer-selective compound promotes RXR association with TAF110 and the corepressor SMRT. Accordingly, binding of this compound induces activity of RXR that is distinct from the activity induced by a pure RXR agonist, such as 9-cis RA. Thus, ligand-induced conformational changes may act as a molecular switch, differentially modulating receptor interactions with coactivators, corepressors, and the basal transcription machinery.

Conclusions and Perspectives

During the last several years, significant progress has been made in our understanding of the molecular mechanism of ligand effects on nuclear receptor functions. The recent results from x-ray crystallography and NMR methods have allowed us to visualize receptor and receptor-DNA complexes and detect conformational alterations imposed by ligand binding.

A variety of new compounds, analogs, and mimetics of known hormones with tissue-selective activities, has been recently described. Insight into their activities has been obtained by a variety of techniques, but primarily by indirect approaches, such as transient transfection experiments or the protease sensitivity assay. It has been postulated that these compounds can diferentially modulate interactions between nuclear receptors and other partner proteins. Although the identities of many of these coregulators remain unknown, it has been proposed that their expression level varies in different tissues. If this model is correct, the identification of these coregulators will provide a very important target for the development of tissue-specific regulators, i.e., new receptor ligands.

For a better understanding of the actual potential of a given hormone in the cell, we believe that a spectrum of receptor interactions and more detailed quantitative analyses of these interactions will be necessary. Parameters such as receptor dimerization constants, DNA-binding thermodynamics and kinetic constants, precise concentrations of receptor, hormones, and interacting factors might potentially be used to generate mathematical models of gene regulation. These models will help to explain how ligand-induced conformational changes, which modulate receptor activities, can generate new spectra of biological activities affecting gene expression.

REFERENCES

Allegretto EA, Shevde N, Zou A, Howell SR, Boehm MF, Hollis BW, and Pike W (1995): Retinoid X receptor acts as a hormone receptor in vivo to induce a key metabolic enzyme for 1,25-dihydroxyvitamin D3. *J. Biol. Chem.* 270: 23906–23909.

Angulo AC, Suto C, Heyman RA, and Ghazal, P (1996): Characterization of the sequences of the human cytomegalovirus enhancer that mediate differential regulation by natural and synthetic retinoids. *Mol Endocrinol.* 10: 781–793.

Apfel CM, Kamber M, Klaus M, Mohr P, Keidel S, and LeMotle PK (1995): Enhancement of HL-60 differentiation by a new class of retinoids with selective activity on retinoid X receptor. *J. Biol. Chem.* 270: 30765–30772.

Arbuckle ND, Dauvois S, and Parker MG (1992): Effects of antioestrogens on the DNA binding activity of oestrogen receptors in vitro. *Nucleic Acids Res.* 20: 3839–3844.

Bagchi MK, Ellison JF, Tsai SY, Edwards DP, Tsai MJ, and O'Malley BW (1988):

Steroid hormone-dependent interaction of human progesterone receptor with its target enhancer element. *Mol. Endocrinol.* 2: 1221–1229.

Bagchi MK, Tsai SY, Tsai MJ, and O'Malley BW (1990): Identification of a functional intermediate in receptor activation in progesterone-dependent cell-free transcription. *Nature* 345: 457–450.

Baniahmad A, Koshne AC, and Renkawitz R (1992): A transferable silencing domain is present in the thyroid hormone receptor, in the v-erbA oncogene product and in the retinoic acid receptor. *EMBO J.* 11: 1015–1023.

Baniahmad A, Steiner C, Köhne AC, and Renkawitz R (1990): Modular structure of a chicken lysozyme silencer: involvement of an unusual thyroid hormone receptor binding site. *Cell* 61: 505–514.

Baulieu E.E. (1989): Contragestion and other clinical applications of RU 486, an antiprogesterone at the receptor. *Science* 245: 1351–1357.

Becker PB, Gloss B, Schmid W, Strahle U, and Shultz G (1986): In vivo protein-DNA interactions in a glucocorticoid response element require the presence of the hormone. *Nature* 324: 686–688.

Beekman JM, Allan GF, Tsai SY, Tsay M-J, O'Malley BW (1993): Transcriptional activation by the estrogen receptor requires a conformational change in the ligand binding domain. *Mol. Endocrinol.* 7: 1266–1274.

Bendik I, and Pfahl M (1995): Similar ligand-induced conformational changes of thyroid hormone receptors regulate homo- and heterodimeric functions. *J. Biol. Chem.* 270: 3107–3114.

Berg OG, Winter RB, and von Hippel PH (1981): Diffusion-driven mechanisms of protein translocation on nucleic acids. 1. Models and theory. *Biochemistry* 20: 629–648.

Bogazzi F, Hudson LD, and Nikodem VM (1994): A novel heterodimerization partner for thyroid hormone receptor. Peroxisome proliferator-activated receptor. *J. Biol. Chem.* 269: 11683–11686.

Bohen, SP, Kralli A, and Yamamoto KR (1995): Hold 'em and fold 'em: chaperones and signal transduction. *Science* 268: 1303–1304.

Bowler J, Lilley TJ, Pittam JD, and Wakeling AE (1989): Novel steroidal pure antiestrogens. *Steroids* 54: 71–99.

Brent GA, Dunn MK, Harney JW, Gulick T, and Larsen PR (1989): Thyroid hormone aporeceptor represses T3-inducible promoters and blocks activity of the retinoic acid receptor. *New Biol.* 1: 329–336.

Brent GA, Williams GR, Harney JW, Forman BM, Samuels HH, and Larsen PR (1992): Capacity for cooperative binding of thyroid hormone (T3) receptor dimers defines wildtype T3 response elements. *Mol. Endocrinol.* 6: 502–514.

Bugge TH, Pohl J, Lonnoy O, and Stunnenberg HG (1992): RXR alpha, a promiscuous partner of retinoic acid and thyroid hormone receptors. *EMBO J.* 11: 1409–1418.

Burnside J, Darling SD, and Chin WW (1990): A nuclear factor that enhances binding of thyroid hormone receptors to thyroid hormone response elements. *J. Biol. Chem.* 265: 2500–2504.

Cao X, Ross FP, Zhang L, MacDonald PN, Chappel J, and Teitelbaum SL (1993): Cloning of the promoter for the avian integrin beta 3 subunit gene and its regulation by 1,25-dihydroxyvitamin D3. *J. Biol. Chem.* 268: 27371–27380.

Carlberg C, Bendik I, Wyss A, Meier E, Sturzenbecker LJ, Grippo JF, and Hunziker (1993): Two nuclear signalling pathways for vitamin D. *Nature* 361: 657–660.

Cavaille's V, Cauvois S, L'Horset F, Lopez G, Hoare S, Kushner PJ, and Parker MG (1995): *EMBO J.* 14: 3741–3751.

Chambon P (1996): A decade of molecular biology of retinoic acid receptors. *FASEB J.* 10: 940–954.

Chen JD, Evans RM (1995): A transcriptional co-repressor that interacts with nuclear hormone receptors. *Nature* 377: 454–457.

Chen JY, Clifford C, Zusi JE, Starrett D, Tortolani J, Ostrowski P, Reczek P, Chambon P, and Gronemeyer H. (1996): *Nature* 382: 819–822.

Chen ZP, Shemshedini L, Durand B, Noy N, Chambon P, and Gronemeyer H (1994): Pure and functionally homogeneous recombinant retinoid X receptor. *J. Biol. Chem.* 269: 25770–25776.

Cheskis B, and Freedman, LP (1994): Ligand modulates the conversion of DNA-bound vitamin D3 receptor (VDR) homodimers into VDR-retinoid X receptor heterodimers. *Mol. Cell. Biol.* 14: 3329–3338.

Cheskis B, Lemon BD, Uskokovic M, Lomedico P, and Freedman LP (1995): Vitamin D3-retinoid X receptor dimerization, DNA binding, and transactivation are differentially affected by analogus of 1,25-dihydroxyvitamin D3. *Mol. Endocrinol.* 9: 1814–1824.

Cheskis B, and Freedman LP (1996): Modulation of nuclear receptor interactions by ligands: kinetic analysis using surface lasmon resonance. *Biochemistry* 35: 3309–3318.

Cheskis B, Karathanasis SK, and Lyttle CR (1997): Estrogen receptor ligands modulate its interaction with DNA. *J. Biol. Chem.* 272: 11384–11391.

Chin WW (1991): Nuclear thyroid hormone receptors. In *Nuclear Hormone Receptors: Molecular Mechanisms, Cellular Functions, Clinical Abnormalities* (ed. Parker MG) pp 79–102, Academic Press, London.

Clifford J, Chiba H, Sobieszezuk D, Metzger D, and Chambon P (1996): RXRalpha-null F9 embryonal carcinoma cells are resistant to the differentiation, anti-proliferative and apoptotic effects of retinoids. *EMBO J.* 15: 4142–4155.

Curtis SW, and Korach KS (1991): Uterine estrogen receptor-DNA complexes: effects of different ERE sequences, ligands, and receptor forms. *Mol. Endocrinol.* 5: 959–966.

Dalman FC, Koenig RJ, Perdew GH, Massa E, and Pratt WB (1990): In contrast to the glucocorticoid receptor, the thyroid hormone receptor is translated in the DNA binding state and is not associated with hsp90. *J. Biol. Chem.* 265: 3615–3618.

Dalman FC, Sturzenbecker LJ, Levin AA, Lucas DA, Perdew GH, Petkovitch M, Chambon P, Grippo F, and Pratt WB (1991): Retinoic acid receptor belongs to a subclass of nuclear receptors that do not form "docking" complexes with hsp90. *Biochemistry* 30: 5605–5608.

Damm K, Thompson CC, and Evans RM (1989): Protein encoded by v-erbA functions as a thyroid-hormone receptor antagonist. *Nature* 339: 593–597.

Darwish HM, and DeLuka HF (1992): Identification of a 1,25-dihydroxyvitamin D3-response element in the 5′-flanking region of the rat calbindin D-9k gene. *Proc. Natl. Acad. Sci. USA* 89: 603–607.

Dauvois S, Danielian PS, White R, and Parker MG (1992): Antiestrogen ICI 164,384 reduces cellular estrogen receptor content by increasing its turnover. *Proc. Natl. Acad. Sci. USA* 89: 4037–4041.

Davis KD, Borrodin TJ, Winkler JD, and Lazar MA (1994): Endogenous retinoid X receptors can function as hormone receptors in pituitary cells. *Mol. Cell. Biol.* 14: 7105–7110.

DeMarzo AM, Nordeen SK, and Edwards DP (1992): Effects of the steroid antagonist RU486 on dimerization of the human progesterone receptor. *Biochemistry* 31: 10491–10501.

DeMarzo AM, Beck CA, Onate SA, and Edwards DP (1991): Dimerization of mammalian progesterone receptors occurs in the absence of DNA and is related to the release of the 90-kDa heat shock protein. *Proc. Natl. Acad. Sci.* USA 88: 72–76.

Demay MB, Gerardi JM, Deluca HF, Kronenberg HM (1990): DNA sequences in the rat osteocalcin gene that bind the 1,25-dihydroxyvitamin D3 receptor and confer responsiveness to 1,25-dihydroxyvitamin D3. *Proc. Natl. Acad. Sci. USA* 87: 369–373.

Desvergne B, Pett, KJ, and Nikodem VM (1991): Functional characterization and receptor binding studies of the malic enzyme thyroid hormone response element. *J. Biol. Chem.* 266: 1008–1013.

Drouin J, Sun YL, Tremblay S, Levender P, Schmidt TJ, Lean A, and Nemert M (1992): Homodimer formation is rate-limiting for high affinity DNA binding by glucocorticoid receptor. *Mol. Endocrinol.* 6: 1299–1309.

Durand B, Saunders M, Leroy P, Leid M, and Chambon P. (1992): All-trans and 9-cis retinoic acid induction of CRABPII transcription is mediated by RAR-RXR heterodimers bound to DR1 and DR2 repeated motifs. *Cell* 71: 73–85.

Durand B, Saunders M, Gaudon C, Roy B, Losson R, and Chambon, P. (1994): Activation function 2 (AF-2) of retinoic acid receptor and 9-cis retinoic acid receptor: presence of a conserved autonomous constitutive activating domain and influence of the nature of the response element on AF-2 activity. *EMBO J.* 13: 5370–5382.

El-Ashry D, Onate S, Nordeen S, and Edwards DP (1989): Human progesterone receptor complexed with the antagonist RU 486 binds to hormone response elements in a structurally altered form. *Mol. Endocrinol.* 3: 1545–1558.

Fawell SE, Lees JA, White R, and Parker MG (1990): Characterization and colocalization of steroid binding and dimerization activities in the mouse estrogen receptor. *Cell* 60: 953–962.

Fawell SE, White R, Hoare S, Sydenham M, Page M, and Parker MG (1990): Inhibition of estrogen receptor-DNA binding by the "pure" antiestrogen ICI 164,384 appears to be mediated by impaired receptor dimerization. *Proc. Natl. Acad. Sci.* USA 87: 6883–6887.

Forman BM, Umesono K, Chen J, and Evans RM (1995): Unique response pathways are established by allosteric interactions among nuclear hormone receptors. *Cell* 81: 541–550.

Forman BM, and Evans RM (1995): Nuclear hormone receptors activate direct, inverted, and everted repeats. *Annals of the New York Academy of Sciences* 761: 29–37.

Freedman LP, Arce V, and Perez R, (1994): DNA sequences that act as high affinity targets for the vitamin D3 receptor in the absence of the retinoid X receptor. *Mol. Endocrinol.* 8: 265–273.

Furlow JD, Murdoch FE, and Gorski J (1993): High affinity binding of the estrogen receptor to a DNA response element does not require. *J. Biol. Chem.* 268: 12519–12525.

Giguere V (1994): Retinoic acid receptors and cellular retinoid binding proteins: complex interplay in retinoid signaling. *Endocr. Rev.* 15: 61–79.

Glass CK (1994): Differential recognition of target genes by nuclear receptor monomers, dimers, and heterodimers. *Endocr. Rev.* 15: 391–407.

Glass CK, and Holloway JM (1990) *Biochim. Biophys. Acta* 1032: 157–176.

Glass CK, Holloway JM, Devary OV, and Rosenfeld MG (1988): The thyroid hormone receptor binds with opposite transcriptional effects to a common sequence motif in thyroid hormone and estrogen response elements *Cell* 54: 313–323.

Gordon MS, and Notides AC (1986): Computer modeling of estrodiol interactions with the estrogen receptor. *J. Steroid Biochem* 25: 177–181.

Graupner G, Wills KN, Tzukerman M, Zhang, X-k, and Pfahl M (1989): Dual regulatory role for thyroid-hormone receptors allows control of retinoic-acid receptor activity. *Nature* 340: 653–656.

Gronemeyer H, and Laudet V (1996): Transcription factors 3: Nuclear Receptors. *Protein profile, Vol. 2,* Academic Press, New York.

Guiochon-Mantel A, Loosfelt H, Lescop P, Sar S, Atger M, Perrot-Applanat M, and Milgrom E (1989): Mechanisms of nuclear localization of the progesterone receptor: evidence for interaction between monomers. *Cell* 57: 1147–1154.

Halachmi S, Marden E, Martin G, MacKay H, Abbondanza C, and Brown M (1994): Estrogen receptor-associated proteins: possible mediators of hormone-induced transcription. *Science* 264: 1455–1458.

Hermann T, Hoffmann B, Zhang X-k, Tran P, and Pfahl M (1992): Heterodimeric receptor complexes determine 3,5,3′-triiodothyronine and retinoid signaling specificities. *Mol. Endocrinol.* 6: 1153–1162.

Hermann T, Hoffmann B, Piedigrafita FJ, Zhang X-k, and Pfahl M (1993): V-erbA requires auxiliary proteins for dominant negative activity. *Oncogene.* 8: 55–65.

Hirst MA, Hinck L, Danielson M, and Reingold GM (1992): Discrimination of DNA response elements for thyroid hormone and estrogen is dependent on dimerization of receptor DNA binding domains. *Proc. Natl. Acad. Sci.* USA 89: 5527–5531.

Horlein AJ, Naar AM, Heinzel T, Torchia J, Gloss B, Kurokawa R, Ryan A, Kamei Y, Soderstrom M, Glass CK, and Rosenfeld MG (1995): Ligand-independent repression by the thyroid hormone receptor mediated by a nuclear receptor co-repressor. *Nature* 377: 397–403.

Horn V, Minucci S, Ogryszko V, Adamson E, Howard B, Levin A, and Ozato K (1996): RAR and RXR selective ligands cooperatively induce apoptosis and neuronal differentiation in P19 embryonal carcinoma cells. *FASEB J.* 10: 1071–1077.

Horwitz KB (1988): Purification, monoclonal antibody production and structural analyses of human progesterone receptors. *J. Steroid Biochem.* 31: 573–578.

Hurd C, Nakao M, Eliezer N, and Moudgill VK (1991): *Molec. Cell. Biochem.* 264: 2203–2211.

Hyder SM, Nawaz Z, Chiappetta C, Yokoyama K, and Stancel GM (1995): The protooncogene c-jun contains an unusual estrogen-inducible enhancer within the coding sequence. *J. Biol. Chem.* 270: 8506–8513.

Ilenchuk TT, and Walters MR (1987): Rat uterine progesterone receptor analyzed by [3H]R5020 photoaffinity labeling: evidence that the A and B subunits are not equimolar. *Endocrinology* 120: 1449–1456.

Kastner P, Leid M, and Chambon P (1994): The role of nuclear retinoic acid receptors in the regulation of gene expression. In *Vitamin A in Health and Disease* (ed. Blomhoff, R.) pp 189–238, Marcel Dekker, New York.

Kastner P, Krust A, Turcotte B, Stropp U, Tora L, Gronemeyer H, and Chambon, P. (1990): Two distinct estrogen-regulated promoters generated transcripts encoding the two functionally different human progesterone receptor forms A and B. *EMBO J* 9: 1603–1614.

Keaveny M, and Stunenberg HG (1995): Retinoic acid receptors. In *Inducible Gene Expression, Vol. 2* (ed. Bauerle PA) pp 187–242, Birkhäuser, Boston.

Kemppainen JA, Lane MV, Sar M, and Wilson EM (1992): Androgen receptor phosphorylation, turnover, nuclear transport, and transcriptional activation. Specificity for steroids and antihormones. *J. Biol. Chem.* 267: 968–974.

Kerner SA, Scott RA, and Pike JW (1989): Sequence elements in the human osteocalcin gene confer basal activation and inducible response to hormonal vitamin D3. *Proc. Natl. Acad. Sci. USA* 86: 4455–4459.

Kersten S, Pan L, and Noy N (1995): On the role of ligand in retinoid signaling: positive cooperativity in the interactions of 9-cis retinoic acid with tetramers of the retinoid X receptor. *Biochemistry* 34: 14263–14269.

Kersten S, Pan L, Chambon P, Gronemeyer H, and Noy N (1995): Role of ligand in retinoid signaling. 9-cis-retinoic acid modulates the ologomeric state of the retinoid X receptor. *Biochemistry* 34 (42): 13717–13721.

Klein-Hitpass K, Tsai SY, Greene GL, Clark JH, Tsai MJ, Argos P, and O'Malley B (1989): Specific binding of estrogen receptor to the estrogen response element. *Mol. Cell. Biol.* 9: 43–49.

Klein-Hitpass L, Cato ACB, Henderson D, and Ryffel G (1991): Two types of antiprogestins identified by their differential action in transcriptionally active extracts from T47D cells. *Nucleic Acid Res.* 19: 1227–1234.

Kliewer SA, Umesono K, Mangelsdorf DJ, and Evans RM (1992): Retinoid X receptor interacts with nuclear receptors in retinoic acid, thyroid hormone and vitamin D3 signalling. *Nature* 355: 446–449.

Klock G, Strahle U, and Schutz G (1987): Oestrogen and glucocorticoid responsive elements are closely related but distinct. *Nature* 329: 734–736.

Koszewski NJ, Reinhardt TA, and Horst RL (1996): *J. Steroid Biochem. Mol. Biol.* 9: 365–384.

Kumar V, and Chambon P. (1988): The estrogen receptor binds tightly to its responsive element as a ligand-induced homodimer. *Cell* 55: 145–156.

Kurokawa R, Soderstorm M, Horlein A, Halachmi S, Brown M, Rosenfeld MG, and Glass CK (1995): Polarity-specific activities of retinoic acid receptors determined by a co-repressor. *Nature* 377: 451–454.

Kurokawa R, DiRenzo J, Boehm M, Sugarman J, Gloss B, Rosenfeld MG, Heyman RA, and Glass CK (1994): Regulation of retinoid signalling by receptor polarity and allosteric control of ligand binding. *Nature* 371: 528–531.

Kurokawa R, Yu VC, Naar A, Kyakumo S, Han Z, Silverman S, Rozenfeld MG, and Glass CK (1993): Differential orientations of the DNA-binding domain and carboxy-terminal dimerization interface regulate binding site selection by nuclear receptor heterodimers. *Genes Dev.* 7: 1423–1435.

Lala DS, Mukherjee R, Schulman IG, Koch SS, Dardashti LJ, Nadzan AM, Croston

GE, Evans RM, and Heyman RA (1996): Activation of specific RXR heterodimers by an antagonist of RXR homodimers. *Nature* 383: 450–453.

Langley E, Zhou Z, and Wilson EM (1995): Evidence for an anti-parallel orientation of the ligand-activated human androgen receptordimer. *J. Biol. Chem.* 270: 29983–29990.

Lazar MA (1993): Thyroid hormone receptors: multiple forms, multiple possibilities. *Endocr. Rev.* 14: 184–193.

Leblane B, and Stunenberg H (1995): 9-cis retinoic acid signaling: changing partners causes some excitement. *Genes Dev.* 9: 1811–1816.

Le Douarin B, Zechel C, Garnier JM, Lutz Y, Tora L, Pierrat B, Heery D, Gronemeyer H, Chambon P, and Losson R (1995): The N-terminal part of TIF1, a putative mediator of the ligand-dependent activation function (AF-2) of nuclear receptors, is fused to B-raf in the oncogenic protein T18. *EMBO J.* 14: 2020–2033.

Lee JW, Ryan F, Swaffield JC, Johnston SA, and Moore DD (1995): Interaction of thyroid-hormone receptor with a conserved transcriptional mediator. *Nature* 374: 91–94.

Lehmann JM, Zhang X-K, Graupner G, Lee, M-O, Hermann T, Hoffmann B, and Pfahl M. (1993): Formation of retinoid X receptor homodimers leads to repression of T3 response: hormonal crosstalk by ligand-induced squelching. *Mol. Cell. Biol.* 13: 7698–7707.

Leid M, Kastner P, Lyons R, Nakshatri H, Saunders M, Zacharewski T, Chen JY, Staub A, Garnier JM, Mader S, and Chambon P (1992): Purification, cloning, and RXR identity of the HeLa cell factor with which RAR or TR heterodimerizes to bind target sequences efficiently. *Cell* 68: 377–395.

Lemon BD, and Freedman LP (1996): Selective effects of ligands on vitamin D3 receptor- and retinoid X receptor-mediated gene activation in vivo. *Mol. Cell. Biol.* 16: 1006–1016.

Leng X, Blanco J, Tsai SY, Ozato, K, O'Malley BW, and Tsai M-J (1994): Mechanisms for synergistic activation of thyroid hormone receptor and retinoid X receptor on different response elements. *J. Biol. Chem.* 269: 31436–31442.

Liao J, Ozono K, Sone T, and Pike JW (1990): Vitamin D receptor interaction with specific DNA requires a nuclear protein and 1,25-dihydroxyvitamin D3. *Proc. Natl. Acad. Sci. USA* 87: 9751–9755.

Liu M, Lee MH, Cohen M, Bommakanti M, and Freedman LP (1996): Transcriptional activation of the Cdk inhibitor p21 by vitamin D3 leads to the induced differentiation of the myelomonocytic cell line U937. *Genes & Dev.* 10: 142–153.

MacDonald PN, Dowd DR, Nakajama S, Galligan MA, Reeder MC, Hausler CA, Ozato K, and Hausler MR (1993): Retinoid X receptors stimulate and 9-cis retinoic acid inhibits 1,25-dihydroxyvitaminD3-activated expression of the rat osteocalcin gene. *Mol. Cell. Biol.* 13(9): 5907–5917.

Mader S, Leroy P, and Chambon P (1993): Multiple parameters control the selectivity of nuclear receptors for their response elements. Selectivity and promiscuity in response element recognition by retinoic acid receptors and retinoid X receptors. *J. Biol. Chem.* 268: 591–600.

Mangelsdorf DJ, Umesono K, and Evans RM (1994): The retinoid receptors. In *The Retinoids: Biology, Chemistry and Medicine* (ed. Sporn MB, Roberts AB, Goodman DS) pp 319–349 Ravens Press, New York.

Mangelsdorf DJ, and Evans RM (1995): The RXR heterodimers and orphan receptors. *Cell* 83: 841–850.

Mangelsdorf DJ, Umesono K, Kliewer SA, Borgmeyer U, Ong ES, and Evans RM (1991): A direct repeat in the cellular retinol-binding protein type II gene confers differential regulation by RXR and RAR. *Cell* 66: 555–561.

Mangelsdorf DJ, Ong ES, Dyck JA, and Evans RM (1990): Nuclear receptor that identifies a novel retinoic acid response pathway. *Nature* 345: 224–229.

Mangelsdorf DJ, Borgmeyer U, Heyman RA, Zhou JY, Ong ES, Oro AE, Kakizuka A, and Evans RM (1992): Characterization of three RXR genes that mediate the action of 9-cis retinoic acid. *Genes & Dev.* 6: 329–344.

Markose EP, Stein JL, Stein GS, and Lain JB (1990): Vitamin D-mediated modifications in protein-DNA interactions at two promoter elements of the osteocalcin gene. *Proc. Natl. Acad. Sci. USA* 87: 1701–1704.

Martinez E, Givel F, and Wahli W (1987): The estrogen-responsive element as an inducible enhancer: DNA sequence requirements and conversion to a glucocorticoid-responsive element. *EMBO J.* 6: 3719–3727.

Medici N, Nigro V, Abbondanza C, Moncharmont B, Molinari, AM, and Puka GA (1991): In vitro binding of the purified hormone-binding subunit of the estrogen receptor to oligonuclotides containing natural or modified sequences of an estrogen-responsive element. *Mol. Endocrin.* 5: 555–563.

Meyer ME, Pornon AJ, Bocquel MT, Chambon P, and Gronemeyer H (1991): Agonistic and antagonistic activities of RU486 on the functions of the human progesterone receptor. *EMBO J.* 12: 3923–3932.

Minucci S, Leid M, Toyama R, Saint-Jeannet J-P, Peterson VJ, Horn V, Ishmael JE, Bhattacharyya N, Dey A, Dawid I, and Ozato K (1997): Retinoid X receptor (RXR) within the RXR-retinoic acid receptor heterodimer binds its ligand and enhances retinoid-dependent gene expression. *Mol. Cell. Biol.* 17: 644–655.

Miyamoto T, Suzuki S, and DeGrot LG (1993): High affinity and specificity of dimeric binding of thyroid hormone receptors to DNA and their ligand-dependent dissociation. *Mol. Endo.* 7: 223–231.

Mullick A, and Katzenellenbogen BS (1986): Antiprogestin-receptor complexes: differences in the interaction of the antiprogestin RU38,486 and the progestin R5020 with the progesterone receptor of human breast cancer cells. *Biochem. Biophis. Res. Commun.* 135: 90–97.

Näär AM, Boutin JM, Lipkin SM, Yu VC, Holloway JM, Glass CK, and Rosenfeld MG (1991): The orientation and spacing of core DNA-binding motifs dictate selective transcriptional responses to three nuclear receptors. *Cell* 65: 1267–1279.

Nagpal S, Friant S, Nakshatri H, and Chambon P. (1993): RARs and RXRs: evidence for two autonomous transactivation functions (AF-1 and AF-2). *EMBO J.* 12: 2349–2360.

Nakshatri H, and Chambon P. (1994): The directly repeated RG(G/T)TCA motifs of the rat and mouse cellular retinol-binding protein II genes are promiscuous binding sites for RAR, RXR, HNF-4, and ARP-1 homo- and heterodimers. *J. Biol. Chem.* 269: 890–902.

Nishikawa J, Kitaura M, Matsumoto M, Imagawa M, and Nishihara T (1994): Difference and similarity of DNA sequence recognized by VDR homodimer and VDR/RXR heterodimer. *Nucl. Acid Res.* 22: 2902–2907.

Noda M, Vogel RL, Craig AM, Prahl JP, DeLuca HF, and Denhardt DT (1990): Identification of a DNA sequence responsible for binding of the 1,25-dihydroxy-vitamin D3 enhancement of mouse secreted phosphoprotein 1 (SPP-1 or osteopontin) gene expression. *Proc. Natl. Acad. Sci. USA* 87: 9995–9999.

Notides AC, Lerner N, and Hamilton DE (1985): in *Molecular Mechanism of Steroid Hormone Action* (ed. Moudgil VK) pp 173–179, De Gruyter, Berlin.

Onrate SA, Tsai SY, Tsai M-J, and O'Malley BW (1995): Sequence and characterization of a coactivator for the steroid hormone receptor superfamily. *Science* 270: 1354–1357.

Ozono K, Liao J, Kerner SA, Scott RA, and Pike JW (1990): The vitamin D-responsive element in the human osteocalcin gene. Association with a nuclear proto-oncogene enhancer. *J. Biol. Chem.* 35: 21881–21888.

Perlmann T, Rangarajan PN, Umesono K, and Evans RM (1993): Determinants for selective RAR and TR recognition of direct repeat HREs. *Genes Dev.* 7: 1411–1422.

Picard D, Dhursheed B, Garabedian MJ, Fortin MG, Lindquist S, and Yamamoto KR (1990): Reduced levels of hsp90 compromise steroid receptor action in vivo. *Nature* 348: 163–168.

Piedrafita FJ, and Pfahl M (1994): Thyroid hormone receptors. Vol. 2 in *Inducible Gene Expression*, (ed. Baeurle P) pp 157–185, Birkhäuser, Boston.

Piedrafita FJ, Bendik I, Ortiz A, and Pfahl M (1995): Thyroid hormone receptor homodimers can function as ligand-sensitive repressors. *Mol. Endocrinol.* 9: 563–578.

Pratt WB (1993): The role of heat shock proteins in regulating the function, folding, and trafficking of the glucocorticoid receptor. *J. Biol. Chem.* 268: 21455–21458.

Redeuilh G, Moncharmont B, Secco C, and Baulieu EE (1987): Subunit composition of the molybdate-stabilized "8–9 S" nontransformed estradiol receptor purified from calf uterus. *J. Biol. Chem.* 262: 6969–6975.

Reese JC, and Katzenellenbogen B (1991): Differential DNA-binding abilities of estrogen receptor occupied with two classes of antiestrogens: studies using human estrogen receptor overexpressed in mammalian cells. *Nucleic Acid Res.* 19: 6595–6602.

Reese JC, and Katzenellenbogen B (1992): Examination of the DNA-binding ability of estrogen receptor in whole cells: implications for hormone-independent transactivation and the actions of antiestrogens. *Mol. Cell. Biol.* 12: 4531–4538.

Saatcioglu F, Deng T, and Karin M (1993): A novel cis element mediating ligand-independent activation by c-ErbA: implications for hormonal regulation. *Cell* 75: 1095–1105.

Sabbah M, Gouilleux F, Sola B, Redeuilih G, and Baulieu EE (1991): Structural differences between the hormone and antihormone estrogen receptor complexes bound to the hormone response element. *Proc. Natl. Acad. Sci. USA* 88: 390–394.

Salomonsson M, Haggblad J, O'Malley BW, and Sitbon GM (1994): The human estrogen receptor hormone binding domain dimerizes independently of ligand activation. *J. Steroid Biochem. Mol. Biol.*, 48: 447–452.

Sartorius CA, Tung L, Takimoto GS, and Horwitz KB (1993): Antagonist-occupied human progesterone receptors bound to DNA are functionally switched to transcriptional agonists by cAMP. *J. Biol. Chem.* 268: 9262–9266.

Schauer M, Chalepakis G, Willmann T, and Beato M (1989): Binding of hormone accelerates the kinetics of glucocorticoid and progesterone receptor binding to DNA. *Proc. Natl. Acad. Sci. USA* 86: 1123–1127.

Schräder M, Müller KM, and Carlberg C (1994): Specificity and flexibility of vitamin D signaling. Modulation of the activation of natural vitamin D response elements by thyroid hormone. *J. Biol. Chem.* 269: 5501–5504.

Schrader M, Nayeri S, Kahlen J-P, Muller KM, and Carlberg C (1995): Natural vitamin D3 response elements formed by inverted palindromes: polarity-directed ligand sensitivity of vitamin D3 receptor-retinoid X receptor heterodimer-mediated transactivation. *Mol. Cell. Biol.* 15: 1154–1161.

Skafar D. (1991): Differences in the binding mechanism of RU486 and progesterone to the progesterone receptor. *Biochemistry* 30: 10829–10832.

Skafar DF (1993): Dimerization of the RU486-bound calf uterine progesterone receptor. *J. Steroid Biochem. Mol. Biol.* 44: 39–43.

Sone T, Kerner SA, and Pike JW (1991): Vitamin D receptor interaction with specific DNA. Association as a 1,25-dihydroxyvitamin D3-modulated heterodimer. *J. Biol. Chem.* 266: 23296–23305.

Sone T, McDonnel DP, O'Malley BW, and Pike JW (1990): Expression of human vitamin D receptor in Saccharomyces cerevisiae. Purification, properties, and generation of polyclonal antibodies. *J. Biol. Chem.* 265: 21997–22003.

Sone T, Ozono K, and Pike JW (1991): A 55-kilodalton accessory factor facilitates vitamin D receptor DNA binding. *Mol. Endocrinol.* 5: 1578–1586.

Tora L, Mullick A, Metzger D, Ponglikitmonkgol M, Park I, and Chambon P (1989): The cloned human oestrogen receptor contains a mutation which alters its hormone binding properties. *EMBO J.* 8: 1981–1986.

Tsai SY, Carlstedt-Duke J, Weigel NL, Dahlman K, Gustafsson J-A, Tsai M-J, and O'Malley BW (1988): Molecular interactions of steroid hormone receptor with its enhancer element: evidence for receptor dimer formation. *Cell* 55: 361–369.

Xing H, and Shapiro DJ (1993): An estrogen receptor mutant exhibiting hormone-independent transactivation and enhanced affinity for the estrogen response element. *J. Biol. Chem.* 268: 23227–23233.

Vegeto E, Shahbaz M, Wen D, Goldman M, O'Malley B, and McDonnel D (1993): Human progesterone receptor A form is a cell- and promoter-specific repressor of human progesterone receptor B function. *Mol. Endocrinol.* 7: 1244–1255.

Wahlstrom GM, Sjîberg M, Andersson M, Nordstrom K, and Vennstrom B (1992) Binding characteristics of the thyroid hormone receptor homo- and heterodimers to consensus AGGTCA repeat motifs. *Mol. Endocrinol.* 6: 1013–1022.

Walter P, Green S, Green G, Krust A, Bornert JM, Jeltsch J-M, Staub A, Jensen E, Scrace G, Waterfield M, and Chambon P. (1985): Cloning of the human estrogen receptor cDNA. *Proc. Natl. Acad. Sci. USA* 82: 7889–7893.

Wang H, Peters GA, Zeng X, Tang M, Ip W, and Khan S (1995): Yeast two-hybrid system demonstrates that estrogen receptor dimerization is ligand-dependent in vivo. *J. Biol. Chem.* 270: 23322–23329.

Weigel NL, Beck CA, Estes PA, Prendergast P, Altman M, Christensen K, and Edwards, DP (1992): *Mol. Endocrinol* 23: 348–357.

Williams GR, Harney JW, Forman BM, Samuels HH, and Brent GA (1991): Oligomeric binding of T3 receptor is required for maximal T3 response. *J. Biol. Chem.* 266: 19636–19644.

Willmann T, and Beato M. (1986): Steroid-free glucocorticoid receptor binds specifically to mouse mammary tumour virus DNA. *Nature* 324: 688–691.

Wrange O, Erikson P, and Perlmann T (1989): The purified activated glucocorticoid receptor is a homodimer. *J. Biol. Chem.* 264: 5253–5259.

Umesono K, Murakami KK, Thompson CC, and Evans RM (1991): Direct repeats as selective response elements for the thyroid hormone, retinoic acid, and vitamin D3 receptors. *Cell* 65: 1255–1266.

Umesono K, and Evans RM (1989): Determinants of target gene specificity for steroid/thyroid hormone receptors. *Cell* 57: 1139–1146.

vom Baur E, Zechel C, Heery D, Heine M, Garnier JM, Vivat V, Le Douarin B, Gronemeyer H, Chambon P, and Losson R (1996): Differential ligand-dependent interactions between the AF-2 activating domain of nuclear receptors and the putative transcriptional intermediary factors mSUG1 and TIF1. *EMBO J.* 15: 110–124.

Yang N, Schule R, Mangelsdorf DJ, and Evans RM (1991): Characterization of DNA binding and retinoic acid binding properties of retinoic acid receptor. *Proc. Natl. Acad. Sci. USA* 88: 3559–3563.

Yen PM, Darling DS, Carter RL, Forgione M, Umeda PK, and Chin WW (1992): Riiodothyronine (T3) decreases binding to DNA by T3-receptor homodimers but not receptor-auxiliary protein heterodimers. *J. Biol. Chem.* 267: 3565–3568.

Yu VC, Delsert C, Andersen B, Holloway JM, Devary OV, Naar AM, Kim SY, Boutin JM, Glass CK, and Rosenfeld MG (1991): RXR beta: a coregulator that enhances binding of retinoic acid, thyroid hormone, and vitamin D receptors to their cognate response elements. *Cell* 67: 1251–1266.

Zechel C, Shen X-O, Chen, J-Y, Chen ZP, Chambon P, and Gronemeyer H (1994): The dimerization interfaces formed between the DNA binding domains of RXR, RAR and TR determine the binding specificity and polarity of the full-length receptors to direct repeats. *EMBO J.* 13: 1425–1433.

Zierold C, Darwish HM, and DeLuca HF (1994): Identification of vitamin D-response element in the rat calcidiol (25-hydroxyvitamin D3) 24-hydroxylase gene. *Proc. Natl. Acad. Sci. USA* 91: 900–902.

Zhang X-k, Hoffman B, Tran P, Graupner G, and Phal M (1992): Retinoid X receptor is an auxiliary protein for thyroid hormone and retinoic acid receptors. *Nature* 355: 441–446.

Zhang X-k, Lehmann JM, Hoffmann B, Dawson MI, Cameron J, Graupner G, Hermann T, and Pfahl M. (1992): Homodimer formation of retinoid X receptor induced by 9-cis retinoic acid. *Nature* 358: 587–591.

Zhang X-k, Wills KN, Hermann T, Graupner G, Tzukerman M, and Pfahl M (1991): Ligand-binding domain of thyroid hormone receptors modulates DNA binding and determines their bifunctional roles. *New Biol.* 3: 169–181.

Zhou Z, Lane MV, Kemppainen JA, French FS, and Wilson EM (1995): *Mol. Endocrin.* 9: 208–218.

6

Molecular Mechanisms of Nuclear Receptor-Mediated Transcriptional Activation and Basal Repression

MILAN K. BAGCHI

INTRODUCTION

The regulation of transcription of eukaryotic protein-coding genes by RNA polymerase II occurs mostly during the initiation and early elongation steps. Initiation of transcription involves the ordered assembly of the polymerase and a set of general initiation factors into a multiprotein initiation complex at the target promoter (Roeder 1991, Zawel and Reinberg 1992, Conway and Conway 1993, Buratowski 1994). The regulation of assembly and function of a transcription initiation complex is determined by the presence of distinct DNA sequence elements in the promoter. Sequence-specific transcription factors bind to these control elements and dramatically enhance or silence transcription by influencing preinitiation complex assembly at the core promoter (Tjian and Maniatis 1994). A precise understanding of the mechanisms underlying these gene regulatory events is one of the major goals of molecular biology.

During the past decades, the signal transduction pathway of nuclear hormone receptors has been studied extensively as a model system to investigate the fundamental mechanisms of transcriptional regulation. These receptors represent a large family of structurally related transcription factors responsible for mediating the physiological action of a variety of hormones, such as steroid and thyroid hormones, vitamin D, and retinoids, all of which play crucial roles in cell growth, morphogenesis, and differentiation (Evans 1988, Tsai and O'Malley 1994, Mangelsdorf et al 1995, Beato et al 1995). The members of the nuclear receptor family share important structural

Molecular Biology of Steroid and Nuclear Hormone Receptors
Leonard P. Freedman, Editor
© 1998 Birkhäuser Boston

features: an amino terminus of variable length, a highly conserved DNA binding domain, and a carboxy-terminal hormone binding domain (Figure 1). Although the hormone-occupied receptors are always localized in the nucleus, many of the receptors are found in the nuclear compartment even in the absence of the cognate ligands. The receptors regulate the expression of specific cellular genes by interacting with distinct DNA sequences, termed hormone response elements, in the target gene promoter (Freedman, 1992, Luisi and Freedman 1995, Glass 1994). It is generally believed that the DNA-bound hormone-receptor complexes modulate the activity of the RNA polymerase II transcription machinery at the target promoter, by interacting either directly or indirectly with one or more of the basal transcription factors. Recent studies in many laboratories indicate that nuclear receptors repress or enhance transcription by interacting with multiple coregulatory factors, which function as signaling intermediates between the receptors and the RNA polymerase II transcription machinery (Horwitz et al 1996). In this review, I will discuss the current models of gene activation and silencing by nuclear receptors, emphasizing the rapidly evolving knowledge of the roles of the coregulatory factors in receptor-mediated signaling.

MECHANISMS OF GENE ACTIVATION BY NUCLEAR HORMONE RECEPTORS

In most instances, a promoter-bound nuclear hormone receptor activates transcription of a target gene in a hormone-dependent manner. *In vivo* gene

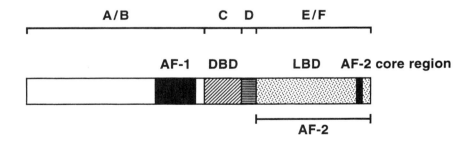

DBD ⟶ DNA binding domain

LBD ⟶ Ligand binding domain

AF-1, AF-2 ⟶ Transactivation Functions 1 & 2

Figure 1.—Structure/Function domains of a typical nuclear hormone receptor.

transfer experiments have revealed that transcriptional activation by these receptors is achieved through autonomous activation functions (AFs): a constitutive AF-1 located in the N-terminal region of the receptor and a hormone-dependent AF-2 located within the C-terminal domain (Gronemeyer 1991). Chambon and co-workers demonstrated that AF-1 and AF-2 domains of estrogen receptor (ER) are functionally distinct and can act independently of each other (Tora 1989, Bocquel 1989, Tasset 1990). Equivalent transactivation domains have been identified in other steroid hormone receptors, such as glucocorticoid receptor (GR) and progesterone receptor (PR), as well as in thyroid hormone receptor (TR) and retinoic acid receptor (RAR) (Hollenberg 1988, Webster 1988, Meyer 1990, Danielian, 1992, Nagpal 1992, Saatcioglu 1993). A central question in receptor-mediated gene regulation is: How does the activation function(s) of the receptor transmit the regulatory signal to the basal transcription machinery ?

A. Interactions of Nuclear Receptor Transactivation Functions with RNA Polymerase II Transcription Machinery

O'Malley and co-workers first explored the mechanism by which a DNA-bound steroid hormone receptor modulates the transcription initiation process at the target promoter (O'Malley et al 1991, Bagchi et al 1992). These investigators utilized a cell-free transcription system reconstituted by combining purified "activated" steroid receptor, such as PR, GR, or ER, with an unfractionated nuclear extract containing RNA polymerase II transcription machinery and a DNA template containing a steroid response element-driven promoter (Klein-Hitpass et al 1990, Bagchi et al 1990, Elliston et al 1990, Tsai et al 1990). The results of the initial studies exploiting this steroid receptor-regulated cell-free system indicated that the receptor stimulates transcriptional initiation by facilitating the formation of a stable preinitiation complex containing RNA polymerase II and other basal transcription factors at the target promoter (Kleinhitpass et al 1990, Bagchi et al 1990).

The formation of a functional preinitiation complex at the TATA-box-containing promoter is a highly ordered process that proceeds through sequential assembly of RNA polymerase II and several other general transcription factors (Figure 2; Buratowski 1989, Zawel and Reinberg 1992, Conway and Conway 1993). The association of TFIID with the TATA box is the first step in the preinitiation complex assembly. The TFIID-TATA complex is then stabilized by binding TFIIA, and the DA-TATA complex is then recognized by TFIIB, producing the DAB-TATA complex (Maldonado et al 1990). The presence of TFIIB is critical for the subsequent recruitment of RNA polymerase II. The association of RNA polymerase II with the DAB complex is mediated by TFIIF (Maldonado et al 1990, Flores et

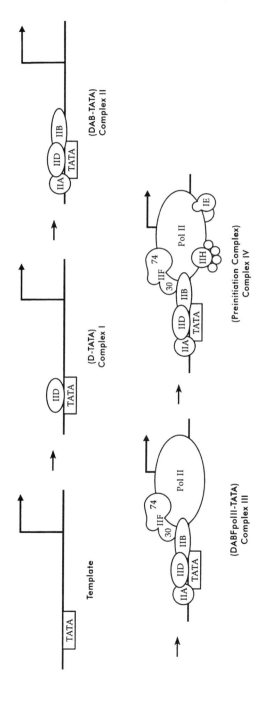

Figure 2. — A model for the assembly of basal RNA polymerase II initiation complexes at the TATA-containing promoter. IIA, IIB, IID, IIE, IIF, IIH, and Pol II denote the transcription factors TFIIB, TFIID, TFIIE, TFIIF, TFIIH, and RNA polymerase II, respectively.

al 1990). The resulting complex, DABpolF, is then recognized by TFIIE and TFIIH (Flores et al 1992, Parvin and Sharp 1993, Goodrich and Tjian 1994). This creates DABpolFEH, which in the presence of ribonucleoside triphosphates, directs a low level of basal RNA synthesis (Goodrich and Tjian 1994). It is reasonable to postulate that the promoter-bound nuclear receptor positively or negatively regulates one or more rate-limiting steps leading to the formation of the functional preinitiation complex.

The receptor may promote transcription initiation by facilitating recognition of promoter by a certain initiation factor(s), or simply by stabilizing the promoter DNA-protein complex once it is formed, or perhaps by influencing both reactions. The receptor may exert its effect by directly interacting with certain of the transcriptional complexes. Alternately, interaction of the receptor with the basal transcriptional machinery might be mediated by a mediator molecule. Since the initial cell-free transcription studies were performed in unfractionated nuclear extracts, the possibility that such mediator factors are indeed present in these nuclear extracts, and participate along with PR in mediating the transcriptional response, could not be excluded. In any case, binding of the receptor or the receptor-mediator complex to an assembling complex may lead to an enhanced recruitment of a downstream basal factor(s). Recently it has been reported that preformed holoenzyme complexes containing RNA polymerase II and all relevant general transcription factors exist in yeast and in higher eukaryotes (Kim et al 1994, Loleske and Young 1994, Ossipow et al 1995). It is conceivable that activators such as steroid hormone receptors can recruit these preformed complexes to the TATA promoter. To understand how nuclear receptor-dependent gene activation occurs, it is crucial to identify the functional target(s) of the receptor during the initiation process.

Many laboratories have attempted to identify the functional target(s) of the activation functions, AF-1 and AF-2, of the nuclear hormone receptors by designing in vitro protein-protein interaction experiments using recombinant affinity-tagged nuclear receptors and radiolabeled basal initiation factors, and vice versa. Studies by Ing et al first revealed that several members of the steroid receptor superfamily, including ER, PR, and COUP-TF, exhibit direct interaction with basal transcription initiation factor TFIIB (Ing et al 1992). This association was observed even when highly purified recombinant receptor and TFIIB were used, suggesting that the interaction occurred independently of additional factors or mediator molecules. Both Baniahmad et al (1993) and Hadzic et al (1995) reported that TR interacts directly with TFIIB in vitro. Blanco et al (1995) observed that vitamin D receptor (VDR) binds TFIIB in a ligand-dependent manner. Collectively, these results indicated that TFIIB is a likely target of the nuclear receptors in the basal initiation complex.

The TATA binding protein (TBP) emerged as another target of nuclear receptor binding. Both AF-1 and AF-2 of ER interact with TBP in vitro

(Sadovsky 1995). Using a yeast two-hybrid assay, Schulman et al (1995) detected specific ligand-dependent interactions between TBP and the LBD of RXR. The AF-2 core region of RXR was found to interact specifically with the conserved carboxy terminal domain of TBP. Recently, Lemon et al (1997) have observed that an RXR-VDR heterodimer facilitates the assembly of TBP-TFIIA, TBP-TFIIA-TFIIB, and TBP-TFIIB complexes on a VDRE-linked promoter, and that the ligand $1,25(OH)_2D_3$ enhanced the formation of these complexes. However, the molecular contacts between VDR and the individual basal factors in these intermediate complexes however remain unknown.

Interactions of nuclear receptors with TBP-associated factors or TAFs in the TFIID complex have also been documented. TR and RXR interact specifically with $TAF_{II}110$ (Schulman et al 1995, Petty et al 1996). Jacq et al (1994) have reported that ER interacts directly with $TAF_{II}30$, a component of the TATA-box binding protein complex TFIID. Although some of the reported interactions between various receptors and either TFIIB or TFIID complex could be functionally significant, in most cases, a correlation between binding activity and hormone-dependent transcriptional activation was not clearly established. For example, the interaction of a nuclear receptor, such as ER, with TBP, TFIIB, or $TAF_{II}30$ occurred in the presence of hormone antagonists that are known to render the receptor transcriptionally inert (Jacq et al 1994, Sadovsky et al 1995). Furthermore, many of the interaction studies are performed under conditions where excess proteins are used *in vitro*, giving rise to the possibility that biologically irrelevant interactions may occur. The true nature of the molecular interactions between the receptor activation domains and the core promoter transcription machinery within the cell that brings about hormone-dependent transcriptional activation of the target promoter therefore remains mostly unresolved.

B. Evidence for the Existence of a Cellular Coactivator(s) That Mediates Transcriptional Activation by Nuclear Receptors

Many sequence-specific transcriptional activators require intermediary cofactors in addition to RNA polymerase II basal transcription machinery to activate gene transcription. These cofactors, termed coactivators, are dispensable for basal transcription but are essential to mediate the response by transactivators (Pugh and Tjian 1990, Pugh and Tjian 1992, Gill and Tjian 1992). The coactivators are envisioned to act as bridging molecules between activation domains and basal transcription machinery (Ptashne 1988). Chambon and co-workers utilized transcriptional interference or "squelching" between different nuclear receptors to suggest that additional cofactors acting in unison with the activated nuclear hormone receptor and the basal

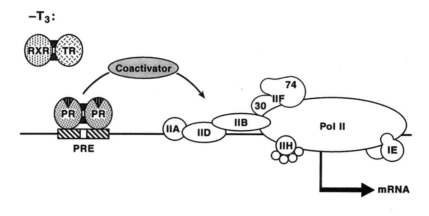

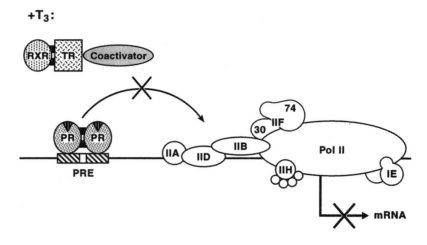

Figure 3.—Ligand-dependent competition between nuclear hormone receptors for a limiting amount of coactivator.

PR, TR, and RXR represent receptors for progesterone, thyroid hormone, and 9-cis retinoic acid, respectively. PRE denotes progesterone response element. A coactivator is depicted as a bridging molecule functionally linking PRE-bound PR homodimers to the basal transcription complex. In the absence of thyroid hormone (T_3), a TR-RXR heterodimer is unable to interact with the coactivator and therefore cannot compete with PR for the limiting amount of coactivator. In the presence of T_3, however, the TR-RXR heterodimer acquires the ability to bind to the coactivator and sequester it, resulting in a suppression of PR-dependent transactivation.

transcription apparatus are necessary for gene activation (Meyer et al 1989 1992). In transcriptional interference experiments, the expression of a given nuclear receptor ("interfering receptor") in a transient transfection system can result in the repression of ligand-induced transcriptional activation of target genes by another member of the superfamily ("activating receptor"). Such mutually antagonistic interactions have been documented between various paired combinations of the steroid receptors ER, GR, PR, and TR (Meyer et al 1989, 1992, Barettino et al 1994, Zhang et al 1996). Conceptually, transcriptional interference occurs as a result of the interaction of the activation domains of nuclear receptors with a common but limiting target protein(s) in their signaling pathways (Figure 3). This phenomenon has therefore been exploited by a number of laboratories to test whether or not two different nuclear receptors share a common target in their gene activation pathways.

Recently Zhang et al (1996) succeeded in mimicking the ligand-dependent transcriptional interference between different nuclear receptors in a cell-free transcription system. In nuclear extracts of T47D cells, addition of T_3-bound recombinant TRβ or its ligand binding domain (LBD) inhibited specifically progesterone-induced transcription. The transcriptional interference was totally dependent on the presence of the hormonal ligand triiodothyronine (T_3). These investigators also observed that T_3-occupied TR LBD could bind to and functionally deplete a soluble cofactor(s) that is critical for transactivation of a progesterone-responsive gene in T47D nuclear extracts, while the basal level of transcription from a minimal TATA promoter, or activated transcription from a control adenovirus major-late promoter, remained unaffected. These results provide strong biochemical evidence in favor of the existence of a common, limiting coactivator molecule that mediates the interaction of ligand-bound PR or TR with the RNA polymerase II transcription machinery, to achieve the activated level of the target gene expression.

The existence of a common coactivator for PR and TR is not surprising, given the presence of highly conserved motifs in the ligand binding domains of these two proteins. Studies by a number of laboratories have identified a 17-amino-acid AF-2 core region containing an amphipathic α-helix, whose main features are well conserved, between all known nuclear receptors that transactivate in a hormone-dependent manner (Figure 4; Danielian et al 1992, Barettino et al 1994, Baniahmad et al, 1995). Point mutations within the AF-2 core domain decrease or abolish ligand-dependent activation, even though the ligand binding, DNA binding, and dimerization properties remain unaffected (Danielian et al 1992, Barettino et al 1994, Baniahmad et al 1995). Furthermore, this domain displays an autonomous activation function when fused to a heterologous DNA binding domain (Barettino et al 1994, Baniahmad et al 1995). Crystal structure analyses of three nuclear receptors, TR, RAR, and RXR, suggest that induction of AF-2 activity upon ligand binding

535	P	L	Y	D	L	L	L	E	M	L	D	A	H	R	L	H	550	hER
748	E	F	P	E	M	L	A	E	I	I	T	N	Q	I	P	K	763	hGR
904	E	F	P	E	M	M	S	E	V	I	A	A	Q	L	P	K	919	hPR
889	D	F	P	E	M	M	A	E	I	I	S	U	Q	V	P	K	904	hAR
955	E	F	P	A	M	L	V	E	I	I	S	D	Q	L	P	K	940	hMR
413	K	L	T	P	L	V	L	E	V	F	G	N	E	I	S	•	427	hVDR
405	S	M	P	P	L	I	Q	E	M	L	E	N	S	E	G	L	420	hRARα
398	S	M	P	P	L	I	Q	E	M	L	E	N	S	E	G	H	413	hRARβ
407	P	M	P	P	L	I	R	E	M	L	E	N	P	E	M	F	422	hRARα
446	P	I	D	T	F	L	M	E	M	L	E	A	P	H	Q	M	461	hRXRα
394	L	F	P	P	L	F	L	E	V	F	E	D	Q	E	V	•	408	hTRα
445	L	L	P	P	L	F	L	E	V	F	E	D	•				456	hTRβ

Figure 4.—Alignment of AF-2 core regions of various nuclear hormone receptors reveals a conserved amphipathic α-helix.

The conserved hydrophobic residues of the α-helix are boxed, and negatively charged residues are shaded.

corresponds to a major conformational change involving the repositioning of the α-helix containing the AF-2 core domain (Bourguet et al 1995, Renaud et al 1995, Wagner et al 1995). These observations have raised the possibility that the AF-2 core domain is involved in creating the surface that interacts directly with a transcriptional coactivator. Consistent with this scenario, a TRβ mutant capable of binding hormone, but lacking six amino acids within the conserved AF-2 core region, failed to squelch PR-mediated transactivation, presumably because of its inability to interact with the common coactivator.

C. Isolation of Candidate Nuclear Receptor Coactivators

Cofactors regulating transcriptional activity of steroid receptors were initially described by genetic studies in yeast, which revealed that one or more nuclear SWI/SNF2 proteins profoundly influence steroid receptor activity. Yoshinaga et al (1992) reported that SWI1, SWI2, and SWI3 proteins control transcriptional activation by GR or ER in yeast. The human homologs of SWI/SNF proteins potentiate transcriptional activation by GR, ER, and RAR (Muchardt and Yaniv 1993, Chiba et al 1994). Although these proteins are present in the RNA polymerase II holoenzyme, the mechanisms by which they enhance transcription remain unclear. It has been proposed that these cofactors stimulate steroid receptor-dependent transcription by relieving the repressive effects of chromatin structure.

During the past two years a number of laboratories have provided strong evidence supporting the existence of intermediary factors or coactivators that mediate the interaction of nuclear hormone receptors with the basal transcription machinery. Using yeast two-hybrid assay or far-Western cloning, several groups have reported the isolation of novel nuclear receptor interacting proteins that may serve as coactivators during hormone-induced transactivation (summarized in Table 1). A genuine coactivator must display the following functional characteristics : (1) it should interact with the receptor in a ligand-dependent manner, (2) overexpression of the receptor interaction domain of this molecule (i.e., truncated coactivator) should interfere with the receptor-mediated gene activation, (3) addition of an excess of this molecule should relieve transcriptional interference by a related nuclear receptor, and (4) receptor-mediated transactivation, but not the basal transcription, should be enhanced upon addition of the coactivator protein to a receptor-regulated gene expression system. So far only a handful of candidate coactivators manage to fulfill these criteria.

Cavailles et al (1994, 1995) and Halachmi et al (1994) have independently reported two nuclear proteins of molecular weights 140 (receptor interacting protein, RIP-140, or estrogen receptor associated protein, ERAP-140) and 160 kDa (RIP-160 or ERAP-160), which interact directly with the hormone-

Table 1.—A Partial List of Candidate Nuclear Receptor Coactivators and Corepressors

Coactivators	Synonyms	Related proteins
hSRC-1 (Onate et al 1995)	p160 (Kamei et al 1996), ERAP 160 (Halachmi et al 1994)	hTIF-2/mGRIP1
hTIF-2 (Voegel et al 1996)	mGRIP1 (Hong et al 1996)	hSRC-1/p160/ERAP 160
ACTR (Chen et al 1997)		hSRC-1/TIF-2
p/CIP (Torchia et al 1997)		hSRC-1/TIF-2
TIF1α (Le Douarin et al 1995)		TIF1β (Le Douarin et al 1996)
RIP 140 (Cavailles et al 1995)		
Trip1 (Lee et al 1995)	ySug1, mSug1 (vom Baur et al 1996)	
CBP (Kwok et al 1994)		p300 (Eckner et al 1994)
Corepressors		
NCoR (Horlein et al 1995)	RIP13 (Lee et al 1995)	SMRT (Chen and Evans 1995)
SMRT	TRAC-2 (Sande and Privalsky, 1996)	NCoR/RIP13 TRAC-1 (Sande and Privalsky)

binding domain of ER in the presence of estrogen but not antiestrogens. The cDNA encoding RIP-140 has been isolated (Cavailles et al 1995). RIP-140 interacts with the conserved AF-2 core region of ER and other nuclear receptors, including RAR, RXR, and TR. Consistent with its proposed role in receptor-mediated transactivation, binding of RIP-140 to ER and RAR is disrupted by point mutations, which inhibit the transcriptional activities of these receptors (Cavailles et al 1995, L'Horset et al 1996). In transient transfection experiments, coexpression of RIP-140 led to only a 2-fold increase in transcriptional activity of the estrogen receptor (Cavailles et al 1995). The reason for this modest effect on receptor-mediated transcription is not clear, but it could be due to already high endogenous levels of RIP-140 in the cell lines tested. RIP-140, when fused to a heterologous DNA binding domain, is able to activate transcription of an appropriate reporter gene, presumably by interacting directly with the basal transcription machinery (L'Horset et al 1996). This property of RIP-140 is also in agreement with its proposed role as a coactivator or bridging protein.

Several groups have recently reported the cloning of the cDNA encoding a 160 kDa nuclear receptor interacting protein. Using yeast two-hybrid screening, Onate et al (1995) reported the isolation of a cDNA encoding a novel coactivator, termed steroid receptor coactivator-1 (SRC-1), which interacts with the LBD of PR in an agonist-dependent manner. This SRC-1 clone encodes a protein of 130 kDa that appears to be an N-terminally truncated version of p160, a 160 kDa protein isolated by Kamei et al (1996) employing far-Western cloning. Chambon and co-workers isolated TIF-2, which is structurally related to, but distinct from, SRC-1 (Voegel et al 1996). Hong et al (1996) independently isolated Grip1, which is identical to TIF-2. In addition, several splice variants of SRC-1 mRNA have been reported (Kamei et al 1996). Two additional proteins, ACTR and p/CIP, which are closely related to SRC-1 or TIF-2, have recently been described (Chen et al 1997, Torchia et al 1997). These results indicate the existence of a novel family of nuclear receptor transcriptional mediators. SRC-1/p160 and TIF-2/Grip1 markedly enhance hormone-induced transactivation by PR, ER, GR, TR, and other nuclear receptors when overexpressed in certain cells in transient transfection assays (Onate et al 1995). A carboxy-terminal nuclear receptor interaction domain of SRC-1 acts as a dominant negative inhibitor of both PR- and TR-mediated transactivation pathways (Onate et al 1995). Furthermore, SRC-1 and TIF-2 relieve transcriptional squelching between different nuclear receptors (Onate et al 1995, Voegel et al 1996). The members of the SRC-1 family are therefore most promising candidate coactivators for nuclear hormone receptors.

Le Douarin et al (1995) isolated a unique cDNA encoding a mouse nuclear protein termed transcription intermediary factor 1α (TIF1α), that interacts with the AF-2 function of a variety of nuclear receptors, such as RAR, RXR, VDR, and ER, in a ligand-dependent manner. Recently they have isolated a structurally related nuclear protein, termed TIF1β (Le Douarin et al 1995). Although the coactivator function of TIF1α or TIFβ could not be demonstrated in transiently transfected cultured cells, it is interesting to note that these proteins contain several conserved domains found in a number of nuclear regulatory proteins such as RING finger and the bromodomain (Le Douarin et al 1995). Using yeast two-hybrid screening, Le Douarin et al (1996) have recently shown that TIF1α interacts with two proteins, mHP1α and mMOD1, which are known to be associated with heterochromatin. On the basis of this observation, these workers have proposed that binding of liganded nuclear receptor to TIF1 might induce chromatin remodeling, converting a transcriptionally "inactive" heterochromatin-like structure to an "active" euchromatin-like structure by triggering the release of HP1 and MOD1. Further studies are clearly required to test this model.

Lee et al (1995a, 1995b) identified a nuclear protein, termed TR-interacting protein 1 (Trip1), that interacts with both TR and RXR in a ligand-dependent fashion. This protein is the human homolog of an essential yeast protein, Sug1 (Lee et al 1995b). von Baur et al (1996) recently reported that the mouse Sug1 interacts with the AF-2 core of a variety of nuclear hormone receptors including ER. Recent data have revealed that Sug1 is a subunit of the 26S proteosome complex that catalyzes the degradation of ubiquitin-conjugated proteins (Rubin et al 1996). The functional significance of Sug1 in ligand-dependent activation by nuclear receptors is therefore unclear.

Recent reports indicate that the cAMP response element-binding protein (CREB) coactivator, CREB-binding protein (CBP), and its homolog P300 may play a mediator role in nuclear receptor-mediated transactivation (Kamei et al 1996, Chakravarti et al 1996, Tso-Pang et al 1996, Hanstein et al 1996). A conserved domain in the N-terminus of CBP/P300 interacts in a ligand-dependent manner with several nuclear receptors including RAR, TR, RXR, and GR (Kamei et al 1996, Chakravarti et al 1996). Furthermore, a C-terminal region of CBP binds independently to the coactivator SRC-1, suggesting that a ternary complex might form between CBP, SRC-1, and the ligand-occupied nuclear receptor at the target promoter (Kamei et al 1996). In transient transfection assays, coexpression of CBP/P300 enhanced ligand-dependent transactivation mediated by either TR, or RAR or RXR. Both Kamei et al (1996) and Chakravarti et al (1996) demonstrated that microinjection of anti-CBP immunoglobulin G inhibits ligand-dependent transcription from retinoid- or glucocorticoid-responsive promoters in intact cells. These observations led to the proposal that CBP/P300 is a common factor required in addition to distinct coactivators for function of diverse transcription factors such as nuclear receptors, CREB, and Jun/Fos. According to this model, CBP serves as an integrator of multiple signaling pathways within the nucleus.

The isolation of additional candidate coactivators has also been reported. Using yeast two-hybrid cloning strategy, Yeh and Chang (1996) identified

ARA_{70}, which interacts with the LBD of AR. Whereas ARA_{70} markedly enhances ligand-dependent transactivation by AR in human prostate cells, it exerts only modest effects on the transcriptional activity of other steroid receptors, such as ER, GR, and PR. These results suggest that ARA_{70} might be specific for AR-mediated transactivation. Imhof and McDonnell (1996) reported that yeast RSP5 and its human homolog hRPF1 potentiate hormone-dependent transactivation by PR and GR without affecting basal transcription. However, more functional characterization of ARA_{70} and RSP5/hRPF1 is needed before they can be considered as genuine coactivators for nuclear receptors.

In an effort to identify specific nuclear proteins that associate with ligand-occupied TR and modulate its transcriptional activity, Fondell et al (1996b) took an interesting approach employing a HeLa-derived cell line that constitutively expresses an epitope-tagged human TRα. Immunopurification of the tagged receptor from hormone-treated cells led to the isolation of at least nine distinct nuclear proteins termed TR-associated proteins (TRAPs). In an *in vitro* system reconstituted from general transcription initiation factors and cofactors, the liganded TR-TRAP complex markedly enhanced transcription from a TRE-linked promoter template. These investigators have proposed that T_3 induces the formation of a multisubunit protein complex *in vivo* containing TR and positive coactivators. The identity and functional roles of individual TRAPs in the receptor-coactivator are currently under investigation. Using a slightly different biochemical approach, Hanstein et al (1996) isolated a group of proteins ranging in size from 30 kDa to 300 kDa from MCF-7 breast cancer cells that associate in a ligand-dependent manner with the LBD of ER. Except for two of these proteins, which have been identified as p300 and SRC-1, the identity and functional activity of the remaining polypeptides remain unknown.

The identification of multiple distinct proteins, SRC-1/p160, TIF-1, RIP-140, TIF-2/Grip1, Sug-1 and CBP, with potential coactivator function raises the intriguing possibility that cell-specific gene regulation by nuclear hormone receptors could be determined by a complex interplay of multiple coactivators with activated receptors that are expressed in that cell. A number of possible scenarios can be considered. For example, differential binding of coactivators to receptor LBD might be a factor determining their activity. Although several candidate coactivators interact with the LBD region containing the conserved amphipathic α-helix, the interaction interface between the receptor and each coactivator may vary, leading to differential activation. Studies by vom Baur et al (1996) have shown that Sug1 and TIF1 display a distinct interaction preference for different nuclear receptors. TR shows a marked preference for Sug1 over TIF1, whereas RXR displays a strong preference for TIF1 over Sug1. RAR and ER interact efficiently with both Sug1 and TIF1 (vom Baur et al 1996). One may also consider the interesting possibility that multiple coactivators binding to the same AF-2 site may modulate each others' activity through competition, and the relative concentrations of different coactivators in a target cell may determine the nature and magnitude of the hormonal response. Moreover, each coactivator may contact the basal transcription machinery differently and therefore may influence the initiation process through different mechanisms. Simultaneous interactions of multiple coactivators with the receptor LBD should also be considered. The work of Kamei et al (1996) suggests that the coactivators CBP and SRC-1 may bind simultaneously to a nuclear receptor, perhaps forming a multiprotein activation complex. In this scenario, the important questions are: How does each individual coactivator function and how do they interact with the basal transcription machinery?

D. Mechanism of Action of a Nuclear Receptor Coactivator: Plausible Models

The details of the molecular interactions between a nuclear receptor and its coactivator remain to be worked out. Recent studies by Jeyakumar et al (1997) demonstrated that a 20-amino-acid peptide containing the AF-2 core domain efficiently inhibits the binding of TR LBD to SRC-1. More importantly, mutations of key acidic (Glu 452, Glu 455, Asp 456) and hydrophobic (Phe 454) amino acids, which impair the ligand-induced transactivation by TR, also abolish the ability of the peptide to interfere with SRC-1 binding to TR. These results provide compelling evidence that the conserved AF-2 core of the receptor is essential for interactions with a nuclear receptor coactivator. Recent crystal structural analyses of the ligand binding domains of TR, RXR, and RAR also suggest that the AF-2 core (helix 12) may play a role in creating the surface for coactivator binding (Bourguet et al 1995, Renaud et al 1995, Wagner et al 1995). In the unoccupied receptor, the helix 12 projects into the solvent (Bourguet et al 1995). In the hormone-occupied receptor, the helix swings back towards the receptor to form a part of the ligand binding cavity (Renaud et al 1995, Wagner et al 1995). The helix packs loosely, with the hydrophobic residues facing inward towards the ligand binding pocket, and the charged residues extending into the solvent. It is conceivable that in the hormone-bound receptor, the helix presents itself as a binding site for the coactivator. As the hydrophobic residues interact with the ligand binding core to stabilize the domain, certain charged residues may be available for direct interaction with the coactivator. Alternatively, the ligand-induced repositioning of helix 12 may trigger a reorganization in other parts of the receptor, creating an interaction surface for the coactivator. In this scenario, the AF-2 core domain may play a critical role by participating in intramolecular interactions with another region(s) of the LBD to help create the proper coactivator binding site.

How does a coactivator provide a functional link between the hormone-bound receptor and the basal transcription machinery? As described above, direct interactions between nuclear hormone receptors and the components of the polymerase II transcription machinery *in vitro* have been documented by a number of laboratories. It is however not clear how the interaction between the ligand-bound receptor and basal transcription factor(s) may eventually lead to the enhancement of the initiation complex formation. One may envision that activation domains such as AF-2, which had been hidden in the unliganded receptor, become unmasked upon ligand binding and available for interaction with the coactivator(s). For steroid receptors such as GR, ER, and PR, ligand-induced release of receptor-associated heat-shock proteins likely precedes coactivator binding (Joab et al 1984, Bagchi et al 1990, 1991). For certain other receptors such as TR and RAR, ligand-dependent displacement of corepressors might be a necessary prerequisite for coactivator action (Horlein et al 1995, Chen and Evans 1995). The coactivator molecule may then function as a physical bridge between the transactivation domain of the receptor and the basal transcription machinery during gene activation. Alternately, ligand-bound nuclear receptors may undergo direct but weak interactions with the basal transcription machinery. In this scenario the role of the coactivator is to promote transcription complex formation by stabilizing the interactions between the receptor and the basal factors. Further studies are clearly necessary to distinguish between these mechanisms.

MECHANISM OF REPRESSION OF BASAL TRANSCRIPTION BY NUCLEAR RECEPTORS

Certain nuclear receptors such as TR and RAR display the dual ability to either activate or repress transcription from genes bearing the cognate hormone response elements. In the absence of hormone, TR or RAR binds to its response element in a ligand-independent manner and functions as a silencer of basal level transcription from the target promoter (Damm et al 1989, Brent et al 1989, Baniahmad et al 1992a). Ligand binding to the receptor releases transcriptional silencing and leads to the activation of target gene expression. The biological significance of transcriptional silencing by unliganded TR or RAR remains unclear. However, recent analyses of transcriptional properties of v-erbA, a mutant TR implicated in oncogenesis by avian erythroblastosis virus and a number of mutant TRs isolated from human patients suffering from a genetic disorder termed generalized resistance to thyroid hormone (GRTH), have shed some light on this question.

The virus-encoded oncoprotein v-erbA is a mutated form of cellular TR that has lost the ability to bind hormonal ligand (Damm et al 1989, Sap et al

1989). It is believed that v-erbA induces neoplastic transformation by arresting normal erythroid differentiation (Gandrillon et al 1989, Schroeder *et al.,* 1990). In cell culture experiments, v-erbA acts as a constitutive silencer of TRE-linked genes and is a dominant negative repressor of the functions of wild-type TRs (Damm et al 1989, Sap et al 1989, Baniahmad et al 1992b). Like TR, v-erbA can potentially form heterodimers with other closely related members of the nuclear receptor superfamily, such as RAR, and interact with their target response elements (Graupner et al 1989, Glass et al 1989). The constitutive silencing activity of v-erbA may, therefore, have far-reaching effects on the expression of at least some RA-responsive genes that regulate cell development and differentiation.

Transcriptional silencing may also contribute to the physiological perturbations associated with the naturally occurring mutations in the gene encoding $TR\beta$ in human patients suffering from GRTH (Refetoff et al 1993). All the GRTH mutations characterized so far result either in failure to bind T_3 or in reduced binding affinity of the hormone. The heterozygous kindreds harbor one mutant and three (1 $TR\beta$ and 2 $TR\alpha$) normal alleles. It appears that the product of the mutant allele interferes efficiently with the function of three normal TR alleles to exhibit dominant negative activity. *In vitro* studies indicate that most of these mutants heterodimerize with RXR and bind to target DNA sites efficiently. Baniahmad et al (1992b), using transient transfection assays, demonstrated that two different GRTH mutants displaying drastically reduced or no T_3 binding activity functioned as constitutive repressors of target genes with strong silencing activity. These results favor the hypothesis that the dominant negative GRTH mutant is not simply an intrinsic transcriptionally nonfunctional receptor. Rather, it is a receptor that has lost the ability to transactivate, but fully retains an active and constitutive silencing function. This viewpoint is consistent with the observation that a homozygous GRTH patient with two mutant (non-hormone-binding but active repressor) $TR\beta$ alleles displayed much more severe clinical symptoms than a homozygous patient who had a complete deletion of the $TR\beta$ gene (Usala et al 1991, Takeda et al 1992).

A. Repression of Basal Transcription by Nuclear Receptors Involves Inhibition of Initiation Complex Assembly

The molecular basis of transcriptional silencing by TR, and the hormonal modulation of this activity have been investigated recently by a number of laboratories. The initial observation that unliganded TR efficiently repressed basal transcription from a TRE-linked minimal promoter containing only a TATA box suggested that the initiation complex assembly process is a potential target of transcriptional silencing by unliganded TR (Damm et al 1989, Sap et al 1989). Consistent with this prediction, recent studies by

Fondell et al (1993) and Tong et al (1995) using a cell-free reconstituted transcription system demonstrated that addition of hormone-free human TRβ during preinitiation complex assembly led to the formation of a transcriptionally inactive complex. In contrast, a fully assembled preinitiation complex is refractory to inhibition by TRβ. These observations suggested that TR functions as a transcriptional repressor by directly influencing an early step leading to the formation of a functional preinitiation complex. The assembling complexes became resistant to repression by TR at a subsequent (undefined) step.

In agreement with this scenario, Tong et al (1995) observed that hormone-free TRβ or TRβ-RXR heterodimer can stably interact with TBP-TFIIB-TATA complex, an early intermediate during initiation complex assembly. These investigators also observed that the binding of TRβ to the TBP-TFIIB-TATA complex is greatly reduced in the presence of T$_3$. In support of these data, Fondell et al (1996a) reported that TRα targets either TBP-TFIIA or TBP-TFIIA-TFIIB-TATA complex for repression of basal transcription. These results led to the proposal that transcriptional silencing by TR involves inhibitory interactions between the receptor and core promoter complexes such as TFIID-TFIIA-TFIIB-TATA (DAB-TATA, Figure 2). These inter-actions may freeze the early intermediate(s) in a non-functional conformation unable to recruit a downstream basal factor(s) such as TFIIF, RNA poly-merase II, TFIIE, or TFIIH. Alternately, TRβ may bind to DAB-TATA and allow it to recruit some or all of the downstream factors. The resulting complex(es), however, enters a nonproductive pathway as the result of an improper configuration. According to this model, ligand binding to TR, which alters receptor conformation, reduces abortive interactions between the receptor and the initiation factors and allows the assembly of productive initiation complexes at the target promoter.

The precise nature of the molecular contacts between TRβ and the individual basal transcription factors remains controversial. Baniahmad et al (1993) demonstrated that the N-terminus of TRβ interacted with the C-terminus of TFIIB in a constitutive manner, while the C-terminal LBD of the receptor recognized the N-terminal region of TFIIB in a hormone-dependent fashion. Previous studies revealed that the C-terminal domain of TFIIB is involved in TFIIB-TBP complex formation at the TATA box, while the N-terminus of TFIIB interacts with TFIIF (Hisatake et al 1993, Ha et al 1993). Taken together, these results raised the interesting possibility that the inter-action of the C-terminus of TRβ with the N-terminus of TFIIB may affect TFIIF binding and subsequent recruitment of RNA polymerase II. Fondell et al (1996a) however, observed recently that TRα can directly interact with TBP as well as TFIIB. Preincubation of TRα with TBP completely alleviated receptor-mediated repression in cell-free transcription extracts, whereas preincubation with TFIIB had no effect. On the basis of these results, Fondell et al (1996a) suggested that unliganded TR interfered with transcription initiation by contacting TBP. The ability of TR to contact TFIIB might be more relevant to a role in activation in than repression.

The work of Baniahmad et al (1993) located the silencing function of TRβ within the C-terminal ligand binding domain between amino acid residues 168–456. Fusion of this fragment of TRβ with the DNA binding domain of an unrelated transcription factor Gal4 resulted in a chimeric protein that bound to a Gal4 target promoter and displayed potent silencing function. These results suggested that the silencing function can be uncoupled from the native DNA binding domain as well as from the specific hormone response element. Two separate C-terminal regions of hTRβ, 168–259 and 260–456, are necessary for gene silencing, and interestingly, these fragments function in *trans* (Baniahmad et al 1993). The 260–456 fragment has been shown to interact with TFIIB. However, this interaction is clearly not sufficient for silencing, and needs the cooperation of the other fragment, 168–259. How multiple C-terminal motifs of TR interact with each other and with the basal transcription machinery to effect silencing is an intriguing question.

B. Roles of Cellular Corepressors in Repression of Basal Transcription by Nuclear Receptors

The initial evidence in favor of the existence of a cellular cofactor mediating TR-dependent silencing was provided by the studies of Casanova et al (1994) and Baniahmad et al (1995), employing chimeric receptors in transient transfection experiments. Casanova et al (1994) demonstrated that coexpression of a LBD peptide (120–410) of chicken TRα and a chimeric protein containing TRα, positioned between a Gal4 DNA binding domain and a herpes simplex virus protein 16 (VP16) activation domain, activated transcription from a Gal4-linked promoter in the absence of thyroid hormone. This transcriptional activation was reversed in the presence of T$_3$. These workers postulated that a cellular inhibitory factor interacted with TR in the chimeric protein to form a receptor-inhibitor complex that suppressed gene activation in the absence of ligand. When the LBD fragment was coexpressed, it competed with the chimeric protein for the limiting amount of inhibitor, and this led to the release of transcriptional inhibition. Binding of ligand to the LBD fragment apparently dissociated the inhibitor-LBD complex. The released inhibitor then bound to the chimeric TR, which itself did not bind to the ligand, to reconstitute gene repression. Casanova et al (1994) suggested that a conserved heptad motif containing leucines (amino acids 365–372) within the LBD of chicken TRα plays a critical role in binding the inhibitory factor.

To analyze the role of the LBD in receptor-mediated silencing, Baniahmad et al (1995) designed competition experiments employing a peptide fragment containing the entire LBD (145–456) of TRβ This peptide, which lacks the

DNA binding domain, did not affect basal RNA synthesis from the TRE-linked promoter when expressed in cultured cells. However, the coexpression of the LBD peptide with full-length TRβ led to a complete reversal of receptor-mediated transcriptional silencing in the absence of thyroid hormone. Similar observations were made by Tong et al (1996) in a cell-free transcription system. These results led to the postulation that the LBD peptide competed with TR for a regulatory molecule, termed a corepressor, that exists in unfractionated nuclear extracts and is essential for efficient receptor-mediated gene repression. The region 168–260 (the D domain) of the LBD was identified as a potential binding site of the putative corepressor (Baniahmad et al 1995, Tong et al 1996). A peptide containing the LBD of retinoic acid receptor (RAR) competed for TR-mediated silencing, suggesting that the RAR LBD may bind to the same corepressor activity as the TR LBD. Interestingly, RAR LBD complexed with its cognate ligand, all-trans retinoic acid, failed to compete for transcriptional silencing by TRβ, indicating that the association of the LBD with the corepressor is ligand-dependent (Tong et al 1996).

Tong et al (1996) also provided strong biochemical evidence supporting the existence of the corepressor activity in the HeLa nuclear extracts. Their studies demonstrated that the silencing activity of TR was greatly reduced in the nuclear extracts preincubated with immobilized, hormone-free GST-LBD fusion proteins, indicating that the corepressor activity was depleted from these extracts through protein-protein interactions with the LBD. Similar treatment with immobilized, hormone-bound GST-LBD, on the other hand, failed to deplete the corepressor activity from the nuclear extracts, indicating that ligand binding to the LBD disrupts its interaction with the corepressor. On the basis of these results, it was proposed that a corepressor binds to the LBD of unliganded TR and critically influences the interaction of the receptor with the basal transcription machinery to promote silencing (Figure 5). Ligand binding to TR results in the release of the corepressor from the LBD and triggers the reversal of silencing by allowing the events leading to gene activation to proceed.

C. Identification of Candidate Nuclear Receptor Corepressors

Recently the cDNAs encoding two distinct but structurally related candidate corepressors, NCoR and SMRT, were isolated by yeast two-hybrid cloning (Table 1; Horlein et al 1995, Chen and Evans 1995, Sande and Privalsky 1996). Horlein et al (1995) reported the identification of NCoR, a novel 270 kDa nuclear protein that binds to the TR or RAR component of DNA-bound TR-RXR or RAR-RXR heterodimers in the absence of T_3 or all-trans-RA, but fails to interact with the ligand-occupied receptors. A recent report suggests that NCoR also acts as a corepressor of an orphan receptor, RevErb, which functions as a constitutive repressor (Zamir et al 1996). The

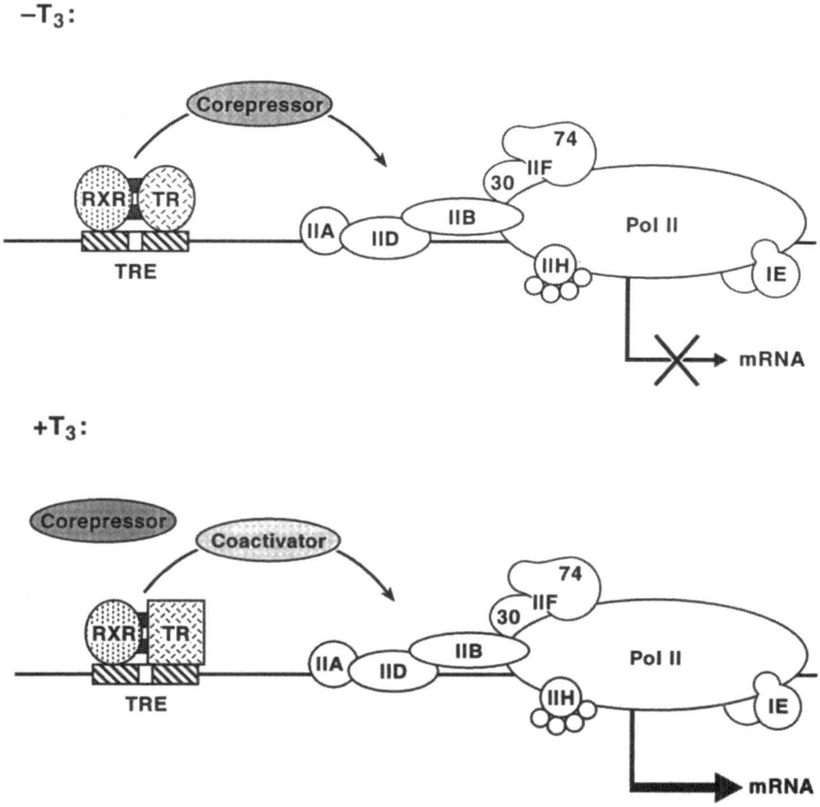

−T$_3$:

+T$_3$:

Figure 5.—A working model for transcriptional repression by TR and its hormone-induced reversal.

In the absence of T$_3$, a corepressor binds to TR and is thought to act as a bridging molecule between the TRE-bound TR-RXR heterodimer and the basal transcription machinery. In the presence of T$_3$, the corepressor is dissociated from TR, allowing the interaction of the heterodimer with a coactivator, which facilitates gene activation.

NCoR interaction domain of TRβ encompasses the hinge region (amino acids 203–230) and the neighboring N-terminal portion (amino acids 230–260) of the LBD. Mutations of conserved amino acids within this domain disrupt NCoR binding as well as basal repression activity of TRβ. Consistent with its role as a mediator of silencing, NCoR, when fused to a heterologous DNA binding domain, can function as a silencer of basal promoter activity. Initial mapping studies indicate that the silencing and receptor interaction domains of the corepressor are physically discrete: extreme N-terminal 300 amino acids of the molecule contain a major repression domain, while the extreme C-terminal 400 amino acids harbor the receptor-interaction domain.

Studies by Kurokawa et al (1995) indicated that the polarity or relative configuration of half-sites of consensus sequence AGGTCA within a retinoid response element can allosterically control the interaction of a RAR-RXR heterodimer with NCoR. A RAR-RXR heterodimer binds to a DR+5 response element consisting of two direct repeats spaced by five base pairs in an asymmetric manner: RXR occupies the upstream half-site and RAR occupies the downstream half-site (Kurokawa et al 1993, Perlmann et al 1993). In the absence of any ligand, the heterodimer interacts with NCoR. Addition of RAR-specific ligand all-*trans*-RA induces dissociation of NCoR bound to the RAR-RXR heterodimers on this element and allows recruitment of coactivators to activate transcription (Kurokawa et al 1995). In contrast, on a DR+1 element in which RAR occupies the upstream half-site, NCoR remains associated with the heterodimer even in the presence of all-*trans*-RA. This failure to release the corepressor results in constitutive repression of a DR+1-linked promoter by RAR-RXR heterodimers. NCoR therefore functions as a polarity-specific corepressor.

Chen and Evans (1995) identified another candidate corepressor, SMRT, which also interacts with the LBD of unliganded TR or RAR. SMRT encodes a polypeptide of relative molecular mass of 168 kDa and is ubiquitously expressed in low quantities. Direct sequence comparison between NCoR and SMRT revealed that these are highly related proteins with an overall amino acid identity of 41% (Chen et al 1996). NCoR and SMRT therefore belong to a new family of receptor regulatory proteins. As is the case with NCoR, association of SMRT with the receptor-DNA complexes is destabilized by ligand. Furthermore, in transient transfection experiments, SMRT can functionally replace the corepressor of TR or RAR. Overexpression of TR or RAR competitors can reverse transcriptional silencing by Gal-TR or Gal-RAR by competition for a limiting amount of corepressor. Coexpression of SMRT reconstitutes the silencing activity of Gal-TR or Gal-RAR. In contrast, the receptor interaction domain of SMRT acts as an antirepressor. A splicing variant, which encodes only the C-terminal receptor interaction domain of SMRT, has been isolated by Sande and Privalsky (1996).

D. Mechanism of Action of a Nuclear Receptor Corepressor: Plausible Models

The molecular interactions between the receptor, the corepressor, and the components of the basal transcription machinery may occur in a number of possible ways. The corepressor may function as a physical link between the promoter-bound receptor and the basal transcription machinery, making contacts with both. As described above, TR may also make direct contacts with the basal transcription machinery, independent of the corepressor. Fondell et al (1993) reported that TRβ is able to repress basal RNA synthesis in a purified, reconstituted transcription system, although the magnitude of this repression is much weaker than to that in an unfractionated system. These results indicate that TR may undergo direct but weak inhibitory interactions with the basal transcription machinery in the absence of the corepressor. In this scenario, the role of the corepressor is to promote repression by stabilizing these interactions between the repressor and the basal factors. Baniahmad et al (1993, 1995) speculated that the binding of the corepressor to the upstream 145–260 region of TRβ may influence the interaction of the downstream 260–456 region with the basal transcription machinery. This model provides a likely mechanism by which multiple C-terminal motifs of TRβ may interact with a specific regulatory factor(s) and the basal transcription machinery to constitute silencing.

Recent studies suggest that the AF-2 core region may influence the interaction of the receptor with corepressor molecules (Schulman et al 1996). Baniahmad et al (1995) described an AF-2 mutant, TR$\beta\Delta C_6$, containing a deletion of six conserved amino acids within the AF-2 core domain. TR$\beta\Delta C_6$ binds ligand and undergoes certain conformational changes in response to ligand binding, but remains a constitutive transcriptional repressor (Leng et al 1993). This is presumably due to a failure to release the corepressor. Therefore, hormone binding is not sufficient to dissociate the corepressor from the AF-2 mutant. Schulman et al (1996) have recently proposed that the AF-2 core domain itself participates in the transition from repressive to active state, by an as yet unknown mechanism. One can speculate that the AF-2 helix helps to maintain a receptor conformation that allows release of corepressor upon hormone binding. The loss of this structure in the TR$\beta\Delta C_6$ mutant prevents corepressor release. Given the crystal structure of TR, however, deletion of six amino acids from the AF-2 core helix is not expected to disrupt the compact bound state of the receptor or its hormone binding cavity, although subtle alterations in the receptor conformation cannot be ruled out (Wagner et al 1995). Alternately, one can imagine that the binding of the coactivator to the AF-2 helix of a ligand-bound receptor serves as the switch that triggers a conformational change, which in turn facilitates corepressor release. Recent observations by Kurokawa et al (1995) however, imply that interactions with the corepressor are dominant over the recruitment of coactivators. Further analyses of the complex interplay between the DNA-bound nuclear receptor and the corepressor or coactivator are therefore needed to discover how these coregulatory molecules modulate the gene regulatory activity of the receptor.

PERSPECTIVES

Major advances have been made during the past few years in elucidating the mechanisms of gene activation and repression by nuclear hormone receptors.

Studies employing *in vivo* gene transfer and *in vitro* cell-free transcription experiments have revealed the existence of novel positive and negative coregulatory factors that are thought to influence the formation of a nuclear-receptor-mediated initiation complex at the target promoter. Despite the remarkable success in identifying candidate coactivators and corepressors, it remains to be established whether any of these factors does in fact perform a critical function during receptor-mediated gene activation or repression. Such insights into the roles of the coactivators and corepressors may become apparent when mice lacking the genes are generated. Another way to evaluate their functional roles would be to exploit the nuclear-receptor-regulated cell-free gene expression systems. An *in vitro* approach employing a transcription system depleted of a coactivator or a corepressor offers a direct means to study regulator functions of these molecules. Some of the future challenges include (1) identification of the rate-limiting step(s) of initiation that is controlled by the receptor-corepressor or receptor-coactivator complexes during repression or activation, (2) identification of the components of the basal RNA polymerase II machinery with which the receptor-coregulator complexes interact, and (3) determination of the functional consequence of such interactions.

An understanding of the role of chromatin structure in regulating the overall levels of receptor-mediated gene expression remains another unattained goal. Recent revelations that candidate nuclear receptor coactivators such as CBP, SRC-1 and ACTR, possess intrinsic histone acetylase activity and that corepressors such as NCoR and SMRT, are associated with histone deacetylase activity, have exciting implications for regulation of transcription through chromatin remodeling (Ogryzko et al 1996, Spencer et al 1997, Chen et al 1997, Heinzel et al 1997 and Nagy et al 1997). The continued development of more sophisticated *in vivo* and *in vitro* nuclear receptor-regulated gene expression systems using appropriate combinations of coactivators, corepressors, basal transcription factors, and DNA templates correctly packaged into chromatin should lead to a better understanding of the molecular modes of action of the receptors in their signal transduction pathways.

ACKNOWLEDGMENTS

I thank Evan Reed for the artwork and Jean Schweis for carefully reading the manuscript.
The author is supported by the grant R01 DK 50257–01 from NIH.

REFERENCES

Bagchi MK, Tsai SY, Weigel NL, Tsai MJ, and O'Malley BW (1990): Regulation of *in vitro* transcription by progesterone receptor: characterization and kinetic studies. *J. Biol. Chem.* 265: 5129–5134.

Bagchi MK, Tsai SY, Tsai M.-J, and O'Malley BW (1990): Identification of a functional intermediate in receptor activation in progesterone-dependent cell-free transcription. *Nature* 345: 547–550.

Bagchi MK, Tsai SY, Tsai M-J, and O'Malley BW (1991): Progesterone enhances gene transcription by receptor free of heat shock proteins hsp90, hsp56, and hsp70. *Mol. Cell Biol.* 11: 4998–5004.

Bagchi MK, Tsai MJ, O'Malley BW, and Tsai SY (1992): Analysis of the mechanism of steroid hormone receptor-dependent gene activation in cell-free systems. *Endocrine Rev.* 13: 525–535.

Baniahmad A, Kohne AC, and Renkawitz R (1992a): A transferable silencing domain is present in the thyroid hormone receptor, in the v-erbA oncogene product and in the retinoic acid receptor. *EMBO J.* 11: 1015–1023.

Baniahmad A, Tsai SY, O'Malley BW, and Tsai M-J (1992b): Kindred S thyroid hormone receptor is an active and constitutive silencer and a repressor for thyroid hormone and retinoic acid responses. *Proc. Natl Acad. Sci. USA* 89: 10633–10637.

Baniahmad A, Ha I, Reinberg D, Tsai SY, Tsai M-J, and O'Malley BW (1993): Interaction of human thyroid hormone receptor β with transcription factor TFIIB may mediate target gene derepression and activation by thyroid hormone. *Proc. Natl. Acad. Sci. USA* 90: 8832–8836.

Baniahmad A, Leng X, Burris TP, Tsai SY, Tsai M-J, and O'Malley BW (1995): The t4 activation domain of the thyroid hormone receptor is required for release of a putative corepressor(s) necessary for transcriptional silencing. *Mol. Cell Biol.* 15: 76–86.

Barettino D, Vivanco Ruiz MM, and Stunnenberg HG (1994) Characterization of the ligand-dependent transactivation domain of thyroid hormone receptor. *EMBO J.* 13: 3039–3049.

Beato M, Herrlich P, and Schutz G. (1995): Steroid hormone receptors: many actors in search of a plot. *Cell* 83: 851–857.

Bocquel MT, Kumar V, Stricker C, Chambon P, Gronemeyer H (1989): The contribution of N- and C-terminal regions of steroid receptors to activation of transcription is both receptor- and cell-specific. *Nucl. Acid Res.* 17: 2581–2595.

Bourguet W, Ruff M, Chambon P, Gronemeyer H, and Moras D (1995): Crystal structure of the ligand-binding domain of the human nuclear receptor RXR-α. *Nature* 375: 377–382.

Brent GA, Dunn MK, Harney JWI, Gulick T, Larsen PR, and Moore DD (1989): Thyroid hormone aporeceptor represses T3 inducible promoters and blocks activity of the retinoic acid receptor. *The New Biologist* 1: 329–336.

Buratowski S, Hahn S, Guarente L, and Sharp PA (1989): Five intermediate complexes in transcription initiation by RNA polymerase II. *Cell* 56: 549–561.

Buratowski S, (1994): The basics of basal transcription by RNA polymersae II. *Cell* 77: 1–3.

Casanova J, Helmer E, Selmi-Ruby S, Qi JS, Au-Fliegner M, Desai-Yajnik V, Koudinova N, Yarm F, Raaka BM, and Samuels HH (1994): Functional evidence for ligand-dependent dissociation of thyroid hormone and retinoic acid receptors from an inhibitory cellular factor. *Mol. Cell Biol.* 14: 5756–5765.

Cavailles V, Dauvois S, Danielian PS, and Parker MG (1994): Interaction of proteins with transcriptionally active estrogen receptors. *Proc. Natl. Acad. Sci. USA.* 91: 10009–10013.

Cavailles V, Dauvois S, L'Horset F, Lopez G, Hoare S, Kushner PJ, and Parker MG (1995): Nuclear factor RIP 140 modulates transcriptional activation by the estrogen receptor. *EMBO J.* 14: 3741–3751.

Chakravarti D, LaMorte VJ, Nelson MC, Nakajima T, Schulman IG, Juguilon H, Montminy M, and Evans RM (1996): Role of CBP/P300 in nuclear receptor signalling. *Nature* 383: 99–103.

Chen JD, and Evans RM (1995): A transcriptional co-repressor that interacts with nuclear hormone receptors. *Nature* 377: 454–457.

Chen JD, Umesono K, and Evans RM (1996): SMRT isoforms mediate repression and anti-repression of nuclear receptor heterodimers. *Proc. Natl. Acad. Sci. USA.* 93: 7567–7571.

Chen H, Lin RJ, Schiltz RL, Chakravarti D, Nash A, Nagy L, Privalsky ML, Nakatani Y, and Evans RM (1997): Nuclear receptor coactivator ACTR is a novel histone acetyltransferase and forms a multimeric activation complex with P/CAF and CBP/p300. *Cell* 90: 569–580.

Chiba H, Muramatsu M, Nomoto A, and Kato H. (1994) Two human homologues of Saccharomyces cerevisiae SNF2/SWI2 and Drosophila BRM genes potentiate transcriptional activation by the glucocorticoid receptor. *EMBO J.* 12: 4279–4290.

Conaway RC, and Conaway JW (1993): General initiation factors for RNA polymerase II. *Ann. Rev. Biochem.* 62: 161–190.

Damm K, Thompson CC, and Evans RM (1989): Protein encoded by v-erbA functions as a thyroid hormone receptor antagonist. *Nature* 339: 593–597.

Danielian PS, White R, Lees JA, and Parker MG (1992): Identification of a conserved region required for hormone dependent transcriptional activation by steroid hormone receptors. *EMBO J.* 11: 1025–1033.

Eckner R, Ewen ME, Newsome D, Gerdes M, DeCaprio JA, Lawrence JB, and Livingstone DM (1994): Molecular cloning and functional analysis of the adenovirus E1A-associated 300-kD protein with properties of a transcriptional activator. *Genes Dev.* 8: 869–884.

Elliston JF, Fawell SE, Klein-Hitpass L, Tsai SY, Tsai MJ, Parker MG, and O'Malley BW (1990): Mechanism of estrogen receptor-dependent transcription in a cell-free system. *Mol. Cell Biol.* 10: 6607–6012.

Evans, RM (1988): The steroid and thyroid hormone receptor superfamily. *Science* 240: 889–895.

Flores O, Ha I, and Reinberg D (1990): Factors involved in specific transcription by mammalian RNA polymerase II: purification and subunit composition of transcription factor IIF. *J. Biol. Chem.* 265: 5629–5634.

Flores O, Lu H, and Reinberg D (1992): Factors involved in specific transcription by mammalian RNA polymerase II: Identification and characterization of factor IIH. *J. Biol. Chem.* 267: 2786–2793.

Fondell JD, Roy AL, and Roeder RG (1993) Unliganded thyroid hormone receptor inhibits formation of a functional initiation complex: Implications for active repression. *Genes and Dev.* 7: 1400–1410.

Fondell JD, Brunel F, Hisatake K, and Roeder RG (1996a): Unliganded thyroid hormone receptor a can target TATA-binding protein for transcriptional repression. *Mol. Cell Biol.* 16: 281–287.

Fondell JD, Ge H, and Roeder RG (1996b): Ligand induction of a transcriptionally active thyroid hormone receptor coactivator complex. *Proc. Natl. Acad. Sci. USA* 93: 8329–8333.

Freedman, L. (1992) Anatomy of the steroid receptor zinc finger region. *Endocrine Rev.* 13: 129–145.

Gandrillon O, Jurdic P, Pain B, Desbois C, Madjar JJ, Moscovici MG, Moscovici C, and Samarut J (1989): Expression of a v-erbA product, an altered nuclear hormone receptor, is sufficient to transform erythrocytic cells in vitro. *Cell* 58: 115–121.

Gill G, and Tjian R (1992): Eukaryotic coactivators associated with the TATA box binding protein. *Current Opin. Genet. Devel.* 2: 236–242.

Glass CK, Lipkin SM, Devary OV, and Rosenfeld MG (1989): Positive and negative regulation of gene transcription by a retinoic acid-thyroid hormone receptor heterodimer. *Cell* 59: 697–708.

Glass CK (1994): Differential regulation of target genes by nuclear receptor monomers, dimers, and heterodimers. *Endocrine Rev.* 15: 391–408.

Goodrich JA, and Tjian R (1994): Transcription factors IIE and IIH and ATP hydrolysis direct promoter clearance by RNA polymerase II. *Cell* 77: 145–156.

Graupner G, Wills KN, Tzukerman M, Zhang X-K, and Pfahl M (1989): Dual regulatory role for thyroid hormone receptors allows control of retinoic acid receptor activity. *Nature* 340: 653–656.

Gronemeyer H (1991) Transcription activation by estrogen and progesterone receptors. *Ann. Rev. Genet.* 25: 89–123.

Ha I, Roberts S, Maldonado E, Sun X, Green MR, and Reinberg D (1993): Multiple functional domains of human transcription factor IIB: Distinct interactions with two general transcription factors and RNA polymerase II. *Genes and Dev.* 7: 1021–1032.

Hadzic E, Desai-Yajnik V, Helmer E, Guo S, Wu S, Koudinova N, Casanova J, Raaka BM, and Samuels HH (1995): A 10-amino-acid sequence in the N-terminal A/B domain of thyroid hormone receptor α is essential for transcriptional activation and interaction with the general transcription factor TFIIB. *Mol. Cell Biol.* 15: 4507–4517.

Halachmi S, Marden E, Martin G, MacKay H, Abbondanza C, and Brown M (1994): Estrogen receptor-associated proteins—possible mediators of hormone-induced transcription. *Science* 264: 1455–1458.

Hanstein B, Eckner R, DiRenzo J, Halachmi S, Liu H, Searcy B, Kurokawa R, and Brown M (1996): p300 is a component of an estrogen receptor coactivator complex. *Proc. Natl. Acad. Sci. USA*, 93: 11540–11545.

Heinzel T, Lavinsky RM, Mullen T-M, Soderstrom M, Laherty CD, Torchia J, Yang WM, Brard G, Ngo SD, Davie JR, Seto E, Eisenman RN, Rose DW, Glass CK, and Rosenfeld MG (1997): A complex containing N-CoR, mSin3, and histone deacetylase mediates transcriptional repression. *Nature* 387: 43–48.

Hisatake K, Roeder RG, and Horikoshi M (1993): Functional dissection of TFIIB domains required for TFIIB-TFIID-promoter complex formation and basal transcription activity. *Nature* 363: 744–747.

Hollenberg SM, and Evans RM (1988): Multiple and cooperative transactivation domains of the human glucocorticoid receptor. *Cell* 55: 899–906.

Hong H, Kohli K, Trivedi A, Johnson DL, and Stallcup MR (1996): GRIP1, a novel mouse protein that serves as a transcriptional coactivator in yeast for the hormone binding domains of steroid receptors. *Proc. Natl. Acad. Sci. USA* 93: 4948–4952.

Horlein AJ, Naar AM, Heinzel T, Torchia J, Gloss B, Kurokawa R, Ryan A, Kamei Y, Soderstrom M, Glass CK, and Rosenfeld MG (1995): Ligand-independent repression by the thyroid hormone receptor mediated by a nuclear receptor co-repressor. *Nature* 377: 397–403.

Horwitz KB, Jackson TA, Bain DL, Richer JK, Takimoto GS, and Tung L (1996) Nuclear receptor coactivators and corepressors. *Mol. Endocrinol.* 10: 1167–1177.

Imhof MO, and McDonnell DP (1996): Yeast RFP5 and its human homolog hRPF1 potentiate hormone-dependent activation of transcription by human progesterone and glucocorticoid receptors. *Mol. Cell Biol.* 16: 2594–2605.

Ing NH, Beekman JM, Tsai SY, Tsai MJ, and O'Malley BW (1992): Members of the steroid receptor superfamily interact with TFIIB (S300-II). *J. Biol. Chem.* 267: 17617–17623.

Jacq X, Brou C, Lutz Y, Davidson I, Chambon P, and Tora L (1994): Human TAFII30 is present in a distinct TFIID complex and is required for transcriptional activation by the estrogen receptor. *Cell* 79: 107–117.

Jeyakumar M, Tanen MR, and Bagchi MK (1997): Analysis of the functional role of steroid receptor coactivator-1 in ligand-induced transactivation by thyroid hormone receptor. *Mol. Endocrinol.* 11: 755–767.

Joab I, Radanyi C, Renoir M, Buchou T, Catelli MG, Binart N, Mester J, and Baulieu EE (1984): Common non-hormone binding component in nontransformed chick oviduct receptors of four steroid hormones. *Nature* 308: 850–853.

Kamei Y, Xu L, Heinzel T, Torchia J, Kurokawa R, Gloss B, Lin SC, Heyman RA, Rose DW, Glass CK, Rosenfeld MG (1996): A CBP integrator complex mediates transcriptional activation and AP-1 inhibition by nuclear receptors. *Cell* 85: 403–414.

Kim YJ, Bjorklund S, Li Y, Sayre MH, and Kornberg RD (1994): A multiprotein mediator of transcriptional activation and its interaction with the C-terminal repeat domain of RNA polymerase II. *Cell* 77: 599–608.

Klein-Hitpass L, Tsai SY, Weigel NL, Allan GA, Riley D, Rodriguez R, Schrader WT, Tsai MJ, and O'Malley BW (1990): The progesterone receptor stimulates cell-free transcription by enhancing the formation of a stable pre-initiation complex. *Cell* 60: 247–257.

Koleske AJ, and Young RA (1994): An RNA polymerase II holoenzyme responsive to activators. *Nature* 368: 466–469.

Kurokawa R, Yu VC, Naar A, Kyakumoto S, Han Z, Silverman S, Rosenfeld MG, and Glass CK (1993): Differential orientations of the DNA-binding domain and carboxy-terminal dimerization interface regulate binding site selection by nuclear receptor heterodimers. *Genes Dev.* 7: 1423–1435.

Kurokawa R, Soderstrom M, Horlein AJ, Halachmi S, Brown M, Rosenfeld MG, and Glass,CK (1995): Polarity-specific activities of retinoic acid receptors determined by a co-repressor. *Nature* 377: 451–454.

Kwok RPS, Lundblad JR, Chrivia JC, Richards JP, Bachinger HP, Brennan RG, Roberts SGE, Green MR, and Goodman RH (1994): Nuclear protein CBP is a coactivator for the transcription factor CREB. *Nature* 370: 223–226.

Le Douarin B, Zechel C, Garnier JM, Lutz Y, Tora L, Pierrat B, Heery D, Gronemeyer H, Chambon P, and Losson R (1995): The N-terminal part of TIF1, a putative mediator of the ligand-dependent activation function (AF-2) of nuclear receptors, is fused to B-raf in the oncogenic protein T18. *EMBO J.* 14: 2020–2033.

Le Douarin B, Nielsen AL, Garnier JM, Ichinose H, Jeanmougin F, Losson R, and Chambon P (1996): A possible involvement of TIF1α and TIF1β in the epigenetic control of transcription by nuclear receptors. *EMBO J.* 15: 6701–6715.

Lee JW, Choi H-S, Gyuris J, Brent R, and Moore DD (1995a): Two classes of proteins dependent on either the presence or absence of thyroid hormone for interaction with the thyroid hormone receptor. *Mol. Endocrinol.* 9: 243–254.

Lee JW, Ryan F, Swaffield JC, Johnston SA, and Moore DD (1995b) Interaction of thyroid hormone receptor with a conserved transcriptional mediator. *Nature* 374: 91–94.

Lemon BD, Fondell JD, and Freedman LP (1997): Retinoid X receptor: Vitamin D3 receptor heterodimers promote stable preinitiation complex formation and direct 1,25-dihydroxyvitamin D3-dependent cell-free transcription. *Mol. Cell Biol.* 17: 1923–1937.

Leng X, Tsai SY, O'Malley BW, and Tsai M-J (1993): Ligand-dependent conformational changes in thyroid hormone and retinoic acid receptors are potentially enhanced by heterodimerization with retinoic X receptor. *J. Steroid Biochem. Mol. Biol.* 46: 643–661.

L'Horset F, Dauvois S, Heery DM, Cavailles V, and Parker MG (1996): RIP-140 interacts with multiple nuclear receptors by means of two distinct sites. *Mol. Cell Biol.* 16: 6029–6036.

Luisi B, and Freedman LP (1995): Dymer, dymer binding tight. *Nature* 375: 359–360.

Maldonado E, Ha I, Cortes P, Weiss L, and Reinberg D (1990): Factors involved in specific transcription by mammalian RNA polymerase II: Role of transcription factors IIA, IID, and IIB during formation of a transcription-competent complex. *Mol. Cell Biol.* 10: 6335–6347.

Mangelsdorf DJ, Thummel C, Beato M, Herrlich P, Schutz G, Umesono K, Blumberg B, Kastner P, Mark M, Chambon P, and Evans RM (1995): The nuclear receptor superfamily: the second decade. *Cell* 83: 835–839.

Meyer ME, Gronemeyer H, Turcotte B, Bocquel MT, Tasset D, and Chambon P (1989) Steroid hormone receptors compete for factors that mediate their enhancer function. *Cell* 57: 433–442.

Meyer ME, Pornon A, Ji J, Bocquel MT, Chambon P, and Gronemeyer H (1990) Agonistic and antagonistic activities of RU486 on the functions of the human progesterone receptor. *EMBO J.* 9: 3923–3932.

Meyer ME, Qurin-Stricker C, Chambon P, and Gronemeyer H (1992): A limiting factor mediates the differential activation of promoters by the human progesterone receptor isoforms. *J. Biol. Chem.* 267: 10882–10887.

Muchardt C, and Yaniv M (1993): A human homologue of Saccharomyces cerevisiae SNF2/SWI2 and Drosophila brm genes potentiates transcriptional activation by the glucocorticoid receptor. *EMBO J.* 12: 4279–4290.

Nagpal S, Saunders M, Kastner P, Durand B, Nakshastri H, and Chambon P (1992): Promoter context- and response element dependent specificity of the transcriptional activation and modulating functions of retinoic acid receptors. *Cell* 70: 1007–1019.

Nagy L, Kao HY, Chakravarti D, Lin RJ, Hassig CA, Ayer DE, Schreiber SL, and Evans RM (1997): Nuclear receptor repression mediated by a complex containing SMRT, mSin3A, and histone deacetylase. *Cell* 89: 373–380.

Ogryzko VV, Schiltz RL, Russanova V, Howard BH, and Nakatani Y (1996): The transcriptional coactivators p300 and CBP are histone acetyltransferases. *Cell* 87: 953–959.

O'Malley BW, Tsai SY, Bagchi MK, Weigel NL, Schrader WT, and Tsai M -J (1991): Molecular mechanism of action of a steroid hormone receptor. *Recent Prog. Hormone Res.* 47: 1–24.

Onate SA, Tsai SY, Tsai MJ, and O'Malley BW (1995): Sequence and characterization of a coactivator for the steroid hormone receptor superfamily. *Science* 270: 1354–1357.

Ossipow V, Tassan, J-P, Nigg EA, and Schibler U (1995): A mammalian RNA polymerase II holoenzyme containing all components required for promoter-specific transcription initiation. *Cell* 83: 137–146.

Parvin JD, and Sharp PA (1993): DNA topology and a minimal set of basal factors for transcription by RNA polymerase II. *Cell* 73: 533–540.

Ptashne M (1988): How do eukaryotic transcriptional activators work? *Nature* 335: 683–689.

Perlmann T, Rangarajan PN, Umesono K, and Evans RM (1993): Determinants for selective RAR and TR recognition of direct repeat HREs. *Genes Dev.* 7: 1411–1422.

Petty KJ, Krimkevich YI, and Thomas D (1996) A TATA-binding protein-associated factor functions as a coactivator for thyroid hormone receptors. *Mol. Endocrinol.* 10: 1632–1645.

Pugh BF, and Tjian R (1990): Mechanism of transcriptional activation by sp1: evidence for coactivators. *Cell* 61: 1187–1197.

Pugh BF, and Tjian R (1992): Diverse transcriptional functions of the multisubunit eukaryotic TFIID complex. *J. Biol. Chem.* 267: 679–682.

Refetoff S, Weiss RE, and Usala SJ (1993): The syndromes of resistance to thyroid hormone. *Endocr. Rev.* 14: 348–399.

Renaud JP, Rochel N, Ruff M, Vivat V, Chambon P, Gronemeyer H, and Moras D (1995): Crystal structure of the RAR-g ligand-binding domain bound to all-trans retinoic acid. *Nature* 378: 681–689.

Roeder RG (1991): The complexities of eukaryotic transcription initiation: regulation of preinitiation complex assembly. *Trends Biochem. Sci.* 16: 402–408.

Rubin DM, Coux O, Wefex I, Hengartner C, Young RA, Goldberg AL, and Finley D (1996): Identification of the gal4 suppressor sug1 as a subunit of the yeast 26S proteosome. *Nature* 379: 655–657.

Saatcioglu F, Bartunek P, Deng T, Zenke M, and Karin M (1993): A conserved C-terminal sequence that is deleted in v-ErbA is essential for the biological activities of c-ErbA (the thyroid hormone receptor). *Mol. Cell Biol.* 13: 3675–3685.

Sadovsky Y, Webb P, Lopez G, Baxter JD, Cavailles V, Parker MG, and Kushner PJ (1995): Transcriptional activators differ in their response to overexpression of TBP. *Mol. Cell Biol.* 15: 1554–1563.

Sande S, and Privalsky ML (1996): Identification of TRACs (T$_3$ receptor-associating cofactors), a family of cofactors that associate with, and modulate the activity of, nuclear hormone receptors. *Mol. Endocrinol.* 10: 813–825.

Sap J, Munoz A, Schmitt J, Stunnenberg H, and Vennstrom B (1989): Repression of transcription mediated at a thyroid hormone response element by the v-erbA oncogene product. *Nature* 340: 242–244.

Schroeder C, Raynoschek C, Fuhrmann U, Damm K, Vennstrom B, and Beug H (1990): The v-erbA oncogene causes repression of erythrocyte-specific genes and an immature, aberrant differentiation phenotype in normal erythroid progenitors. *Oncogene* 5: 1445–1453.

Schulman IG, Chakravarti D, Juguilon H, Romo A, and Evans RM (1995): Interactions between the retinoid X receptor and a conserved region of the TATA-binding protein mediate hormone-dependent transactivation. *Proc. Natl. Acad. Sci. USA,* 92: 8288–8292.

Schulman IG, Juguilon H, and Evans RM (1996): Activation and repression by nuclear hormone receptors: Hormone modulates an equillibrium between active and repressive states. *Mol. Cell Biol.* 16: 3807–3813.

Spencer TE, Jenster G, Burcin MM, Allis CD, Zhou J, Mizzen CA, Mckenna NJ, Onate SA, Tsai SY, Tsai MJ, and O'Malley BW (1997): Steroid receptor coactivator-1 is a histone acetyltransferase. *Nature* 389: 194–197.

Takeda K, Sakurai A, DeGroot LJ, and Refetoff S. (1992) Recessive inheritance of thyroid hormone resistance caused by complete deletion of the protein-coding region of the thyroid hormone receptor-β gene. *J. Clin. Endocrinol. Metab.* 74: 49–55.

Tasset D, Tora L, Fromental C, Scheer E, and Chambon P (1990): Distinct classes of transcriptional activation domains function by different mechanisms. *Cell* 62: 1177–1187.

Tjian R, and Maniatis T (1994): Transcriptional activation: a complex puzzle with few easy pieces. *Cell* 77: 5–8.

Tong G-X, Tanen MR, and Bagchi MK (1995): Ligand modulates the interaction of thyroid hormone receptor b with the basal transcription machinery. *J. Biol. Chem.* 270: 10601–10611.

Tong G-X, Jeyakumar M, Tanen MR, and Bagchi MK (1996): Transcriptional silencing by unliganded thyroid hormone receptor b requires a soluble corepressor that interacts with the ligand-binding domain of the receptor. *Mol. Cell Biol.* 16: 1909–1920.

Tora L, White J, Brou C, Tasset D, Webster NJG, Scheer E, and Chambon P (1989): The human estrogen receptor has two independent nonacidic transcriptional activation functions. *Cell* 59: 477–487.

Torchia J, Rose DW, Inostroza J, Kamei Y, Westin S, Glass CK, and Rosenfeld MG (1997): The transcriptional co-activator p/CIP binds CBP and mediates nuclear–receptor function. *Nature* 387: 677–684.

Tsai M-J, and O'Malley BW (1994): Molecular mechanisms of action of steroid/thyroid receptor superfamily members. *Ann. Rev. Biochem.* 63: 451–486.

Tsai SY, Srinivasan G, Allan GF, Thompson EB, O'Malley BW, and Tsai M-J (1990): Recombinant human glucocorticoid receptor induces transcription of hormone-responsive genes in vitro. *J. Biol. Chem.* 265: 17055–17061.

Tso-Pang Y, Ku G, Zhou N, Scully R, and Livingstone DM (1996): The nuclear hormone receptor coactivator SRC-1 is a specific target of p300. *Proc. Natl. Acad. Sci. USA* 93: 10626–10631.

Usala SJ, Menke JB, Watson TL, Wondisford FE, Weintraub BD, Berard J, Bradley WEC, Ono S, Mueller OT, and Bercu BB (1991): A homozygous deletion in the c-erbAβ thyroid hormone receptor gene in a patient with generalized thyroid hormone resistance: isolation and characterization of the mutant receptor. *Mol. Endocrinol.* 5: 327–335.

Voegel JJ, Heine, MJS, Zechel C, Chambon P, and Gronemeyer H (1996): TIF2, a

160kDa transcriptional mediator for the ligand-dependent activation function AF-2 of nuclear receptors. *EMBO J.* 15: 3667–3675.

vom Baur E, Zechel C, Heery D, Heine MJS, Garnier JM, Vivat V, Le Douarin B, Gronemeyer H, Chambon P, and Losson R (1996): Differential ligand-dependent interactions between the AF-2 activating domain of nuclear receptors and the putative transcriptional intermediary factors mSUG1 and TIF1. *EMBO J.* 15: 110–124

Wagner RL, Apriletti JW, McGrath ME, West BL, Baxter JD, and Fletterick RJ (1995): A structural role for hormone in the thyroid hormone receptor. *Nature* 378: 690–697.

Webster NJG, Green S, Jin JR, and Chambon P (1988): The hormone binding domains of the estrogen and glucocorticoid receptors contain an inducible transcription activation function. *Cell* 54: 199–207.

Yeh S, and Chang C (1996): Cloning and characterization of a specific coactivator, ARA$_{70}$, for the androgen receptor in human prostatic cells. *Proc. Natl. Acad. Sci. USA* 93: 5517–5521.

Yoshinaga SK, Peterson CL, Herskowitz I, and Yamamoto KR (1992): Roles of SWI1, SWI2, and SWI3 proteins for transcriptional enhancement by steroid receptors. *Science* 258: 1598–1604.

Zamir I, Harding HP, Atkins GB, Horlein AJ, Glass CK, Rosenfeld MG, and Lazar MA (1996): A nuclear hormone receptor corepressor mediates trancriptional silencing by receptors with distinct repression domains. *Mol. Cell Biol.* 16: 5458–5465.

Zawel L, and Reinberg D, (1992): Advances in RNA polymerase II transcription. *Current Opin. Cell Biol.* 4: 488–495.

Zhang X, Jeyakumar M, and Bagchi MK (1996): Ligand-dependent cross-talk between steroid and thyroid hormone receptors. *J. Biol. Chem.* 271: 14825–14833.

7

Transcriptional Cross-Talk by Steroid Hormone Receptors

PETER HERRLICH AND MARTIN GÖTTLICHER

INTRODUCTION

In the enormous signal flow between the cell's environment and the nucleus, mechanisms must exist that integrate the fluxes, which balance out opposing influences and permit an ordered realization of a genetic program. In addition, any stimulus inducing signaling to the nucleus must subsequently be "turned off" or limited by a negative control step. These goals of integration and control can be achieved on many levels: (1) at the cell surface (e.g., growth factor receptor antagonists: Hannum et al 1990, Schweitzer et al 1995), (2) during signal transduction (e.g., PKC inhibits the EGF receptor: King and Cooper 1986, Lin et al 1986; proline-directed protein kinase activation is mostly transient and obviously subject to negative regulation), and (3) ultimately in the nucleus, where incoming signals are converted into gene expression. The cross-talk between transcription factors to be covered here represents an example of this latter class.

Among the host of stimuli cells are exposed to, hormones form an important set of signal molecules. Interestingly, they exert both stimulatory and inhibitory influences on cells and on complex processes in the organism. Inhibitory actions of steroid hormones and retinoids have been known for several decades and have a long history of application: the anti-inflammatory action of glucocorticoids, the antiproliferative and anti-tumor-promotive function of many hormones (including the hormone-like vitamin D, retinoids and glucocorticoids) (Baxter and Forsham 1972, Belman and Troll 1972, Viaje et al 1977, Lotan 1980, Gudas et al 1994), the effect of the negative

Molecular Biology of Steroid and Nuclear Hormone Receptors
Leonard P. Freedman, Editor
© 1998 Birkhäuser Boston

feedback loops of hormones on their producing glands and the glands further upstream in the hormone hierarchy, the induction of apoptosis (Wyllie 1980, Ucker 1987). With the recognition that hormones act through nuclear receptors and that nuclear receptors are ligand-dependent transcription factors selecting specific target genes, the notion has found support that inhibitory influences of hormones are also the result of induced changes of gene expression (for a recent series of reviews on the nuclear receptor superfamily see Beato et al 1995, Kastner et al 1995, Mangelsdorf and Evans 1995, Mangelsdorf et al 1995). The inhibitory outcome of hormone action on a cellular function could be generated by hormone-receptor-dependent synthesis of an inhibitory gene product. Alternately, under certain conditions, the hormone receptors could act as transcriptional repressors.

THE NEGATIVE ACTIONS OF HORMONES

For the sake of analysis it may be useful to classify the negative actions of hormones and retinoids.

Apoptosis can be induced by hormones under specific conditions. As the best-studied example, T lymphocytes go into apoptosis upon treatment with glucocorticoid hormone or retinoids (Wyllie 1980, Ucker 1987). The target cells may be in proliferation like the expanding T cell clone upon antigenic stimulation of a T cell line, or seemingly noncycling like the thymocytes.

Cell-cycle inhibition prevails in the action of glucocorticoids or retinoids on skin, of retinoids on certain cancers (Belman and Troll 1972, Viaje et al 1977), and of dioxins, nonphysiologic ligands of the dioxin receptor, on cells in culture (Weiss et al 1996). Cell cycle inhibition accompanies many differentiation-inducing actions of glucocorticoid hormone, e.g., on adipocytes (Williams et al 1991) and on mammary epithelium (Doppler et al 1989).

Repression of transcription of individual genes was first discovered in the mid-1980s (Medcalf et al 1986, Frisch and Ruley 1987). Now classic examples are the repression of synthesis of proopiomelanotropic hormone in the pituitary (Tremblay et al 1988), of metalloproteases in cultured primary fibroblasts and HeLa cells (first presented at Matrix Metalloproteinase Conference, Sandestin Beach, 1989, see Jonat et al 1990, Jonat et al 1992) and (*in vivo*) in osteoblasts (Tuckermann, unpublished), of interleukins and their receptors (reviewed in Baeuerle and Henkel 1994), and of ICAM-1 in T lymphocytes (van de Stolpe et al 1993, van de Stolpe et al 1994, Caldenhoven et al 1995). Several steroid hormones and the retinoids exert these effects.

It is not *a priori* clear that all negative effects, including the induction of apoptosis, the interference with the cell cycle, and inhibition of specific genes, should be considered consequences of one and the same principle.

Are Nuclear Receptors Involved, and If So, Do They Need To Be Liganded?

Several cell lines possess no functional nuclear receptor, e.g., CV-1, COS, glucocorticoid-receptor-deficient S49 and Jurkat T cell lines. These cells cannot activate receptor-dependent target promoters. They also cannot repress AP-1 or NFκB-dependent genes, and cannot be induced by the appropriate hormone to arrest the cell cycle or to go into apoptosis. All these deficiencies in response to a given hormone can be overcome by reintroducing a cDNA expression clone coding for the corresponding nuclear receptor (Gehring and Tomkins 1974, Sibley and Tomkins 1974, Coffino et al 1975, Jonat et al 1990, Yang-Yen et al 1990, Auphan et al 1995, Helmberg et al 1995).

Mice with a glucocorticoid receptor gene disruption die soon after birth from lung failure (Cole et al 1995). Their thymocytes cannot be induced by glucocorticoids to apoptose. Thus, this negative action of glucocorticoids is mediated by their nuclear receptor.

Steroid hormone receptors require ligand binding for both transactivating and inhibiting properties because they are complexed with inactivating chaperones in the nonliganded state. Ligand is required for release of the chaperones and for further activation steps (Beato et al 1995).

In conclusion, liganded receptors are required for the inhibitory hormone actions. Yet, how is a positive transcription factor converted to an inhibitory molecule?

Cross-Talk Between Nuclear Receptors and Other Transcription Factors: A Physiologic Mode of Regulation

An interesting aspect of the negative regulation has been discovered: steroid hormone receptors and the transcription factors AP-1 (Jun-Fos-ATF-family) or NFκB mutually interfere with their transactivating functions (Jonat et al 1990, Lucibello et al 1990, Yang-Yen et al 1990, Schüle et al 1990, Shemshedini et al 1991, Touray et al 1991, Ray and Prefontaine 1994, Scheinman et al 1995b, Caldenhoven et al 1995, Stein and Yang 1995). This type of cross-talk is the subject of this chapter. Other types of cross-talks (Felli et al 1991, Kutoh et al 1992, Stauber et al 1992, Guathier et al 1993, Imai et al 1993, Sylvester and Scholer 1994, Stein and Yang 1995, Minucci et al 1996) may follow similar or identical rules. The nuclear receptors can positively or negatively act on AP-1- and NFκB-dependent promoters that carry no *bona fide* DNA element for the nuclear receptor. Elevated AP-1 or NFκB activity in turn can inhibit nuclear receptor-dependent transcription from promoters that carry only hormone response elements. Unfortunately, many studies have been performed with nonphysiological levels of transiently over-expressed factors. A small set of data with endogenous components is however available: e.g., dexamethasone strongly inhibited AP-1 (Fos/Jun) dependent collagenase transcription in primary skin fibroblasts and in HeLa cells (Jonat et al 1990, Jonat et al 1992), plasminogen activator RNA was reduced (Medcalf et al 1986), endogenous TNFα induced NFκB activity was inhibited by dexamethasone in HeLa cells (Scheinman et al 1995b); in the whole mouse, glucocorticoids inhibited the NFκB activity induced in thymus and lymph nodes by injection of antibodies to CD3 (Auphan et al 1995), putatively elevated Fos activity in mice carrying the lethal albino mutation inhibited the glucocorticoid-induced transcription of tyrosine aminotransfer-ase in the liver (Kelsey and Schütz 1993). Activation of glucocorticoid receptor function in retinal nerve cells in chick embryos depends on the disappearance or downmodulation of Jun (Berko-Flint et al 1994). Thus, mutual interference of nuclear receptors and other transcription factors can be taken as a physiologic mechanism of transcriptional control.

TOWARDS A MECHANISM OF MUTUAL CROSS-TALK

Inhibition of the function of one transcription factor by another could occur at steps preceeding or following its activation. In the past, research concen-trated on the exploration of linear signaling chains; e.g., a hormone binds to and activates a specific receptor, which then finds its cognate promoter element, and, in the case of its *bona fide* activity, turns on transcription. The literature of the last few years has, however, revealed the existence of protein shuttling between compartments (e.g., of Jun and JNK between nucleus and cytoplasm), and of numerous interconnections between puta-tively "unrelated" signaling pathways (e.g., trimeric G proteins couple their seven-membrane-spanning peptide hormone receptors to the EGF receptor and Src) (Daub et al 1996, Luttrell et al 1996). These new directions of research should caution us about assuming *a priori* that nuclear receptors as transcription factors have to exert their effect through transcription in the nucleus. So far there is no evidence that nuclear receptors block protein kinase cascades that activate the transcription factors AP-1 or NFκB (Figure 1A). Interestingly however, both the estrogen receptor and the dioxin receptor, a ligand-dependent transcription factor not related to the steroid receptor family, activate proline-directed protein kinases, seemingly without prior protein synthesis (Ignar-Trowbridge et al 1996, Migliaccio et al 1996, Weiss et al, unpublished 1996). Furthermore, recently a reduction, perhaps indirect, of

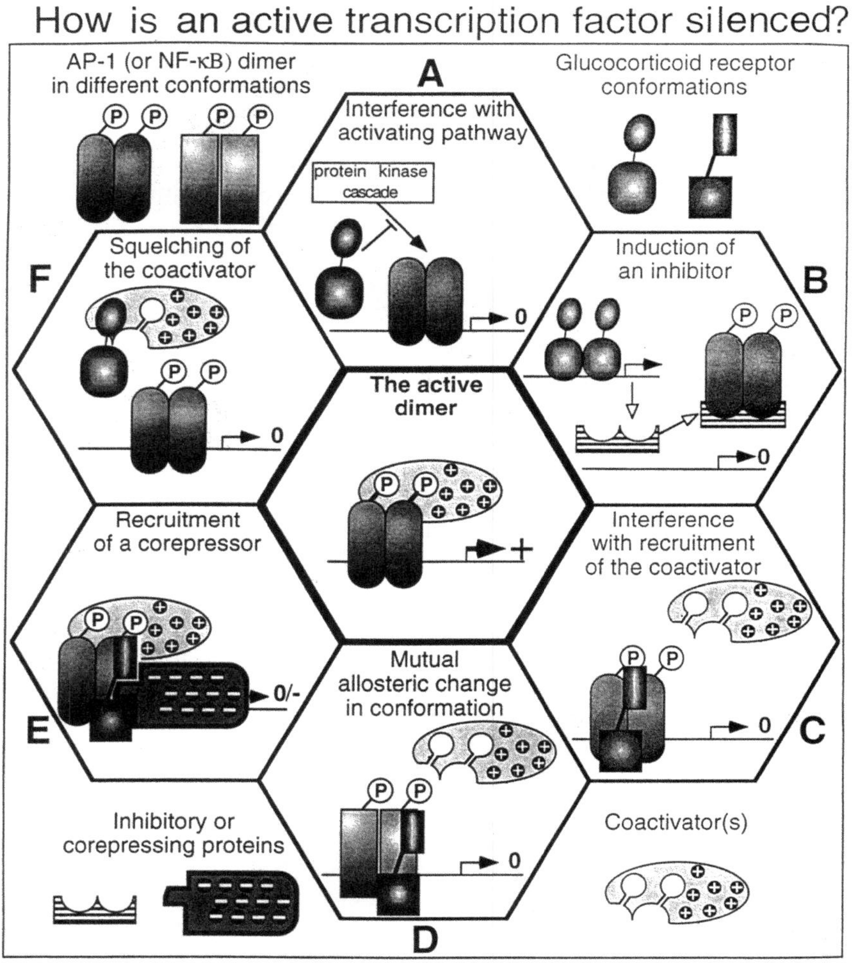

How is an active transcription factor silenced?

Figure 1.—Schematic drawing of the possibilities of how a nuclear receptor could repress the activity of AP-1 or NK-κB transcription factors. The symbols used are explained in the corners of the drawing, and details of the proposed mechanisms are discussed in the text.

S6 kinase by glucocorticoids has been reported (Monfar and Blenis 1996). In the case of AP-1 inhibition by glucocorticoid receptor, phorbol-ester dependent activation and phosphorylation of Jun and Fos was not inhibited by glucocorticoids, nor was Fos expression (Jonat et al 1990), suggesting that interference did not occur prior to AP-1 activation.

PROTEIN-PROTEIN INTERACTIONS AS THE KEY TO TRANSCRIPTIONAL CROSS-TALK

If the nuclear receptor does not interfere with the upstream activating pathways (e.g., of AP-1) and does apparently not need a proper DNA element in the promoter to be inhibited, it may act through the increased synthesis of an inhibitory protein (Figure 1B). Major arguments against this hypothesis are (1) the kinetics of the reaction (inhibition is detectable within 15 minutes), (2) the cycloheximide resistance, (3) the fact that the hormone concentrations required for repression are lower than those for nuclear receptor dependent transcription (Jonat et al 1990), and (4) the fact that the inhibitory and transactivating functions of nuclear receptors can be dissociated by choice of ligands and mutation of the receptor (see below). This statement does not mean that no inhibitor could be induced. An inhibitor of NFκB, IκBα, is indeed inducible by glucocorticoid hormone, both in cultured cells and in lymphocytes within the organism (Auphan et al 1995, Scheinman et al 1995a). Accordingly, the induced inhibition of NFκB bandshift activity of Jurkat cell extracts was cycloheximide-sensitive (Auphan et al 1995). By appropriate choice of conditions, however, NFκB activity can be repressed in an osteo-blast cell line and in fibroblasts without an induction of IκBα (Stein and Yang 1995, Heck et al submitted). Moreover, elevated IκBα levels do not suffice to counteract NFκB activity (Heck et al 1997), suggesting that other modes of interference with NFκB must be operative. These data would be compatible with the cycloheximide sensitivity reported for Jurkat cells if existence of a labile protein in the repression pathway were assumed.

The features of negative transcriptional cross-talk—the fact that the inhibiting factor does not need DNA contact, the rapid cycloheximide-resistant kinetics, and the fact that inhibition is mutual—support the idea that cross-talk of nuclear receptors with AP-1 or NFκB is established by protein-protein interaction. A nuclear receptor/AP-1 or NFκB complex with or without additional partner proteins would conceal the transactivating functions of both transcription factors (Figure 1C and D). This hypothesis is compatible with the following findings. Genomic footprints of the collagenase promoter revealed no major changes upon dexamethasone dependent inhibi-tion of transcription. In particular, the decisive AP-1 site remained occupied, as if the interfering glucocorticoid receptor were tethered to the DNA-bound AP-1 (König et al 1992). The methodology could not rule out the possibility that the AP-1 subunits were exchanged without altering the pattern of protected guanosines. Unfortunately, genomic footprints of an NFκB site in the repressed condition have not been reported. Retinoic-acid-dependent downregulation of Oct3/4, interestingly, correlated with loss of the *in vivo* footprint (Minucci et al 1996). The hormone-activated glucocorticoid recep-tor gave rise to an *in vivo* footprint over the hormone response element of the tyrosine aminotransferase gene promoter, irrespective of the state of inhibition by phorbol ester treatment of hepatoma cells (presumably through activated AP-1) (Reik et al 1994). Attempts to detect interactions *in vitro* have yielded controversial results. Some authors have seen such interactions by coprecipitation with or without cross-linking, or by GST-pulldowns (e.g., (Jonat et al 1990, Schüle et al 1990, Yang-Yen et al 1990, Caldenhoven et al 1995, Scheinman et al 1995b). The *in vitro* complexes in some of these studies seemed unable to bind to DNA. Another laboratory failed to detect direct *in vitro* interaction between nuclear receptor and AP-1 (Lucibello et al 1990).

DISSOCIATION OF TRANSACTIVATION AND TRANSREPRESSION

Many members of the nuclear receptor superfamily are involved in mutual interference with AP-1 and/or NFκB, e.g., the progesterone and androgen receptors (Shemshedini et al 1991), the estrogen receptor (Stein and Yang 1995, Heck et al 1997), and the retinoid receptors (Schüle et al 1990, Fanjul et al 1994, Chen et al 1995). Interference with AP-1 and NFκB by the glucocorticoid receptor has been most thoroughly studied. It would be attractive to postulate a common principle of cross-talk by all members of the nuclear receptor superfamily, much as transactivation depends on highly conserved motif(s) in all nuclear receptors (Barettino et al 1994).

Interestingly, the ability of nuclear receptors to repress transcription can be dissociated from its transactivating function by mutation of the receptor, or by the choice of ligand. Mutants in the D-loop of the glucocorticoid receptor DNA binding domain impair receptor dimerization and binding to the palindromic hormone response element. Nevertheless, they inhibit AP-1 and NFκB (Heck et al 1994, and Heck et al 1997). This finding suggests that repression of AP-1 and NFκB does not require receptor dimerization. Similarly, a mutation of the carboxy terminal side of the second zinc finger (LS7) inhibits perfectly but is severely reduced in transactivation (Godowski et al 1989, Helmberg et al 1995, Heck & Kullmann, unpublished). Other mutations of the DNA binding domain, e.g., point mutants converting to the sequence of the DNA binding domain of the mineralocorticoid receptor, affect transrepression with less or little effect on transactivation (Pearce and Yamamoto 1993, Heck et al 1994, Starr et al 1996). Thus, receptor mutations suggest that repression and transactivation depend, at least in part, on different molecular properties of the receptor.

Transactivating and repressing functions of nuclear receptors are also dissociated by the choice of the ligands. Retinoid acid receptors can be arrested in either a transactivating form or an AP-1 repressing form (Fanjul et al 1994, Chen et al 1995). Dissociating ligands exist also for the glucocorticoid receptor (Heck et al 1994, and Heck et al 1997). The inevitable conclusion is that ligands determine the conformation of the nuclear receptor and that there are transactivating and repressing conformations.

DETERMINANTS OF NUCLEAR RECEPTOR CONFORMATION

In addition to ligand, other determinants appear to dictate receptor conformation. The effects of the DNA elements and of the composition of the cross-talk protein partner have both been reported to influence conformation. The existence of negative hormone response elements, from which glucocorticoid receptor intrinsically inhibits transcription, (Sakai et al 1988, Drouin et al 1989, Cairns et al 1993, Drouin et al 1993) is still debated. The role of the interaction of the glucocorticoid receptor with other transcription factors acting on these promoters has not been sufficiently explored. The composition of AP-1 can influence the function of the glucocorticoid receptor tethered to it. Jun homodimers formed in F9 cells by Jun overexpression can synergize with the glucocorticoid receptor at an AP-1-site-dependent promoter while Jun-Fos heterodimers are repressed (Teurich and Angel 1995). This AP-1 subunit dependent change from synergy to repression was first observed at a so-called composite element isolated from the promoter of the proliferin gene (Diamond et al 1990, Miner and Yamamoto 1992). The hypothesis is that in the composite element the receptor also contacts DNA, but it is not converted into an active conformation as it would be by a proper palindromic glucocorticoid response element (Yamamoto et al 1993, Lefstin et al 1994, Starr et al 1996). The authors propose that AP-1, if present as Jun homodimer on the composite response element, may substitute for the role of the DNA response element by protein-protein interaction, whereas the Jun/Fos heterodimer cannot. Such allosteric modulation can be proposed also for the cross-talk on simple AP-1 sites (Figure 1D) (Teurich and Angel 1995).

FACETS OF THE MECHANISM

The mechanism of transactivation by DNA-binding factors requires the participation of additional molecules that appear to bridge to the assembly of factors at the start site of transcription (basal machinery), or to cause acetylation of histones and thereby perhaps the opening of chromatin (Figure 1, central cartoon). These so-called coactivators bind to the transactivation domains of transcription factors: e.g., CBP to the phosphorylated form of CREB (Chrivia et al 1993, Kwok et al 1994, Parker et al 1996) or BOB to Oct-2 (Zwilling et al 1997). The ligand-dependent AF-2 transactivation domain of nuclear receptors also contacts putative coactivators (p140, p160, TIF-2). Their overexpression enhances ligand-induced transactivation. AF-2 is part of helix 12 of the ligand binding domain of retinoid X receptor (RXR) α (Bourguet et al 1995). Comparisons of the structures of the liganded retinoic acid receptor (RAR) γ and thyroid hormone receptor (TR) (Renaud et al 1995, Wagner et al 1995) suggest that ligand binding causes a relocalization of helix 12 and major additional changes in structure to permit interaction with the coactivators.

One may postulate that repression of transcription also requires conformational changes that recruit another set of partner proteins, corepressors (Figure 1E). These may cause types of interaction with the basal machinery different from those of coactivators, or may induce deacetylation of chromatin. Such factors have been found for unliganded TR and RXR (Chen and Evans 1995, Hörlein et al 1995, Kurokawa et al 1995): N-CoR or SMRT. In the cross-talk between nuclear receptors and other factors, there is no convincing evidence for repression below basal-level transcription. Nevertheless, it is feasible that the principle of cross-talk involves corepressors.

Alternately, the function of the transactivation domains of both partners in the negatively cross-talking complex could be reduced. One possibility proposed early is a mechanism of squelching: competition for a limiting cofactor (Figure 1F), (Shemshedini et al 1991). This hypothesis has received recent revival because of the finding that CBP and its family member p300 are probably used as cofactors by many transcription factors including those participating in negative cross-talk: CREB (Chrivia et al 1993, Imai et al 1993, Kwok et al 1994, Parker et al 1996), Jun, Fos, and STATs (Bhattacharya et al 1996, Stöcklin et al 1996). CBP by overexpression improves transcription. It may cooperate with or "integrate" the activities of several transcription factors, may itself exhibit acetylase activity (Bannister and Kouzarides 1996), or may act on a histone acetylase (p/CAF; Yang et al 1996). More telling than overexpressions are experiments with microinjection of neutralising antibodies against CBP/p300, which interfere with ligand-dependent transcription by RAR, RXR, and GR (Chakravarti et al 1996, Kamei et al 1996), and with the function of myoD and E47 (Eckner et al 1996). Although both CBP and p300 are abundant proteins, it may well be that they exist in different functional pools and that they are limiting. Support for the assumption that CBP levels can be limiting has come from the finding of a human disease with apparent haploid insufficiency for CBP (Rubenstein-Taybi Syndrome; Petrij et al 1995).

Nevertheless, the CBP/p300 hypothesis suffers from a number of unexplained and inconsistent observations.

The c-fos promoter is not repressed by nuclear receptors (Jonat et al 1990), although this would be expected as CBP acts as coactivator at the c-fos promoter (Janknecht and Hunter 1996, Janknecht and Nordheim 1996).

The cross-talk interactions between GR and Fos/Jun have been mapped to the bZip region (Lucibello et al 1990, Schüle et al 1990, Schüle and Evans 1991): (1) a Jun-bZip-GalTAD or Jun-bZip-GHF1TAD is repressed by GR (Teurich and Angel 1995, and unpublished), (2) the JunbZip fragment suffices to repress GR (Schüle et al 1990). The CBP hypothesis could be salvaged if CBP also interacted with the bZip region. There is no such evidence as yet, however.

Dissociating mutations and ligands (which poison transactivation, but not the transrepressing cross-talk) would need to create receptors that, despite their failure to activate transcription, squelch CBP/p300—which does not sound likely.

The most efficient inhibitor of AP-1 activity, the glucocorticoid receptor, is the poorest binder of CBP/p300 (Chakravarti et al 1996, Kamei et al 1996).

Non-phosphorylatable c-Jun mutants and v-Jun, which do not bind CBP, are just as efficient inhibitors of glucocorticoid receptor transactivation as wild-type c-Jun (Claret & Karin, cited in Karin et al 1997).

In conclusion, a role of CPB/p300 in the inhibitory cross-talk of nuclear receptors has to be further clarified. Whether CBP participates or not, one or several additional partners in the protein-protein interaction during cross-talk are needed to explain some of the observations, such as the observation that tissue-specific components determine whether cross-talk between nuclear receptors and AP-1 is inhibitory or synergistic (Shemshedini et al 1991, Maroder et al 1993, Bubulya et al 1996).

CAN ALL NEGATIVE EFFECTS BE LINKED TO INFLUENCES ON TRANSCRIPTION?

A transcription factor should exert its influences through transcription. Nevertheless it is theoretically possible that the liganded nuclear receptor activates or inactivates by protein-protein contact regulators of the cell cycle or of the apoptotic pathway. Interestingly, a nuclear receptor binding protein was identified as Rap-46 (Zeiner and Gehring 1995, Cato, unpublished), a member of a family of proteins that also complex with Bcl-2. It is not yet known what this link means in physiological terms.

So far the induction of apoptosis and the inhibition of cell cycle progression in G1 by nuclear receptors appear to correlate with transcriptional activities. There are possibly more than one apoptotic pathways, which would explain the seemingly contradictory results in the literature. For instance, apoptosis in lymphoid cells induced by glucocorticoid hormone has been reported to require hormone-dependent gene expression (Caron-Leslie et al 1991, Dieken and Miesfeld 1992, Shi et al 1992), while other researchers have induced lymphoid apoptosis using GR mutants defective in transactivation (Helmberg et al 1995, Kullmann, unpublished).

PERSPECTIVES

We have argued in the introduction that negative control is important for an organism in various ways. In additional support of a physiological significance of the transcriptional cross-talk, there is good correlation between the efficiency of ligands as anti-inflammatory agents and as inhibitors of AP-1 and NFκB (Heck et al 1997). An ultimate documentation of the role of the negative cross-talk in an organism must come from mice carrying dissociating point mutations. To create such mice by site-directed mutagenesis and gene replacement in the germ line will be the next important goal in exploring the molecular function of nuclear receptors.

ACKNOWLEDGMENTS

The authors thank Stefanie Heck, Peter Angel and Andrew Cato for stimulating discussions. The Fonds der Chemie Schen Industrie (PH) and the Deutsche Forschungsgemeinschaft (MG) are acknowledged for financial support.

REFERENCES

Auphan N, Di Donato JA, Rosette C, Helmberg A, and Karin M (1995): Immunosuppression by glucocorticoids: Inhibition of NF-kappa B activity through induction of I kappa B synthesis. *Science* 270: 286–290.

Baeuerle PA, and Henkel T (1994): Function and activation of NF-kappa B in the immune system. *Annu. Rev. Immunol.* 12: 141–179.

Bannister AJ, and Kouzarides T (1996): The CBP co-activator is a histone acetyltransferase. *Nature* 384: 641–643.

Barettino D, Vivanco Ruiz MM, and Stunnenberg HG (1994): Characterization of the ligand-dependent transactivation domain of thyroid hormone receptor. *Embo. J.* 13: 3039–3049.

Baxter JD, and Forsham PH (1972): Tissue effects of glucocorticoids. *Am. J. Med.* 53: 573–589.

Beato M, Herrlich P, and Schütz G (1995): Steroid hormone receptors: many actors in search of a plot. *Cell* 83: 851–857.

Belman S, and Troll W (1972): The inhibition of croton oil-promoted mouse skin tumorigenesis by steroid hormones. *Cancer Res.* 32: 450–454.

Berko-Flint Y, Levkowitz G, and Vardimon L (1994): Involvement of c-Jun in the control of glucocorticoid receptor transcriptional activity during development of chicken retinal tissue. *Embo. J.* 13: 646–654.

Bhattacharya S, Eckner R, Grossman S, Oldread E, Arany Z, D'Andrea A, and Livingston DM (1996): Cooperation of Stat2 and p300/CBP in signalling induced by interferon-alpha. *Nature* 383: 344–347.

Bourguet W, Ruff M, Chambon P, Gronemeyer H, and Moras D (1995): Crystal structure of the ligand-binding domain of the human nuclear receptor RXR-alpha. *Nature* 375: 377–382.

Bubulya A, Wise SC, Shen XQ, Burmeister LA, and Shemshedini L (1996): c-Jun can mediate androgen receptor-induced transactivation. *J. Biol. Chem.* 271: 24583–24589.

Cairns C, Cairns W, and Okret S (1993): Inhibition of gene expression by steroid hormone receptors via a negative glucocorticoid response element: evidence for the involvement of DNA-binding and agonistic effects of the antiglucocorticoid/anti-progestin RU486. *DNA Cell Biol.* 12: 695–702.

Caldenhoven E, Liden J, Wissink S, Van de Stolpe A, Raaijmakers J, Koenderman L, Okret S, Gustafsson JA, and Van der Saag PT (1995): Negative cross-talk between RelA and the glucocorticoid receptor: a possible mechanism for the antiinflammatory action of glucocorticoids. *Mol. Endocrinol.* 9: 401–412.

Caron-Leslie LM, Schwartzman RA, Gaido ML, Compton MM, and Cidlowski JA (1991): Identification and characterization of glucocorticoid-regulated nuclease(s) in lymphoid cells undergoing apoptosis. *J. Steroid Biochem. Mol. Biol.* 40: 661–671.

Chakravarti D, LaMorte VJ, Nelson MC, Nakajima T, Schulman IG, Juguilon H, Montminy M, and Evans RM (1996): Role of CBP/p300 in nuclear receptor signalling. *Nature* 383: 99–103.

Chen JD, and Evans RM (1995): A transcriptional co-repressor that interacts with nuclear hormone receptors. *Nature* 377: 454–457.

Chen JY, Penco S, Ostrowski J, Balaguer P, Pons M, Starrett JE, Reczek P, Chambon P, and Gronemeyer H (1995): RAR-specific agonist/antagonists which dissociate transactivation and AP1 transrepression inhibit anchorage-independent cell proliferation. *Embo. J.* 14: 1187–1197.

Chrivia JC, Kwok RP, Lamb N, Hagiwara M, Montminy MR, and Goodman RH (1993): Phosphorylated CREB binds specifically to the nuclear protein CBP. *Nature* 365: 855–859.

Coffino P, Bourne HR, and Tomkins GM (1975): Mechanism of lymphoma cell death induced by cyclic AMP. *Am. J. Pathol.* 81: 199–204.

Cole TJ, Blendy JA, Monaghan AP, Krieglstein K, Schmid W, Aguzzi A, Fantuzzi G, Hummler E, Unsicker K, and Schütz G (1995): Targeted disruption of the glucocorticoid receptor gene blocks adrenergic chromaffin cell development and severely retards lung maturation. *Genes Dev.* 9: 1608–1621.

Daub H, Weiss FU, Wallasch C, and Ullrich A (1996): Role of transactivation of the EGF receptor in signalling by G-protein-coupled receptors. *Nature* 379: 557–560.

Diamond MI, Miner JN, Yoshinaga SK, and Yamamoto KR (1990): Transcription factor interactions: selectors of positive or negative regulation from a single DNA element. *Science* 249: 1266–1272.

Dieken ES, and Miesfeld RL (1992): Transcriptional transactivation functions localized to the glucocorticoid receptor N terminus are necessary for steroid induction of lymphocyte apoptosis. *Mol. Cell Biol.* 12: 589–597.

Doppler W, Groner B, and Ball, RK (1989): Prolactin and glucocorticoid hormones synergistically induce expression of transfected rat beta-casein gene promoter constructs in a mammary epithelial cell line. *Proc. Natl. Acad. Sci. USA* 86: 104–108.

Drouin J, Sun YL, Chamberland M, Gauthier Y, De Lean A, Nemer M, and Schmidt TJ (1993): Novel glucocorticoid receptor complex with DNA element of the hormone-repressed POMC gene. *Embo. J.* 12: 145–156.

Drouin J, Trifiro MA, Plante RK, Nemer M, Eriksson P, and Wrange O (1989): Glucocorticoid receptor binding to a specific DNA sequence is required for hormone-dependent repression of pro-opiomelanocortin gene transcription. *Mol. Cell Biol.* 9: 5305–5314.

Eckner R, Yao TP, Oldread E, and Livingston DM (1996): Interaction and functional collaboration of p300/CBP and bHLH proteins in muscle and B-cell differentiation. *Genes. Dev.* 10: 2478–2490.

Fanjul A, Dawson MI, Hobbs PD, Jong L, Cameron JF, Harlev E, Graupner G, Lu XP, and Pfahl M (1994): A new class of retinoids with selective inhibition of AP-1 inhibits proliferation. *Nature* 372: 107–111.

Felli MP, Vacca A, Meco D, Screpanti I, Farina AR, Maroder M, Martinotti S, Petrangeli E, Frati L, and Gulino A (1991): Retinoic acid-induced down-regulation of the interleukin-2 promoter via cis-regulatory sequences containing an octamer motif. *Mol Cell Biol* 11: 4771–4778.

Frisch SM, and Ruley HE (1987): Transcription from the stromelysin promoter is induced by interleukin-1 and repressed by dexamethasone. *J. Biol. Chem.* 262: 16300–16304.

Gehring U, and Tomkins GM (1974): A new mechanism for steroid unresponsiveness: loss of nuclear binding activity of a steroid hormone receptor. *Cell* 3: 301–306.

Godowski PJ, Sakai DD, and Yamamoto KR (1989): In: *DNA-Protein Interactions in Transcription. UCLA Symposia on Molecular and Cellular Biology*, J. D. Gralla, ed. New York: Alan R. Liss, pp. 197–210.

Guathier JM, Bourachot B, Doucas V, Yaniv M, and Moreau-Gachelin F (1993): Functional interference between the Spi-1/PU.1 oncoprotein and steroid hormone or vitamin receptors. *Embo. J.* 12: 5089–5096.

Gudas LJ, Sporn MB, and Roberts AB (1994): In: *The Retinoids*, M.B. Sporn, A.B. Roberts and D.S. Goodman, eds. New York: Raven Press, pp. 443–520.

Hannum CH, Wilcox CJ, Arend WP, Joslin FG, Dripps DJ, Heimdal PL, Armes LG, Sommer A, Eisenberg SP, and Thompson RC (1990): Interleukin-1 receptor antagonist activity of a human interleukin-1 inhibitor. *Nature* 343: 336–340.

Heck S, Kullmann M, Gast A, Ponta H, Rahmsdorf HJ, Herrlich P, and Cato AC (1994): A distinct modulating domain in glucocorticoid receptor monomers in the repression of activity of the transcription factor AP-1. *Embo. J.* 13: 4087–4095.

Heck S. Bender K. Kullmann M, Göttlicher M, Herrlich P, and Cato ACB (1997): IκBα independent downregulation of NK-κB activity by glucocorticoid receptor. *Embo. J.* 16: 4698–4707.

Helmberg A, Auphan N, Caelles C, and Karin M (1995): Glucocorticoid-induced apoptosis of human leukemic cells is caused by the repressive function of the glucocorticoid receptor. *Embo. J.* 14: 452–460.

Hörlein AJ, Näär AM, Heinzel T, Torchia J, Gloss B, Kurokawa R, Ryan A, Kamei Y, Soderstrom M, Glass CK, and et al (1995): Ligand-independent repression by the thyroid hormone receptor mediated by a nuclear receptor co-repressor. *Nature* 377: 397–404.

Ignar-Trowbridge DM, Pimentel M, Parker MG, McLachlan JA, and Korach KS (1996): Peptide growth factor cross-talk with the estrogen receptor requires the A/B domain and occurs independently of protein kinase C or estradiol. *Endocrinology* 137: 1735–1744.

Imai E, Miner JN, Mitchell JA, Yamamoto KR, and Granner DK (1993): Glucocorticoid receptor-cAMP response element-binding protein interaction and the response of the phosphoenolpyruvate carboxykinase gene to glucocorticoids. *J. Biol. Chem.* 268: 5353–5356.

Janknecht R, and Hunter T (1996): Transcription. A growing coactivator network [news]. *Nature* 383: 22–23.

Janknecht R, and Nordheim A (1996): Regulation of the c-fos promoter by the ternary complex factor Sap-1a and its coactivator CBP. *Oncogene* 12: 1961–1969.

Jonat C, Rahmsdorf HJ, Park KK, Cato AC, Gebel S, Ponta H, and Herrlich P (1990): Antitumor promotion and antiinflammation: down-modulation of AP-1 (Fos/Jun) activity by glucocorticoid hormone. *Cell* 62: 1189–2104.

Jonat C, Stein B, Ponta H, Herrlich P, and Rahmsdorf HJ (1992): Positive and negative regulation of collagenase gene expression. *Matrix Suppl.* 1: 145–155.

Kamei Y, Xu L, Heinzel T, Torchia J, Kurokawa R, Gloss B, Lin SC, Heyman RA, Rose DW, Glass CK, and Rosenfeld MG (1996): A CBP integrator complex mediates transcriptional activation and AP-1 inhibition by nuclear receptors. *Cell* 85: 403–414.

Kastner P, Mark M, and Chambon P (1995): Nonsteroid nuclear receptors: What are genetic studies telling us about their role in real life? *Cell* 83: 859–869.

Kelsey G, and Schütz G (1993): Lessons from lethal albino mice. *Curr. Opin. Genet. Dev.* 3: 259–264.

King CS, and Cooper JA (1986): Effects of protein kinase C activation after epidermal growth factor binding on epidermal growth factor receptor phosphorylation. *J. Biol. Chem.* 261: 10073–10078.

König H, Ponta H, Rahmsdorf HJ, and Herrlich P (1992): Interference between pathway-specific transcription factors: glucocorticoids antagonize phorbol ester-induced AP-1 activity without altering AP-1 site occupation in vivo. *Embo. J.* 11: 2241–2246.

Kurokawa R, Söderstrom M, Hörlein A, Halachmi S, Brown M, Rosenfeld MG, and Glass CK (1995): Polarity-specific activities of retinoic acid receptors determined by a co-repressor. *Nature* 377: 451–454.

Kutoh E, Stromstedt PE, and Poellinger L (1992): Functional interference between the ubiquitous and constitutive octamer transcription factor 1 (OTF-1) and the glucocorticoid receptor by direct protein-protein interaction involving the homeo subdomain of OTF-1. *Mol. Cell Biol.* 12: 4960–4969.

Kwok RP, Lundblad JR, Chrivia JC, Richards JP, Bachinger HP, Brennan RG, Roberts SG, Green MR, and Goodman RH (1994): Nuclear protein CBP is a coactivator for the transcription factor CREB. *Nature* 370: 223–226.

Lefstin JA, Thomas JR, and Yamamoto KR (1994): Influence of a steroid receptor DNA-binding domain on transcriptional regulatory functions. *Genes. Dev.* 8: 2842–2856.

Lin CR, Chen WS, Lazar CS, Carpenter CD, Gill GN, Evans RM, and Rosenfeld, MG (1986): Protein kinase C phosphorylation at Thr 654 of the unoccupied EGF receptor and EGF binding regulate functional receptor loss by independent mechanisms. *Cell* 44: 839–848.

Lotan R (1980): Effects of vitamin A and its analogs (retinoids) on normal and neoplastic cells. *Biochim. Biophys. Acta.* 605: 33–91.

Lucibello FC, Slater EP, Jooss KU, Beato M, and Muller R (1990): Mutual transrepression of Fos and the glucocorticoid receptor: involvement of a functional domain in Fos which is absent in FosB. *Embo. J.* 9: 2827–2834.

Luttrell LM, Hawes BE, van Biesen T, Luttrell DK, Lansing TJ, and Lefkowitz RJ (1996): Role of c-Src tyrosine kinase in G protein-coupled receptor- and Gbetagamma subunit-mediated activation of mitogen-activated protein kinases. *J. Biol. Chem.* 271: 19443–19450.

Mangelsdorf DJ, and Evans RM (1995): The RXR heterodimers and orphan receptors. *Cell* 83: 841–850.

Mangelsdorf DJ, Thummel C, Beato M, Herrlich P, Schütz G, Umesono K, Blumberg B, Kastner P, Mark M, Chambon P, and et al (1995): The nuclear receptor superfamily: the second decade. *Cell* 83: 835–839.

Maroder M, Farina AR, Vacca A, Felli MP, Meco D, Screpanti I, Frati L, and Gulino A (1993): Cell-specific bifunctional role of Jun oncogene family members on glucocorticoid receptor-dependent transcription. *Mol. Endocrinol.* 7: 570–584.

Medcalf RL, Richards RI, Crawford RJ, and Hamilton, J. A (1986): Suppression of urokinase-type plasminogen activator mRNA levels in human fibrosarcoma cells and synovial fibroblasts by anti-inflammatory glucocorticoids. *Embo. J.* 5: 2217–2222.

Migliaccio A, Di Domenico M, Castoria G, de Falco A, Bontempo P, Nola E, and Auricchio F (1996): Tyrosine kinase/p21ras/MAP-kinase pathway activation by estradiol-receptor complex in MCF-7 cells. *Embo. J.* 15: 1292–1300.

Miner JN, and Yamamoto KR (1992): The basic region of AP-1 specifies glucocorticoid receptor activity at a composite response element. *Genes. Dev.* 6: 2491–2501.

Minucci S, Botquin V, Yeom YI, Dey A, Sylvester I, Zand DJ, Ohbo K, Ozato K, and Scholer HR (1996): Retinoic acid-mediated down-regulation of Oct3/4 coincides with the loss of promoter occupancy in vivo. *Embo. J.* 15: 888–899.

Monfar M, and Blenis J (1996): Inhibition of p70/p85 S6 kinase activities in T cells by dexamethasone. *Mol. Endocrinol.* 10: 1107–1115.

Parker D, Ferreri K, Nakajima T, La Morte VJ, Evans R, Koerber SC, Hoeger C, and Montminy MR (1996): Phosphorylation of CREB at Ser-133 induces complex formation with CREB-binding protein via a direct mechanism. *Mol. Cell Biol.* 16: 694–703.

Pearce D, and Yamamoto KR (1993): Mineralocorticoid and glucocorticoid receptor activities distinguished by nonreceptor factors at a composite response element. *Science* 259: 1161–1165.

Petrij F, Giles RH, Dauwerse HG, Saris JJ, Hennekam RC, Masuno M, Tommerup N, van Ommen GJ, Goodman RH, Peters DJ, et al (1995): Rubinstein-Taybi syndrome caused by mutations in the transcriptional co-activator CBP. *Nature* 376: 348–351.

Ray A, and Prefontaine KE (1994): Physical association and functional antagonism between the p65 subunit of transcription factor NF-kappa B and the glucocorticoid receptor. *Proc. Natl. Acad. Sci. USA* 91: 752–756.

Reik A, Stewart AF, and Schütz G (1994): Cross-talk modulation of signal transduction pathways: Two mechanisms are involved in the control of tyrosine aminotransferase gene expression by phorbol esters. *Mol. Endocrinol.* 8: 490–497.

Renaud JP, Rochel N, Ruff M, Vivat V, Chambon P, Gronemeyer H, and Moras D (1995): Crystal structure of the RAR-gamma ligand-binding domain bound to all-trans retinoic acid. *Nature* 378: 681–689.

Sakai DD, Helms S, Carlstedt-Duke J, Gustafsson JA, Rottman FM, and Yamamoto KR (1988): Hormone-mediated repression: a negative glucocorticoid response element from the bovine prolactin gene. *Genes. Dev.* 2: 1144–1154.

Scheinman RI, Cogswell PC, Lofquist AK, and Baldwin AS, Jr (1995a): Role of transcriptional activation of I kappa B alpha in mediation of immunosuppression by glucocorticoids. *Science* 270: 283–286.

Scheinman RI, Gualberto A, Jewell CM, Cidlowski JA, and Baldwin AS, Jr (1995b): Characterization of mechanisms involved in transrepression of NF-kappa B by activated glucocorticoid receptors. *Mol. Cell Biol.* 15: 943–953.

Schüle R, and Evans RM (1991): Cross-coupling of signal transduction pathways: Zinc finger meets leucine zipper. *Trends Genet.* 7: 377–381.

Schüle R, Rangarajan P, Kliewer S, Ransone LJ, Bolado J, Yang N, Verma IM, and Evans RM (1990): Functional antagonism between oncoprotein c-Jun and the glucocorticoid receptor. *Cell* 62: 1217–1226.

Schweitzer R, Howes R, Smith R, Shilo BZ, and Freeman M (1995): Inhibition of Drosophila EGF receptor activation by the secreted protein Argos. *Nature* 376: 699–702.

Shemshedini L, Knauthe R, Sassone-Corsi P, Pornon A, and Gronemeyer H (1991): Cell-specific inhibitory and stimulatory effects of Fos and Jun on transcription activation by nuclear receptors. *Embo. J.* 10: 3839–3849.

Shi Y, Glynn JM, Guilbert LJ, Cotter TG, Bissonnette RP, and Green DR (1992): Role for c-myc in activation-induced apoptotic cell death in T cell hybridomas. *Science* 257: 212–214.

Sibley CH, and Tomkins, G. M (1974): Isolation of lymphoma cell variants resistant to killing by glucocorticoids. *Cell* 2: 213–220.

Starr DB, Matsui W, Thomas JR, and Yamamoto KR (1996): Intracellular receptors use a common mechanism to interpret signaling information at response elements. *Genes. Dev.* 10: 1271–1283.

Stauber C, Altschmied J, Akerblom IE, Marron JL, and Mellon PL (1992): Mutual cross-interference between glucocorticoid receptor and CREB inhibits transactivation in placental cells. *New Biol.* 4: 527–540.

Stein B, and Yang MX (1995): Repression of the interleukin-6 promoter by estrogen receptor is mediated by NF-kappa B and C/EBP beta. *Mol. Cell Biol.* 15: 4971–4979.

Stöcklin E, Wissler M, Gouilleux F, and Groner B (1996): Functional interactions between Stat5 and the glucocorticoid receptor. *Nature* 383: 726–728.

Sylvester I, and Scholer HR (1994): Regulation of the Oct-4 gene by nuclear receptors. *Nucleic Acids Res.* 22: 901–911.

Teurich S, and Angel P (1995): The glucocorticoid receptor synergizes with Jun homodimers to activate AP-1-regulated promoters lacking GR binding sites. *Chem. Senses* 20: 251–255.

Touray M, Ryan F, Jaggi R, and Martin F (1991): Characterisation of functional inhibition of the glucocorticoid receptor by Fos/Jun. *Oncogene* 6: 1227–1234.

Tremblay Y, Tretjakoff I, Peterson A, Antakly T, Zhang CX, and Drouin J (1988): Pituitary-specific expression and glucocorticoid regulation of a proopiomelanocortin fusion gene in transgenic mice. *Proc. Natl. Acad. Sci. USA* 85: 8890–8894.

Ucker DS (1987): Cytotoxic T lymphocytes and glucocorticoids activate an endogenous suicide process in target cells. *Nature* 327: 62–64.

van de Stolpe A, Caldenhoven E, Raaijmakers JA, van der Saag PT, and Koenderman L (1993): Glucocorticoid-mediated repression of intercellular adhesion molecule-1 expression in human monocytic and bronchial epithelial cell lines. *Am. J. Respir. Cell Mol. Biol.* 8: 340–347.

van de Stolpe A, Caldenhoven E, Stade BG, Koenderman L, Raaijmakers JA, Johnson JP, and van der Saag PT (1994): 12-O-tetradecanoylphorbol-13-acetate- and tumor necrosis factor alpha-mediated induction of intercellular adhesion molecule-1 is inhibited by dexamethasone. Functional analysis of the human intercellular adhesion molecular-1 promoter. *J. Biol. Chem.* 269: 6185–6192.

Viaje A, Slaga TJ, Wigler M, and Weinstein, I. B (1977): Effects of antiinflammatory agents on mouse skin tumor promotion, epidermal DNA synthesis, phorbol ester-induced cellular proliferation, and production of plasminogen activator. *Cancer Res.* 37: 1530–1536.

Wagner RL, Apriletti JW, McGrath ME, West BL, Baxter JD, and Fletterick, RJ (1995): A structural role for hormone in the thyroid hormone receptor. *Nature* 378: 690–697.

Weiss C, Kolluri SK, Kiefer F, and Göttlicher M (1996): Complementation of Ah receptor deficiency in hepatoma cells: negative feedback regulation and cell cycle control by the Ah receptor. *Exp. Cell Res.* 226: 154–163.

Williams PM, Navre M, and Ringold GM (1991): Glucocorticoid induction of the adipocyte clone 5 gene requires high cell density. *Mol. Endocrinol.* 5: 615–618.

Wyllie AH (1980): Glucocorticoid-induced thymocyte apoptosis is associated with endogenous endonuclease activation. *Nature* 284: 555–556.

Yamamoto KR, Pearce D, Thomas JR, and Miner JN (1993): Combinatorial regulation at a mammalian composite element. In: *Transcriptional Regulation*, S. McKnight and K. R. Yamamoto, eds. Cold Spring Harbour, NY: Cold Spring Harbour Laboratory Press, pp. 3–32.

Yang XJ, Ogryzko VV, Nishikawa J, Howard BH, and Nakatani Y (1996): A p300/CBP-associated factor that competes with the adenoviral oncoprotein E1A. *Nature* 382: 319–324.

Yang-Yen HF, Chambard JC, Sun YL, Smeal T, Schmidt TJ, Drouin J, and Karin M (1990): Transcriptional interference between c-Jun and the glucocorticoid receptor: mutual inhibition of DNA binding due to direct protein-protein interaction. *Cell* 62: 1205–1215.

Zeiner M, and Gehring U (1995): A protein that interacts with members of the nuclear hormone receptor family: identification and cDNA cloning. *Proc. Natl. Acad Sci USA* 92: 11465–11469.

Zwilling S, Dieckmann A, Pfisterer P, Angel P, and Wirth T (1997): Inducible expression of coactivator BOB.1/OBF.1 in T cells. *Science* 277: 221–225.

8

Chromatin and Steroid-Receptor-Mediated Transcription

CATHERINE E. WATSON AND TREVOR K. ARCHER

I. INTRODUCTION

The normal growth and development of multicellular organisms requires precise spatial and temporal regulation of gene expression. For a subset of genes, steroid and nuclear hormone receptors act to modulate gene transcription by binding to hormone-responsive elements (HREs), usually present within the 5′-flanking region of the gene. Within the eukaryotic nucleus these regulatory DNA elements are associated with histone and nonhistone proteins to form chromatin. Consequently, a thorough mechanistic understanding of receptor-mediated transcription must reflect the fact that receptors, in concert with other transcription factors, act on genes highly organized as chromatin.

This review will focus on the interaction of steroid hormone receptors with their target genes assembled as chromatin, and will emphasize recent work with the mouse mammary tumor virus long terminal repeat (MMTV LTR) model system in mouse and human cells. We will provide a brief introduction to chromatin and steroid receptors, as well as a description of the MMTV system. This will be followed by a detailed discussion of experimental findings that argue for a central role for chromatin structure in steroid-receptor-mediated gene transcription. A more comprehensive description of individual steriod and nuclear hormone receptors can be found in accompanying chapters in this volume.

Molecular Biology of Steroid and Nuclear Hormone Receptors
Leonard P. Freedman, Editor
© 1998 Birkhäuser Boston

A. Chromatin

In eukaryotic cells, DNA is tightly associated with proteins into a highly organized structure know as chromatin (Wolffe 1995, van Holde 1988). The building block of chromatin is the nucleosome, which consists of DNA (146 bp) wrapped around a protein core (Kornberg 1974). This core is composed of an octamer of histones: two each of H2A, H2B, H3, and H4 (Arents and Moudrianakis 1993, Richmond et al 1984). More specifically, the core is assembled from an H3/H4 tetramer bounded by two H2A/H2B dimers (Arents et al 1991). The resulting wedge-shaped complex has a path of positive charges around its outer perimeter with which the negatively charged DNA interacts (Arents and Moudrianakis 1993). The DNA wraps around the octamer $1\frac{3}{4}$ turns, entering and leaving near the same location. It is at this site that the fifth histone, H1, has traditionally been thought to interact with the DNA (van Holde 1988, Ali and Singh 1987). However, recent experiments with the 5S rRNA gene suggest that H1 may adopt specific and asymmetric binding with the nucleosome (Hayes 1996, Pruss et al 1996). The positioning of histone H1 and its interaction with the DNA between nucleosomes have implicated it in the higher-order packaging of the nucleosomes that eventually leads to the fully condensed chromosome (Zlatanova and van Holde 1992, Roth and Allis 1992).

During the last decade, studies have demonstrated that these DNA/histone interactions, which are important for the compaction and organization of the genome, may also play a critical role in gene regulation (Felsenfeld 1992). Strong support for this concept hase come from genetic experiments in *Saccharomyces cerevisiae* where deletion of one or more core histone genes led to the constitutive expression of a number of previously inducible genes (Clark-Adams et al 1988, Han and Grunstein 1988, Grunstein 1991). Numerous *in vitro* studies indicate that when DNA is assembled into nucleosomes, transcription factors are generally unable to bind their sites and gene expression is repressed (Workman and Buchman 1993, Hayes and Wolffe 1992). In some instances transcription is elevated, presumably as a result of the compaction of promoter DNA, which brings distant enhancers closer to the proximal promoter and the basal transcription machinery (Wolffe 1994).

The ability of histones to compact DNA in the nucleus raises the possibility of an additional level of complexity in the control of gene expression in eukaryotes (Wolffe 1995, Felsenfeld 1992). Within chromatin, regions of DNA that are highly condensed are often transcriptionally silent, with transcription originating from more loosely packaged or disordered regions of the genome. This "disorder" is often reflected in an increased sensitivity or hypersensitivity to a variety of nucleases and chemical probes (Almer and Hörz 1986, Zaret and Yamamoto 1984, Richard-Foy and Hager 1987, McGhee et al 1981, Wu 1980). These hypersensitive sites may be constitutive, or may be inducible by a variety of agents, including steroid hormones (Yamamoto 1985).

B. The LTR of MMTV

MMTV is a B-type retrovirus that integrates into the host genome as a double-stranded DNA provirus (Panganiban 1985). When the MMTV LTR is stably introduced into tissue culture cells it adopts a highly ordered structure of six phased nucleosomes (Figure 1) (Richard-Foy and Hager 1987). The nucleosomal array is acquired whether this promoter is maintained as part of a bovine papilloma virus (BPV) episomal vector or integrated into the genome (Archer et al 1989, Richard-Foy and Hager 1987). This predictable chromatin conformation has made the MMTV promoter a valuable tool for studying the role of chromatin during transcriptional activation (Archer and Mymryk 1995, Hager et al 1993).

The MMTV LTR is activated by glucocorticoids, resulting in a dramatic increase in transcription and the appearance of a nuclease hypersensitive region in the proximal portion of the LTR (Young et al 1977, Zaret and Yamamoto 1984). Approximately 200bp of the MMTV proximal promoter is required to confer hormone-regulated transcription to a chimeric construct containing a heterologous sequence (Chandler et al 1983, Huang et al 1981). *In vitro* binding studies have demonstrated that the glucocorticoid receptor (GR) interacts specifically with the MMTV LTR in the region -200 to -100 (Chandler et al 1983). Within this region, four GR-protected areas are detected during DNaseI footprinting experiments (Payvar et al 1981). These sites share the hexanucleotide motif TGTTCT, and are termed HREs as they have also been shown to mediate progesterone, androgen, and mineralcorticoid receptor responsiveness (Cato et al 1987, Gowland and Buetti 1989, Otten et al 1988). The proximal promoter also possesses binding sites for other transcription factors, including nuclear factor 1 (NF1), octamer transcription factor (OTF), and TATA-binding protein (TBP) (Cordingley and Hager 1988, Cavallini et al 1988, Brüggemeier et al 1991, Rosenfeld and Kelly 1986). NF1 binds to a site adjacent to the proximal HRE, and this interaction is necessary for hormone-activated transcription (Buetti et al 1989, Buetti and Kühnel 1986). The two OTF binding sites are required for basal transcription (Brüggemeier et al 1991, Buetti 1994).

Precise mapping of the phased nucleosomes within the MMTV LTR places the second nucleosome (Nuc-B) over the entire HRE region as well as part of the NF1 binding site (Fragoso et al 1995, Richard-Foy and Hager 1987). The first nucleosome (Nuc-A) is positioned over the TATA-box and the transcription start site, while the OTF sites are located in the linker region between Nuc-A and Nuc-B (Figure 1). This organization suggests that the phasing of these nucleosomes may restrict access of some transcription factors to their cognate

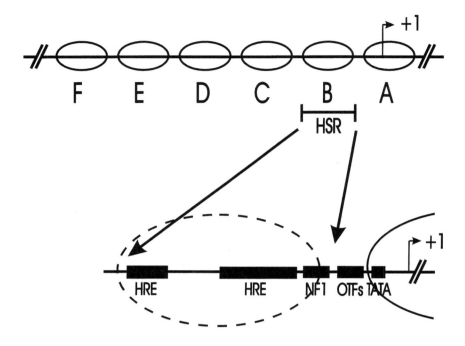

Figure 1.—Schematic representation of the nucleosomal organization of the MMTV LTR.
The MMTV LTR reproducibly adopts a phased array of six nucleosomes (A-F) *in vivo* with transcription initiating at the A nucleosome. A hormone-inducible hypersensitive region (HSR) coincides with the position of nucleosome B. Transcription factor binding sites within the proximal portion of the MMTV LTR include those for the hormone-responsive elements (HREs), nuclear factor 1 (NF1), octamer transcription factors (OTFs), and the TATA-box.

binding sites and thus provide a mechanism by which the MMTV promoter chromatin structure might influence transcription (Archer et al 1989).

II. Steroid Receptor Modulation of Transcription and Chromatin

A. The Glucocorticoid Receptor

1. *Activation of the MMTV LTR*

Glucocorticoid-induced transcription has been studied extensively in GR+ mouse breast cancer cells engineered to stably maintain a BPV minichromosome containing the MMTV LTR attached to reporter genes (Ostrowski et al 1983, Ostrowski et al 1984). In the absence of glucocorticoids, transcription from the LTR is minimal, with negligible transcription factor loading on the promoter (Lee and Archer 1994, Archer et al 1992, Cordingley et al 1987). Upon dexamethasone (Dex; a synthetic glucocorticoid) addition, transcription from the MMTV LTR is dramatically increased, and the DNA encompassed by Nuc-B becomes hypersensitive to many reagents (Richard-Foy and Hager 1987, Zaret and Yamamoto 1984, Archer et al 1991, Archer et al 1992). For example, hormone-induced cells exhibit a substantial increase in the accessibility of restriction enzyme sites within Nuc-B (Archer and Mymryk 1995). These results suggest that Nuc-B acquires a more "open" conformation during promoter activation. Congruent with the alteration of Nuc-B chromatin structure, nuclear factor loading studies detected enhanced NF1, OTF, and TBP binding to the Dex-induced promoter (Lee and Archer 1994). These results formed the basis of a hypothesis that linked the specific chromatin structure of the promoter to glucocorticoid regulation of transcription. This hypothesis suggests that in the absence of hormone the precise positioning of Nuc-A and -B leads to the exclusion of the ubiquitous transcription factors necessary for transcription from the promoter. Hormone-activated GR would modify Nuc-B such that NF1 and the other transacting factors could bind, and thus initiate the formation of a transcription initiation complex (Figure 2) (Hager et al 1993, Archer 1993).

Support for this model has come from a variety of *in vitro* and *in vivo* experiments (Perlmann and Wrange 1988, Piña et al 1990, Archer et al 1991). In particular, although NF1 has a strong affinity for its site on naked DNA, *in vitro* reconstitution studies using either di- or mono-nucleosomes and purified transcription factors have shown that NF1 is incapable of binding in a nucleosomal context (Archer et al 1991). In contrast, GR, which has substantially lower affinity for its site in free DNA, was able to bind to its cognate sites assembled as chromatin. In subsequent experiments, hydroxylradical footprinting of Nuc-B suggested that only two of the four HREs are in a location favorable to receptor binding (Piña et al 1990). However, while *in vivo* hormone-dependent binding of NF1 is readily observed, similar *in vivo* footprinting studies have been unable to reproducibly detect GR bound to the HREs in the presence of hormone (Lee and Archer 1994, Truss et al 1995, Archer et al 1992, Cordingley et al 1987). This result suggests that GR interacts with its cognate sites in a "hit and run" manner. Thus it can be argued that NF1 and GR are representative of transcription factors to which the nucleosomal structure can be viewed as "opaque" (prevents binding) or "transparent" (allows binding), respectively. Implicit in these observations is the view that remodeling of Nuc-B, upon receptor binding, is necessary for transcription factor access (Cordingley et al 1987, Archer et al 1989).

To assess the contribution of chromatin structure to glucocorticoid regulation of MMTV transcription, an experimental model was developed in which

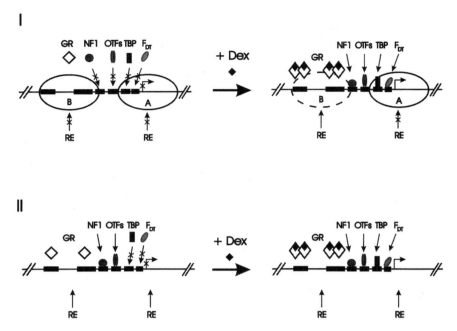

Figure 2.—A comparison of stable and transient MMTV templates in GR+/pr− cells. An overview of differences between stable chromatin (**I**) and transiently transfected templates (**II**) with respect to transcription factor binding and restriction enzyme (RE) accessibility is depicted. Stably maintained copies of the MMTV LTR adopt a phased array of nucleosomes, while transiently transfected copies do not. **I**: Transcription factor loading (NF1, OTF, TBP, and F_{DT}) onto the promoter is dexamethasone- (Dex) dependent. Addition of Dex alters the conformation of Nuc-B (B), but not of Nuc-A (A), allowing RE access within this region. **II**: Transcription factors NF1 and OTF are constitutively bound to their sites within the promoter. Dex is required for TBP and F_{DT} binding, and for activated transcription. The lack of organized nucleosomes on the DNA allows constitutive access by REs to sites in both Nuc-B and Nuc-A.

MMTV reporter plasmids are transiently transfected into cells that stably maintain the MMTV promoter (Archer et al 1992, Archer 1993). The transient and stable copies of the MMTV promoter are distinguished by different reporter genes and diagnostic restriction enzyme sites, which allows both function and structure to be independently monitored (Archer 1993, Archer and Mymryk 1995). Extensive characterization of the transiently transfected MMTV DNA revealed that it is not assembled into the same phased array of nucleosomes seen for the stable chromatin copy (Archer et al 1992). In addition, the DNA is hypersensitive to nucleases in both the presence and absence of hormone (Lee and Archer 1994, Archer et al 1992). Thus, there are fundamental differences between stably-maintained and transiently-expressed copies of the MMTV LTR (Figure 2), with the former more closely resembling the chromatin conformation of an MMTV provirus. This distinction between stable chromosomal and transient templates has been used to examine the role of chromatin on hormone-induced transcription from the MMTV LTR (Archer et al 1994a, Lee and Archer 1994, Archer et al 1992). Consistent with the hypothesis that phased nucleosomes exclude NF1 and the other transcription factors, the transient templates exhibited constitutive binding of NF1 and the OTFs (Lee and Archer 1994). These experiments argue for the GR-mediated remodeling of the MMTV as a prerequisite for NF1 and OTF binding (Lee and Archer 1994, Archer et al 1992, Cordingley et al 1987).

Previous experiments indicated that the binding of NF1 and the OTFs was critical for the GR activation of the MMTV promoter, yet transient templates that had these proteins bound constitutively remained hormone inducible. This apparent paradox was resolved by further *in vivo* footprinting of the transient templates. It was established that while NF1 and the OTFs exhibited constitutive loading onto the promoter, the binding of TBP remained hormone inducible, as seen for the stable chromatin template. In addition a second factor, designated factor downstream of TBP (F_{DT}), was shown to exhibit hormone-dependent binding to the transient templates (Lee and Archer 1994). This set of experiments suggested that it was possible to further subdivide those transcription factors that were "opaque" to binding nucleosomes. The first group, represented by NF1 and OTF, was able to bind its recognition elements, providing that the chromatin structure did not render these sites inaccessible. The second group, represented by TBP and F_{DT}, required additional activities associated with GR to be brought to the promoter, presumably by protein-protein interaction GR and/or associated factors (see section **IV**). Taken together, the data suggested a dual role for hormone-bound GR in the activation of transcription from the MMTV promoter. Not only is GR required to bind to its HRE and disrupt Nuc-B, but it is also required to recruit TBP and F_{DT} to form the preinitiation complex (PIC) (Lee and Archer 1994).

Analysis of the activation profile of the MMTV promoter revealed that, while it is maximally induced 1 hour after hormone treatment, the promoter subsequently becomes refractory to GR activation upon continuous hormone treatment (Lee and Archer 1994, Archer et al 1994a). This decline in promoter activation following hormone stimulation is characteristic and unique for templates assembled as chromatin, as it is not observed when the same sequence is transiently transfected into these cells. In fact, this "silencing" occurs despite the presence of relevant transcription factors in the nucleus during the refractory phase, as these proteins were able to interact with transiently introduced MMTV DNA (Lee and Archer 1994). Consequently, it was argued that these factors were prevented from binding to their recognition sequences by the reformation of Nuc-B, resulting in the inhibition of transcription (Lee and Archer 1994). These observations strengthen the view that while chromatin structure is vital to the understanding of transcriptional activation by GR, it also provides a mechanism for the cessation of transcription after prolonged hormone stimulation. In the future it will be important to understand the precise changes in chromatin structure that occur upon extended exposure to hormone, and the mechanism by which GR is prevented from remodeling the MMTV promoter.

2. Activation of the Tyrosine Aminotransferase (TAT) Gene

The role of chromatin structure in GR-mediated transcription has also been extensively studied using the rat tyrosine aminotransferase (TAT) gene (Crettaz et al 1988, Schmid et al 1987). TAT is a liver-specific gene that is induced by glucocorticoids acting through two glucocorticoid response elements (GREs) found at approximately -2500 and -5400 in the 5′-flanking region of the TAT gene (Schütz et al 1986). The proximal GRE confers hormone-dependent transcription to a heterologous gene in transfection studies (Grange et al 1989). The distal GRE acts synergistically with the proximal element to increase GR-mediated expression by 3–4 fold (Grange et al 1989). Unlike the proximal GRE, the distal element fails to function independently as a hormone response element. The region surrounding the proximal GRE also contains binding sites for additional transcription factors that function in concert with the GRE to mediate hormone induction (Strähle et al 1988).

Similar to the MMTV LTR, the TAT promoter adopts a phased array of nucleosomes over the promoter, and DNaseI studies have shown a rapid induction of a glucocorticoid-induced hypersensitive site at the proximal GRE (Grange et al 1989, Reik et al 1991). Micrococcal nuclease studies have demonstrated a disruption of the nucleosomal structure upon hormone administration. These results have been interpreted to indicate that hormone treatment and subsequent receptor binding lead to a displacement of two nucleosomes in the TAT promoter (Reik et al 1991). The displacement of the repressive nucleosomes would then allow transcription factors, such as HNF5, to access their binding sites and enhance transcription. Interestingly, the binding site for the liver-specific DNA-binding protein, HNF5, overlaps the GRE, yet DNaseI footprinting detected hormone-dependent binding of HNF5 within 10 minutes of Dex treatment (Rigaud et al 1991). This suggests that glucocorticoid treatment rapidly disrupts the nucleosomal structure of the region in a "hit and run" fashion similar to that proposed for MMTV.

B. The Progesterone Receptor

Like glucocorticoids, progestins have been shown to increase transcription from the MMTV LTR during transient transfection studies (Cato et al 1987). *In vitro* DNA-binding studies have established that both receptors bind to the HREs in nearly identical orientations (Gowland and Buetti 1989, Otten et al 1988, Cato et al 1987). Biochemical reconstitution experiments demonstrated that like GR, the progesterone receptor (PR) bound to an HRE in chromatin (Piña et al 1990, Pham et al 1992b). However, the position of an HRE in a nucleosome was shown to profoundly affect the affinity of PR for its site (Pham et al 1992b). Further, when the chicken PR was transiently expressed in mouse breast cells that contain a functional GR and multiple copies of a chromosomal MMTV construct, an unanticipated difference was observed. Although Dex efficiently stimulated transcription from the LTR, progesterone failed to induce restriction enzyme hypersensitivity in Nuc-B, or to increase mRNA production from the promoter (Smith et al 1993, Archer et al 1994b). Interestingly, the transiently transfected PR was capable of transactivating transiently transfected MMTV templates within the same cells. These unexpected results prompted a closer examination of the PR's interaction with MMTV chromatin in cells that express only the PR, and in cells that express both the PR and the GR.

1. Progesterone Receptor Action in Cells That Express Only PR

In order to investigate the ability of the PR to induce hormone-activated transcription from nucleosomal templates, gr−/PR+ human T47D breast cancer cells were stably transfected with the MMTV LTR driving a chloramphenicol acetyl transferase (CAT) reporter (Mymryk et al 1995). Although CAT assays established that R5020 (a synthetic progestin) induced transcription from these stable templates, the induction was significantly less than that observed with the GR+/pr− mouse cells treated with Dex (Mymryk et al 1995). This difference in induction was attributed to an increase in the basal transcription from the promoter in these gr−/PR+ cells. Micrococcal nuclease studies demonstrated an identically positioned array of nucleosomes in GR+/pr− and gr−/PR+ cell lines (Mymryk et al 1995). Closer examination of the chromatin structure of the promoter revealed no restriction enzyme hypersensitivity upon R5020 addition. Instead, substantial cutting was observed in the absence of hormone, suggesting that the chromatin was constitutively in a relaxed or "open" conformation (Mymryk et al 1995). In this way, the chromatin structure of the MMTV promoter in these cells is similar to that seen in the transiently transfected templates (Archer et al 1992). However, in this case, the adjacent nucleosomes, such as Nuc-A, were resistant to restriction enzyme cleavage, suggesting that the "open" chromatin conformation was confined to the Nuc-B region.

Consistent with the "open" state of the Nuc-B chromatin, *in vivo* footprinting demonstrated significant NF1 and OTF binding in both the presence and absence of progesterone (Mymryk et al 1995). This result suggested that these factors may contribute to the elevated basal transcription that is characteristic of these gr−/PR+ cells. Interestingly, although *in vivo* footprinting experiments failed to detect GR binding to the promoter (even in the presence of Dex), the PR was bound to the promoter in the presence and absence of R5020 (Mymryk et al 1995). This difference in receptor binding may be due inpart to the location of the two unliganded receptors. While GR is found primarily in the cytosol prior to glucocorticoid stimulation (Picard and Yamamoto 1987), unliganded PR is already in the nucleus (Guiochon-Mantel et al 1989). Alternatively this may reflect inherent differences between the way GR and PR interact with chromatin (Archer et al 1994b).

The constitutive binding of PR and NF1 and the constitutively "open" conformation of Nuc-B in gr−/PR+ cells predict that this nucleosomal structure is dependent on PR binding. To address this question, progesterone antagonist that either allowed DNA binding (type II) or prevented DNA binding (type I) while blocking PR activation was utilized (Bocquel et al 1993, Klein-Hitpass et al 1991). Upon the addition of a type I antagonist, the promoter assumed a "closed" or repressed chromatin structure (Mymryk and Archer 1995). The closing of Nuc-B chromatin coincided with the loss of transcription factor loading and inactivation of the promoter. In contrast, addition of a type II antagonist inhibited transcription, yet maintained the hypersensitive character of the promoter (Mymryk and Archer 1995). These findings demonstrated that the "open" conformation of the promoter was a result of PR binding and was maintained by the receptor with or without exogenous ligand. The dynamic nature of receptor-mediated chromatin remodeling was illustrated by reversibly cycling Nuc-B between "open" and "closed" conformations, using consecutive short-term treatments of agonist and antagonist (Mymryk and Archer 1995). These observations allowed further refinement of the model for the receptor-mediated induction of chromatin structure and transcriptional activation. PR complexed with a type II antagonist demonstrated that a change in chromatin conformation is necessary, but not sufficient, to induce assembly of a PIC, and hence transactivation.

2. Progesterone Receptor Action in Cells that Express Both PR and GR

To further investigate the (role) steroid receptor status plays in the transcriptional activation of MMTV, an expression vector containing the rat GR gene was stably transfected into human gr−/PR+ breast cancer cells, T47D, along with a construct containing the MMTV LTR driving a luciferase reporter gene (Nordeen et al 1989). The resulting cell line, T47D/A1-2, had comparable levels of the PR-A, PR-B, and newly introduced GR (Nordeen et al 1989). However, while these GR+/PR+ cells showed substantial Dex-induced transcription, little R5020 activation was observed. In contrast, transiently transfected MMTV templates demonstrated considerable

hormone-inducible transcription by both agonists, suggesting a clear difference in the interaction of PR with the two classes of MMTV templates (Archer
et al 1994b). This was not completely unexpected, as it confirmed the previous
experiments where PR had been introduced by transient transfection and
failed to activate the chromatin template (Archer et al 1994a).

As is consistent with the observed promoter activity from the chromatin
templates, *in vivo* restriction enzyme cleavage of Nuc-B revealed that while
Dex induced hypersensitivity, R5020 did not. In addition, transcription factor
loading onto the promoter was apparent only during glucocorticoid not
progestin, stimulation (Archer et al 1994b). These results suggested that, as
with the transient transfection of PR into GR+/pr− mouse cells, PR was
ineffective at activating MMTV from a chromatin template in the presence of
GR. Thus, the differences in chromatin structure and hormone-mediated
response observed in the gr−/PR+ and GR+/PR+ human cells suggested
that the receptor complement of a cell is important for chromatin organization, and ultimately for gene expression (Archer et al 1995).

Somewhat surprisingly, when the reciprocal experimental manipulation
was employed (stable introduction of PR into GR+/pr− cells), PR activated a
chromatin template (Smith et al 1993). The reasons for these divergent results
are not entirely clear, but overt differences in the experimental systems, such
as variations between mouse C127 and human T47D cells (including altered
expression of the PR isoforms) and differences in the MMTV DNA used to
make the cell lines, may be responsible (Smith et al 1993, Archer et al 1994a).
In addition, potential differences in the chromatin structure between these
human and mouse cell lines remain to be evaluated.

In T47D/A1-2 cells that express both GR and PR, coadministration of Dex
and R5020 resulted in a profound reduction in the ability of Dex to induce
chromatin remodeling and subsequent transcription (Archer et al 1994b).
This suggested that while PR itself is a nonproductive activator of the MMTV
promoter in chromatin, it can effectively interpose itself in the GR-mediated
activation pathway. It should be possible to further dissect PR's inhibitory
mechanism using PR antagonists that target specific steps in the hormone
activation pathway (Bocquel et al 1993, Klein-Hitpass et al 1991). As an
example, experiments using type I or type II antagonist may establish whether
PR must be active or merely able to bind DNA to inhibit GR remodeling of
chromatin and transcription.

C. The Estrogen Receptor

Early experiments with estrogen-induced genes demonstrated a clear correlation between hormone-mediated transcription and a change in chromatin
conformation (Weintraub 1983). Experiments on the chicken vitellogenin
gene revealed estrogen-dependent DNaseI hypersensitive sites whose appear-

ance parallelled increases in mRNA levels (Folger et al 1983, Burch and Weintraub 1983). In addition, developmental and sex-specific hypersensitive sites were detected in the estrogen-responsive apoVLDLII gene (Hache and Deeley 1988). Similar studies using the chicken ovalbumin gene also reported estrogen-mediated changes in both the DNaseI and micrococcal nuclease patterns during hormone-activated transcription (Bellard et al 1986, Bloom and Anderson 1982).

Building on these observations, recent studies have used the estrogen-responsive rat prolactin (rPRL) gene as a model system to study hormone-induced chromatin changes. Micrococcal nuclease experiments examining the rPRL promoter on stably maintained BPV-based minichromosomes found that the DNA was organized as a phased array of nucleosomes (Seyfred and Gorski 1990). The spacing of the nucleosomes within the promoter places the estrogen response unit (ERU), which is located 1700 bp upstream of the transcriptional start site, in a nucleosome-free zone (Willis and Seyfred 1996). This positioning is important since the estrogen receptor (ER) is unable to bind to the rPRL HRE in a nucleosomal context. Consistent with these observations, only cell lines that normally express rPRL, such as rat pituitary cells, exhibited phased nucleosomes over the promoter. In nonexpressing embryonic fibroblasts, a random arrangement of nucleosomes was seen over this region (Willis and Seyfred 1996). This finding implies that the rPRL promoter, sequence alone is insufficient to establish the phased array of nucleosomes and maintain the ERU in an accessible conformation. It is therefore likely that additional cell-specific factors are required for the expression of the rPRL gene.

Although the nucleosome phasing of the rPRL promoter is not affected by estrogen or antiestrogen treatment, hormone-dependent DNaseI and S1 nuclease hypersensitivity was observed around the transcriptional start site (Willis and Seyfred 1996, Seyfred and Gorski 1990). Such findings suggested an interaction between the ERU and the proximal promoter despite the presence of almost 2 kb of intervening DNA. This was confirmed using an *in vivo* nuclear ligation assay which detected a dramatic increase in the association of the ERU and the proximal promoter in estrogen-treated cells (Cullen et al 1993). Therefore, the chromatin structure of the rPRL promoter appears to bring these two regions into close proximity in an estrogen-dependent manner (Cullen et al 1993).

A similar spatial reorganization by nucleosomes has also been documented in the *Xenopus* vitellogenin B1 gene (Schild et al 1993). In this promoter, a single nucleosome separates the estrogen response element (ERE) and the proximal promoter. Nucleosome reconstitution studies, in concert with promoter deletions, demonstrated that the ability of the intervening DNA to form a nucleosome, and thus bring the ERE and transcription start site closer, was absolutely required for the estrogen-induced activation of the gene (Schild et al 1993). Subsequent binding studies determined that neither ER nor NF1 could bind to their specific elements within the vitellogenin B1 promoter when those sites were located within a nucleosome (Schild et al 1993). These observations imply that the position of the nucleosomes is crucial for organizing the spatial arrangement of transcription factors within the promoter, and is critical for estrogen-responsiveness.

Recently, investigators have used yeast as a model system to investigate the mechanism of hormone-induced chromatin remodeling. By transferring ER and a synthetic promoter containing multiple EREs into *S. cerevisiae*, estrogen-inducible gene activation can be reconstituted in yeast (Kladde et al 1996, Pham et al 1992a, Pham et al 1991b). Interestingly, at least two EREs were required to achieve the estrogen-mediated, chromatin disruption required for transcription activation (Pham et al 1991b). In this reconstituted system, the interaction of either estrogens or antiestrogens with ER enabled the receptor complex to bind the ERE and effect similar chromatin changes. However, the antiestrogen-bound ER was unable to subsequently induce transcription, and was therefore a nonproductive complex (Pham et al 1991a). In addition, both transcriptional activating domains of ER operated synergistically to disrupt chromatin and transactivate the promoter, although only the action of the C-terminal domain was hormone-dependent (Pham et al 1992a).

The ability of this type of reconstituted estrogen-responsive system to detect hormone-induced changes to chromatin has been augmented by *in vivo* expression of a Dam methylase (Kladde et al 1996). Since yeast does not contain endogenous DNA methylase, the introduction of a copper-inducible gene for SssI DNA methyltransferase into yeast has created a powerful system with which to investigate transcription factor/chromatin interactions inside a living cell. When this system was used, changes in methylation patterns definitively showed that ER binding to the ERE is a hormone-dependent event. However, when the ER was overexpressed, binding of unliganded receptor was detected (Kladde et al 1996).

D. The Androgen Receptor

Androgen-mediated gene expression also produces detectable changes in chromatin conformation. The mouse sex-limited protein (*Slp*) gene is inducible in a tissue- and sex-specific manner. DNaseI experiments located male-specific hypersensitive sites 2 kb upstream of the transcription start site in *Slp*-expressing tissues (Hemenway and Robins 1987). Closer examination of this region revealed that an ancient provirus had integrated at this site, conferring androgen responsiveness to the *Slp* gene (Stavenhagen and Robins 1988). This enhancer contains a consensus HRE that is functional in transient transfection studies, but binds weakly to the androgen receptor (AR) *in vitro* (Adler et al 1991). In addition, two adjacent, degenerate HREs augment the response to the first, but cannot function independently. Sites for numerous other transcription factors are also present proximate to the HREs and are required for maximal hormone responsiveness (Adler et al 1991). As expected, *in vitro* DNaseI footprinting experiments demonstrated hormone-dependent protein binding to the HRE and non-receptor sites (Scarlett and Robins 1995). However, little AR interaction with the HRE was detected *in vivo* in the presence or absence of hormone. This observation suggested that, like the "hit and run" scenario proposed for GR binding to the MMTV LTR and TAT promoter, androgen-mediated binding of the AR is transient. In contrast, the octamer transcription factor footprint was detected *in vivo* in a sex-specific manner, suggesting that AR requires OTF for proper regulation of the *Slp* gene (Scarlett and Robins 1995).

E. The Thyroid Hormone Receptor

Many thyroid hormone (T_3)-inducible genes possess multiple hormone-inducible DNaseI hypersensitive sites (Spindler et al 1989, Usala et al 1988, Jump et al 1990). In rat pituitary tumor cells, which express the growth hormone gene, thyroid hormone induction produced three hypersensitive sites in the region -200 to $+150$. These sites mapped to the thyroid response element (TRE), the TATA-box, and the first intron (Spindler et al 1989). In the malic enzyme gene, T_3 induced two DNaseI-hypersensitive sites in the proximal promoter at -310 and -50 (Usala et al 1988). Treatment of rats with thyroid hormone also rapidly induced the expression of the *S14* gene in the liver and lactating mammary glands. Chromatin isolated from these two tissues revealed multiple changes in the DNaseI cleavage pattern in the *S14* proximal promoter (Jump et al 1987). The appearance of these three hypersensitive sites correlated with *S14* gene activation (Jump et al 1990). A protein (P1) that may be responsible for one of the sites in the *S14* gene has been identified. The activity of P1 was inversely proportional to *S14* mRNA levels, suggesting that it functions as a repressor of thyroid receptor (TR) activation (Wong et al 1993).

   In addition to mediating thyroid hormone activation of genes, the unliganded TR, which is often found constitutively bound to chromatin *in vivo*, can also repress gene expression (see Chapter 6). This premature binding may passively block access of transcription factors to their cognate sites, thus inhibiting gene activation and PIC formation (Fondell et al 1993, Glass et al 1988). This type of receptor-mediated repression of transcription contrasts with that observed with MMTV in gr−/PR+ cells. With PR, unliganded binding of the receptor resulted in an "open" nucleosome, which allowed non-hormone-dependent transcription factors to bind constitutively to their sites within the MMTV promoter (Mymryk et al 1995).

   TR can also repress transcription by forming inactive heterodimers with other receptors (Glass et al 1989). This repressive action of TR was demonstrated on the TRβA promoter (Wong et al 1995). Here, the TRE was bound constitutively by an unliganded TR/RXR heterodimer. This complex repressed expression of the TRβA gene in the absence of hormone. However, T_3 addition resulted in a change in the micrococcal nuclease pattern and a dramatic increase in transcription. The change in the nucleosomal organization upon hormone induction was not dependent on transcription, since addition of transcription inhibitors did not affect the hormone-dependent chromatin disruption. Interestingly, repression by the unliganded heterodimer was augmented if the receptors were present during nucleosome formation, suggesting that the complex affects replication-coupled chromatin assembly (Wong et al 1995).

F. The Vitamin D Receptor

Although hormone-mediated binding of the vitamin D receptor has not been extensively studied in a chromatin context, it is clear that this interaction does alter chromatin structures. For example, osteocalcin, a bone-specific vitamin D-responsive protein, is expressed in osteoblasts during differentiation. As little as 0.1 kb of the proximal promoter of the osteocalcin gene was needed to produce minimal basal transcription in transient transfections (Frenkel et al 1996). This construct was inactive in a stable/chromatin context, and required at least 0.72 kb for maximal basal and vitamin D-induced transcription. This larger fragment of the promoter contained hormone-dependent DNaseI hypersensitive sites. These sites corresponded to the proximal promoter and vitamin D response element of the gene, and were not present in non-osseous cells that do not express osteocalcin (Montecino et al 1994). These results suggested that receptor-mediated remodeling of the chromatin structure plays an important role in the vitamin D responsiveness of the osteocalcin gene.

III. Cofactors

As argued above, it is clear that steroid and nuclear receptors play an integral role in altering chromatin structure and transcription activation. Recent experiments suggest that while the receptor is an important catalyst for these activities, it is not solely responsible for either activity. Genetic and biochemical experiments have described a number of additional proteins that interact with members of the steroid receptor family (for a review, see Horwitz et al 1996; see also Chapter 6). These include coactivators such as SRC-1 (Onate et al 1995), RIP-140 (Cavaillès et al 1995), ERAP-160 (Halachmi et al

1994), SPT6 (Baniahmad et al 1995), TIF1 (Le Douarin et al 1995), and GRIP 170 (Eggert et al 1995), and corepressors such as NCoR (Chen and Evans 1995), SMRT (Hörlein et al 1995), and calreticulin (Dedhar et al 1994).

In general, coactivators are fairly promiscuous as to which receptor they interact with (Horwitz et al 1996). For example, hSRC-1 was isolated by the yeast two-hybrid system using hPR as bait (Onate et al 1995). However, SRC-1, and its homolog ERAP160, not only increased transcription by the PR in a hormone-dependent fashion, but also enhanced transcription by GR, ER, TR, and RXR (Onate et al 1995, Halachmi et al 1994). This lack of specificity among coactivators, and their presence in most tissues, suggests that they play a universal role in transcriptional activation. The ubiquitous function of these coactivators is also supported by observations that they interact with the basal transcription machinery (Hong et al 1996). An important caveat is that although many of these proteins have been shown to interact with the receptors and components of the PIC, they have not been tested within the context of chromatin. It will therefore be important to determine how coactivators and corepressors interact with chromatin-bound steroid receptors and ultimately affect transcription regulation.

In contrast to the coactivators described above, recent experiments have described a number of multiprotein complexes that may function as steroid receptor cofactors. These include the SWI/SNF, RSC, and NURF complexes, which have been shown to mediate chromatin remodeling (Imbalzano et al 1994, Côté et al 1994, Cairns et al 1996b, Tsukiyama et al 1995). The most extensively studied coactivator of steroid-induced chromatin changes is the SWI/SNF complex, first identified in yeast by its involvement in mating-type switching and sucrose fermentation, and its mammalian homolog hBRG1 (Kingston et al 1996, Laurent and Carlson 1992, Peterson and Herskowitz 1992). SWI/SNF was later shown to be involved in transcription regulation in yeast by remodeling chromatin and allowing transcription factor access to nucleosomal DNA (Côté et al 1994, Imbalzano et al 1994, Peterson and Tamkun 1995, Travers 1992). Subsequent experiments demonstrated that GR-mediated activation of transcription was lost in yeast strains where one or more of the SWI proteins was disrupted (Yoshinaga et al 1992). The link to receptor-mediated changes in chromatin was enhanced when a subunit of the SWI/SNF complex, Swi3p, was shown to physically interact with GR *in vitro* (Yoshinaga et al 1992). More recent studies have demonstrated that the yeast protein Swp73 (SNF12) and its mammalian homolog BAF60a are also required for GR activation of transcription (Cairns et al 1996a, Wang et al 1996a, Wang et al 1996b, Miller et al 1996). Another yeast protein, SPT6, is also important in gene regulation by steroid hormones. SPT6, which binds to the ligand binding domain of ER *in vitro*, was found associated with areas of extensive chromatin remodeling and enhanced ER-mediated transcription in yeast and mammalian cells (Baniahmad et al 1995, Botvin and Winston 1996).

In addition to novel proteins isolated for their interaction with the steroid receptors and the chromatin remodeling proteins, recent experiments have linked the receptors with a number of previously characterized and studied proteins (Katzenellenbogen et al 1996, Horwitz et al 1996). A particularly intriguing finding is the potential involvement of the Rb tumor suppressor gene and CBP/p300 in SWI/SNF-mediated GR activation (Kamei et al 1996, Singh et al 1995). While Rb interacts with the human homologs of SWI/SNF, it is also dependent on this association to potentiate GR action (Singh et al 1995). CBP/p300, like SNF12, is a target for the adenovirus oncoprotein E1A, and interestingly, has also been linked to targeted acetylation of chromatin (Ogryzko et al 1996, Miller et al 1996). These observations suggest that the steroid receptors may function at the juctions of important cellular pathways, the activation of which may be modulated by receptor::chromatin interactions.

IV. SUMMARY

Steroid and nuclear receptors are important regulators of eukaryotic gene expression, which act in part by remodeling chromatin. While this super-family of transcription factors shares extensive structural homology, they appear to function through distinct mechanisms involving numerous auxiliary proteins. An extensive body of work has been compiled regarding transcriptional activation of genes assembled as chromatin by the glucocorticoid, progesterone, and estrogen receptors. As a result, we are now able to propose models of receptor-activated gene transcription that reflect the additional regulatory constraints imposed by chromatin (Figure 3).

When the HRE is located within the nucleosomal DNA, the binding of the activated receptor to its site apparently recruits other protein factors, such as SWI/SNF, which can help modulate chromatin structure. In cases such as the GR activation of the TAT gene, this transformation appears to cause a loss of the nucleosome(s). However, the receptor-associated chromatin remodeling machinery may not need to completely displace the nucleosome, but rather, as with MMTV promoter, may convert it into a more relaxed or "open" conformation. In both scenarios, this conformational change eliminates the repressive nature of the "closed" nucleosome, allowing transcription factors unrestricted access to their sites within the promoter. Once the transcription factors are able to bind, they and the receptors can recruit other factors required for assembly of a functional PIC to initiate transcription.

This mechanism of remodeling nucleosomes to activate a repressed promoter is in contrast to that found in a number of ER-activated genes. In the vitellogenin and rPRL genes, the nucleosomes seem to enhance rather than repress transcription. In these cases, the nucleosomal structure of the promoters is required to bring distant parts of the promoter into closer proximity.

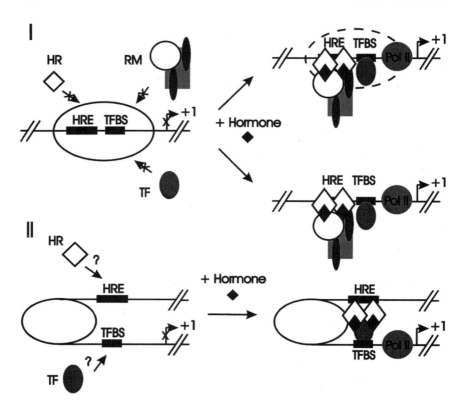

Figure 3.—Models for steroid receptor-mediated activation of transcription.
I). In the absence of hormone, the position of the nucleosome excludes transcription factor(s) (TF) from binding their cognate sites (TFBS), and transcription is repressed. Addition of hormone allows receptor (HR) dimerization, DNA binding, and interaction with chromatin remodeling machinery (RM). The result of these interactions can either be an alteration in nucleosome conformation, as with the MMTV, LTR, or displacement of the nucleosome, as with the TAT gene. The nucleosome remodeling now allows TF binding and recruitment of RNA polymerase II (PolII) and associated factors to achieve activated transcription. **II).** The wrapping of the DNA around the nucleosome(s) brings distant regulatory elements within the promoter into closer proximity, as in the vitellogenin and rPRL genes. Hormone addition permits interaction between the activated receptor and other TFs, resulting in assembly of a preinitiation complex.

This chromatin arrangement allows the interaction of the activated receptor with transcription factors in the proximal promoter, to help recruit the basal transcription machinery.

These models provide an outline of the mechanistic process, and they are necessarily general in their presentation. Future experiments will be necessary, to define the roles of the various activators and repressors as well as the contribution of higher-order chromatin structure. Finally, posttranslational modification of the receptors, activator proteins, and histones will undoubtedly have a profound impact on steroid-receptor-mediated activation of transcription from gene promoters assembled as chromatin.

ACKNOWLEDGMENTS

We wish to thank the following people for critical reading of the manuscript: Bonnie Deroo, Christy Fryer, Kim Kenward, Krista McAllister and Wendi Rodrigueza. Work in the authors' laboratory was supported by grants from the NCIC, MRC, LRCC, and LHSC to Trevor K. Archer.

REFERENCES

Adler AJ, Scheller A, Hoffman Y, and Robins DM (1991): Multiple components of a complex androgen-dependent enhancer. *Mol. Endo.* 5: 1587–1596.

Ali Z, and Singh N. (1987): Binding of linker histones to core nucleosome. Journal of *Biological Chemistry* 262: 12989–12993.

Almer A, and Hörz W. (1986): Nuclease hypersensitive regions with adjacent positioned nucleosomes mark the gene boundaries of the PHO5/PHO3 locus in yeast. *EMBO J.* 5: 2681–2687.

Amero SA, Kretsinger RH, Moncrief ND, Yamamoto KR, and Pearson WR (1992): The origin of nuclear receptor proteins: A single precursor distinct from other transcription factors. *Mol. Endocrinol.* 6: 3–7.

Archer TK, Cordingley MG, Marsaud V, Richard-Foy H, and Hager GL (1989): Steroid transactivation at a promoter organized in a specifically-positioned array of nucleosomes. In *Steroid/Thyroid Hormone Receptor Family and Gene Regulation*. JA Gustafsson, H. Eriksson, and J. Carlstedt-Duke, eds. Berlin: Birkhauser Verlag AG, pp. 221–238.

Archer TK, Cordingley MG, Wolford RG, and Hager GL (1991): Transcription factor access is mediated by accurately positioned nucleosomes on the mouse mammary tumor virus promoter. *Mol. Cell. Biol.* 11: 688–698.

Archer TK, Lefebvre P, Wolford RG, and Hager GL (1992): Transcription factor loading on the MMTV promoter: A bimodal mechanism for promoter activation. *Science* 255: 1573–1576.

Archer TK (1993): Nucleosomes modulate access of transcription factor to the MMTV promoter *in vivo* and *in vitro*. *Ann. NY Acad. Sci.* 684: 196–198.

Archer TK, Lee, H.-L, Cordingley MG, Mymryk JS, Fragoso G, Berard DS, and Hager GL (1994a). Differential steroid hormone induction of transcription from the mouse mammary tumor virus promoter. *Mol. Endo.* 8: 568–576.

Archer TK, Zaniewski E, Moyer M, and Nordeen SK (1994b). The differential capacity of glucocorticoids and progestins to alter chromatin structure and induce gene expression in human breast cancer cells. *Mol. Endo.* 8: 1154–1162.

Archer TK, Fryer CJ, Lee, H.-L, Zaniewski E, Liang T, and Mymryk JS (1995):

Steroid hormone receptor status defines the MMTV promoter chromatin structure *in vivo. J. Steroid Biochem. Mol. Biol.* 53: 421–429.

Archer, T.K, and Mymryk JS (1995): Modulation of transcription factor access and activity at the MMTV promoter *in vivo*. In *The Nucleus*. AP Wolffe, ed. Greenwich: JAI Press Inc., pp. 123–150.

Arents G, Burlingame RW, Wang, B.-C, Love WE, and Moudrianakis EN (1991): The nucleosomal core histone octamer at 3.1 Å resolution: A tripartite protein assembly and a left-handed superhelix. *Proc. Natl. Acad. Sci. USA* 88: 10148–10152.

Arents G, and Moudrianakis EN (1993): Topography of the histone octamer surface: Repeating structural motifs utilized in the docking of nucleosomal DNA. *Proc. Natl. Acad. Sci. USA* 90: 10489–10493.

Baniahmad C, Nawaz Z, Baniahmad A, Gleeson, M.AG, Tsai, M.-J, and O'Malley BW (1995): Enhancement of human estrogen receptor activity by SPT6: A potential coactivator. *Mol. Endo.* 9: 34–43.

Bellard M, Dretzen G, Bellard F, Kaye JS, Pratt-Kaye S, and Chambon, P (1986): Hormonally induced alterations of chromatin structure in the polyadenylation and transcription termination regions of the chicken ovalbumin gene. *EMBO J.* 5: 567–574.

Bloom KS, and Anderson JN (1982): Hormonal regulation of the conformation of the ovalbumin gene in chick oviduct chromatin. *J. Biol. Chem.* 257: 13018–13027.

Bocquel MT, Ji J, Ylikomi T, Benhamou B, Vergezac A, Chambon P, and Gronemeyer H. (1993): Type II antagonists impair the DNA binding of steroid hormone receptors without affecting dimerization. *J. Steroid. Biochem. Mol. Biol.* 45: 205–215.

Botvin A, and Winston, F (1996): Evidence that Spt6p controls chromatin structure by a direct interaction with histones. *Science* 272: 1473–1476.

Brüggemeier U, Kalff M, Franke S, Scheidereit C, and Beato, M (1991): Ubiquitous transcription factor OTF-1 mediates induction of the MMTV promoter through synergistic interaction with hormone receptors. *Cell* 64: 565–572.

Buetti E, Kühnel B, and Diggelmann, H (1989): Dual function of a nuclear factor I binding site in MMTV transcription regulation. *Nucl. Acids Res.* 17: 3065–3078.

Buetti, E (1994): Stably integrated mouse mammary tumor virus long terminal repeat DNA requires the octamer motifs for basal promoter activity. *Mol. Cell. Biol.* 14: 1191–1203.

Buetti E, and Kühnel, B (1986): Distinct sequence elements involved in the glucocorticoid regulation of the mouse mammary tumor virus promoter identified by linker scanning mutagenesis. J. Mol. Biol. 190: 379–389.

Burch, J.B, and Weintraub, H (1983): Temporal order of chromatin structural changes associated with activation of the major chicken vitellogenin gene. *Cell* 33: 65–76.

Cairns BR, Levinson RS, Yamamoto KR, and Kornberg RD (1996a). Essential role of Swp73p in the function of yeast Swi/Snf complex. *Genes Dev.* 10: 2131–2144.

Cairns BR, Lorch Y, Li Y, Zhang M, Lacomis L, Erdjument-Bromage H, Tempst P, Du J, Laurent B, and Kornberg RD (1996b). RSC, an essential, abundant chromatin-remodeling complex. *Cell* 87: 1249–1260.

Cato, A.CB, Henderson D, and Ponta, H (1987): The hormone response element of the mouse mammary tumour virus DNA mediates the progestin and androgen induction of transcription in the proviral long terminal repeat region. *EMBO J.* 6: 363–368.

Cavaillès V, Dauvois S, L'Horset F, Lopez G, Hoare S, Kushner PJ, and Parker MG (1995): Nuclear factor RIP140 modulates transcriptional activation by the estrogen receptor. *EMBO J.* 14: 3741–3751.

Cavallini B, Huet J, Plassat JL, Sentenac A, Egly JM, and Chambon, P (1988): A yeast activity can substitute for the HeLa cell TATA box factor. *Nature* 334: 77–80.

Chandler VL, Maler BA, and Yamamoto KR (1983): DNA sequences bound specifically by glucocorticoid receptor in vitro render a heterologous promoter hormone responsive in vivo. *Cell* 33: 489–499.

Chen, J.D, and Evans RM (1995): A transcriptional co-repressor that interacts with nuclear hormone receptors. *Nature* 377: 454–457.

Clark-Adams CD, Norris D, Osley MA, Fassler JS, and Winston, F (1988): Changes in histone gene dosage alter transcription in yeast. *Genes Dev.* 2: 150–159.

Cordingley MG, Riegel AT, and Hager GL (1987): Steroid-dependent interaction of transcription factors with the inducible promoter of mouse mammary tumor virus in vivo. *Cell* 48: 261–270.

Cordingley, M.G, and Hager GL (1988): Binding of multiple factors to the MMTV promoter in crude and fractionated nuclear extracts. *Nucl. Acids Res.* 16: 609–628.

Côté J, Quinn J, Workman JL, and Peterson CL (1994): Stimulation of GAL4 derivative binding to nucleosomal DNA by the yeast SWI/SNF complex. *Science* 265: 53–60.

Crettaz M, Muller-Weiland D, and Kahn CR (1988): Transcriptional and posttranscriptional regulation of tyrosine aminotransferase by insulin in rat hepatoma cells. *Biochemistry* 27: 495–500.

Cullen KE, Kladde MP, and Seyfred MA (1993): Interaction between transcription regulatory regions of prolactin chromatin. *Science* 261: 203–206.

Dedhar S, Rennie PS, Shago M, Leung Hagesteijn, C.-Y, Yang H, Filmus J, Hawley RG, Bruchovsky N, Cheng H, Matusik RJ, and Giguère V (1994): Inhibition of nuclear hormone receptor activity by calreticulin. *Nature* 367: 480–483.

Eggert M, Möws CC, Tripier D, Arnold R, Michel J, Nickel J, Schmidt S, Beato M, and Renkawitz R (1995): A fraction enriched in a novel glucocorticoid receptor-interacting protein stimulates receptor-department transcription *in vitro. J. Biol. Chem.* 270: 30755–30759.

Evans RM (1988): The steroid and thyroid hormone receptor superfamily. *Science* 240: 889–895.

Felsenfeld G (1992): Chromatin as an essential part of the transcriptional mechanism. *Nature* 355: 219–224.

Folger K, Anderson JN, Hayward MA, and Shapiro DJ (1983): Nuclease sensitivity and DNA methylation in estrogen regulation of Xenopus laevis vitellogenin gene expression. *J. Biol. Chem.* 258: 8908–8914.

Fondell JD, Roy AL, and Roeder RG (1993): Unliganded thyroid hormone receptor inhibits formation of a functional preinitiation complex: Implications for active repression. *Genes Dev.* 7: 1400–1410.

Fragoso G, John S, Roberts MS, and Hager GL (1995): Nucleosome positioning on the MMTV LTR results from the frequency-biased occupancy of multiple frames. *Genes Dev.* 9: 1933–1947.

Frenkel B, Montecino M, Green J, Aslam F, Desai R, Banerjee C, Stein JL, Lian JB, and Stein GS (1996): Basal and vitamin D-responsive activity of the rat osteocalcin promoter in stably transfected osteosarcoma cells: Requirement of upstream sequences for control by the proximal regulatory domain. *Endocrinology* 137: 1080–1088.

Glass CK, Holloway JM, Devary OV, and Rosenfeld MG (1988): The thyroid hormone receptor binds with opposite transcriptional effects to a common sequence motif in thyroid hormone and estrogen responsive elements. *Cell* 54: 313–323.

Glass CK, Lipkin SM, Devary OV, and Rosenfeld MG (1989): Positive and negative regulation of gene transcription by a retinoic acid-thyroid hormone receptor heterodimer. *Cell* 59: 697–708.

Gowland, P.L, and Buetti E (1989): Mutations in the hormone regulatory element of mouse mammary tumor virus differentially affect the response to progestins, androgens, and glucocorticoids. *Mol. Cell. Biol.* 9: 3999–4008.

Grange T, Roux J, Rigaud G, and Pictet R (1989): Two remote glucocorticoid responsive units interact cooperatively to promote glucocorticoid induction of rat tyrosine aminotransferase gene expression. *Nucl. Acids Res.* 17: 8695–8709.

Green S, Kumar V, Krust A, Walter P, and Chambon P (1986): Structural and functional domains of the estrogen receptor. *Cold Spring Harbor Symp. Quant. Biol.* 51: 751–758.

Green S, and Chambon P (1986): A superfamily of potentially oncogenic hormone receptors [news]. *Nature* 324: 615–617.

Green S, and Chambon P (1988): Nuclear receptors enhance our understanding of transcription regulation. *Trends Genet.* 4: 309–314.

Grunstein M (1990): Histone function in transcription. *Annu. Rev. Cell. Biol.* 6: 643–678.

Guiochon-Mantel A, Loosfelt H, Lescop P, Sar S, Atger M, Perrot-Applanat M, and Milgrom E (1989): Mechanisms of nuclear localization of the progesterone receptor: evidence for interaction between monomers. *Cell* 57: 1147–1154.

Hache, R.J, and Deeley RG (1988): Organization, sequence and nuclease hypersensitivity of repetitive elements flanking the chicken apoVLDLII gene: Extended sequence similarity to elements flanking the chicken vitellogenin gene. *Nucl. Acids Res.* 16: 97–113.

Hager GL, Archer TK, Fragoso G, Bresnick EH, Tsukagoshi Y, John S, and Smith CL (1993): Influence of chromatin structure on the binding of transcription factors to DNA. *Cold Spring Harbor Symp. Quant. Biol.: DNA and Chromosomes* 58: 63–71.

Halachmi S, Marden E, Martin G, MacKay H, Abbondanza C, and Brown M (1994): Estrogen receptor-associated proteins: Possible mediators of hormone-induced transcription. *Science* 264: 1455–1458.

Han M, and Grunstein M (1988): Nucleosome loss activates yeast downstream promoters in vivo. *Cell* 55: 1137–1145.

Hayes JJ (1996): Site-directed cleavage of DNA by a linker histone-Fe(II) EDTA conjugate: Localization of a globular domain binding site within a nucleosome. *Biochemistry* 35: 11931–11937.

Hayes, J.J, and Wolffe AP (1992): The interaction of transcription factors with nucleosomal DNA. *BioEssays* 14: 597–603.

Hemenway C, and Robins DM (1987): DNase I-hypersensitive sites associated with expression and hormonal regulation of mouse C4 and Slp genes. *Proc. Natl. Acad. Sci. USA* 84: 4816–4820.

Hong H, Kohli K, Trivedi A, Johnson DL, and Stallcup MR (1996): GRIP1, a novel mouse protein that serves as a transcriptional coactivator in yeast for the hormone binding domains of steroid receptors. *Proc. Natl. Acad. Sci. USA* 93: 4948–4952.

Horwitz KB, Jackson TA, Bain DL, Richer JK, Takimoto GS, and Tung L (1996): Nuclear receptor coactivators and corepressors. *Mol. Endo.* 10: 1167–1177.

Hörlein AJ, Näär AM, Heinzel T, Torchia J, Gloss B, Kurokawa R, Ryan A, Kamei Y, Söderström M, Glass CK, and Rosenfeld MG (1995): Ligand-independent repression by the thyroid hormone receptor mediated by a nuclear receptor co-repressor. *Nature* 377: 397–404.

Huang AL, Ostrowski MC, Berard D, and Hager GL (1981): Glucocorticoid regulation of the Ha-MuSV p21 gene conferred by sequences from mouse mammary tumor virus. *Cell 27(2 Pt 1)*, 245–255.

Imbalzano AN, Kwon H, Green MR, and Kingston RE (1994): Facilitated binding of TATA-binding protein to nucleosomal DNA. *Nature* 370: 481–485.

Jump DB, Wong NC, and Oppenheimer JH (1987): Chromatin structure and methylation state of a thyroid hormone-responsive gene in rat liver. *J. Biol. Chem.* 262: 778–784.

Jump DB, Bell A, and Santiago V (1990): Thyroid hormone and dietary carbohydrate interact to regulate rat liver S14 gene transcription and chromatin structure. *J. Biol. Chem.* 265: 3474–3478.

Kamei Y, Xu L, Heinzel T, Torchia J, Kurokawa R, Gloss B, Lin, S.-C, Heyman RA, Rose DW, Glass CK, and Rosenfeld MG (1996): A CBP integrator complex mediates transcriptional activation and AP-1 inhibition by nuclear receptors. *Cell* 85: 403–414.

Katzenellenbogen JA, O'Malley BW, and Katzenellenbogen BS (1996): Tripartite steroid hormone receptor pharmacology: Interaction with multiple effector sites as a basis for the cell- and promoter-specific action of these hormones. *Mol. Endocrinol.* 10: 119–131.

Kingston RE, Bunker CA, and Imbalzano AN (1996): Repression and activation by multiprotein complexes that alter chromatin structure. *Genes Dev.* 10: 905–920.

Kladde MP, Xu M, and Simpson RT (1996): Direct study of DNA-protein interactions in repressed and active chromatin in living cells. *EMBO J.* 15: 6290–6300.

Klein-Hitpass L, Cato, A.CB, Henderson D, and Ryffel GU (1991): Two types of antiprogestins identified by their differential action in transcriptionally active extracts from T47D cells. *Nucleic Acids Res.* 19: 1227–1234.

Kornberg RD (1974): Chromatin structure: A repeating unit of histones and DNA. *Science* 184: 868–871.

Laurent, B.C, and Carlson M (1992): Yeast SNF2/SWI2, SNF5, and SNF6 proteins function coordinately with the gene-specific transcriptional activators Gal4 and bicoid. *Genes Dev.* 6: 1707–1715.

Le Douarin B, Zechel C, Garnier, J.-M, Lutz Y, Tora L, Pierrat B, Heery D, Gronemeyer H, Chambon P, and Losson R (1995): The N-terminal part of TIF1, a putative mediator of ligand-dependent activation function (AF-2) of nuclear receptors, is fused to B-raf in the oncogenic protein T18. *EMBO J.* 14: 2020–2033.

Lee, H.-L, and Archer TK (1994): Nucleosome-mediated disruption of transcription factor-chromatin initiation complexes at the mouse mammary tumor virus long terminal repeat *in vivo. Mol. Cell. Biol.* 14: 32–41.

Mangelsdorf DJ, Thummel C, Beato M, Herrlich P, Schütz G, Umesono K, Blumberg B, Kastner P, Mark M, Chambon P, and Evans RM (1995): The nuclear receptor superfamily: The second decade. *Cell* 83: 835–839.

McGhee JD, Wood WI, Dolan M, Engel JD, and Felsenfeld G (1981): A 200 base pair region at the $5'$ end of the chicken adult beta-globin gene is accessible to nuclease digestion. *Cell* 27: 45–55.

Miller ME, Cairns BR, Levinson RS, Yamamoto KR, Engel DA, and Smith MM (1996): Adenovirus E1A specifically blocks SWI/SNF-dependent transcriptional activation. *Mol. Cell. Biol.* 16: 5737–5743.

Montecino M, Pockwinse S, Lian J, Stein G, and Stein J (1994): DNase I hypersensitivity sites in promoter elements associated with basal and vitamin D dependent transcription of the bone-specific osteocalcin gene. *Biochemistry* 33: 348–353.

Mymryk JS, Berard D, Hager GL, and Archer TK (1995): Mouse mammary tumor virus chromatin in human breast cancer cells is constitutively hypersensitive and exhibits steroid hormone-independent loading of transcription factors *in vivo*. *Mol. Cell. Biol.* 15: 26–34.

Mymryk, J.S, and Archer TK (1995): Dissection of progesterone receptor-mediated chromatin remodeling and transcriptional activation *in vivo*. *Genes Dev.* 9: 1366–1376.

Nordeen SK, Kühnel B, Lawler-Heavner J, Barber DA, and Edwards DP (1989): A quantitative comparison of dual control of a hormone response element by progestins and glucocorticoids in the same cell line. *Mol. Endo.* 3: 1270–1278.

Ogryzko VV, Schiltz RL, Russanova V, Howard BH, and Nakatani Y (1996): The transcriptional coactivators p300 and CBP are histone acetyltransferases. *Cell* 87: 953–959.

Onate SA, Tsai SY, Tsai, M.-J, and O'Malley BW (1995): Sequence and characterization of a coactivator for the steroid hormone receptor superfamily. *Science* 270: 1354–1357.

Ostrowski MC, Richard-Foy H, Wolford RG, Berard DS, and Hager GL (1983): Glucocorticoid regulation of transcription at an amplified, episomal promoter. *Mol. Cell. Biol.* 3: 2045–2057.

Ostrowski MC, Huang AL, Kessel M, Wolford RG, and Hager GL (1984): Modulation of enhancer activity by the hormone responsive regulatory element from mouse mammary tumor virus. *EMBO J.* 3: 1891–1899.

Otten AD, Sanders MM, and McKnight GS (1988): The MMTV LTR promoter is induced by progesterone and dihydrotestosterone but not by estrogen. *Mol. Endo.* 2: 143–147.

Panganiban AT (1985): Retroviral DNA integration. *Cell* 42: 5–6.

Payvar F, Wrange O, Carlstedt-Duke J, Okret S, Gustafsson JA, and Yamamoto KR (1981): Purified glucocorticoid receptors bind selectively in vitro to a cloned DNA fragment whose transcription is regulated by glucocorticoids in vivo. *Proc. Natl. Acad. Sci. USA* 78: 6628–6632.

Perlmann T, and Wrange O (1988): Specific glucocorticoid receptor binding to DNA reconstituted in a nucleosome. *EMBO J.* 7: 3073–3079.

Peterson, C.L, and Herskowitz I (1992): Characterization of the yeast SWI1, SWI2, and SWI3 genes, which encode a global activator of transcription. *Cell* 68: 573–583.

Peterson, C.L, and Tamkun JW (1995): The SWI-SNF complex: A chromatin remodeling machine? *Trend Biochem. Sci.* 20: 143–146.

Pham TA, Elliston JF, Nawaz Z, McDonnell DP, Tsai, M.-J, and O'Malley BW (1991a). Antiestrogen can establish nonproductive receptor complexes and alter chromatin structure at target enhancers. *Proc. Natl. Acad. Sci. USA* 88: 3125–3129.

Pham TA, Hwung Y, McDonnell DP, and O'Malley BW (1991b). Transactivation functions facilitate the disruption of chromatin structure by estrogen receptor derivatives *in vivo*. *J. Biol. Chem.* 266: 18179–18187.

Pham TA, Hwung YP, Santiso-Mere D, McDonnell DP, and O'Malley BW (1992a). Ligand-dependent and -independent function of the transactivation regions of the human estrogen receptor in yeast. *Mol. Endo.* 6: 1043–1050.

Pham TA, McDonnell DP, Tsai, M.-J, and O'Malley BW (1992b). Modulation of progesterone receptor binding to progesterone response elements by positioned nucleosomes. *Biochemistry* 31: 1570–1578.

Picard D, and Yamamoto KR (1987): Two signals mediate hormone-dependent nuclear localization of the glucocorticoid receptor. *EMBO J.* 6: 3333–3340.

Piña B, Brüggemeier U, and Beato M (1990): Nucleosome positioning modulates accessibility of regulatory proteins to the mouse mammary tumor virus promoter. *Cell* 60: 719–731.

Pruss D, Bartholomew B, Persinger J, Hayes JJ, Arents G, Moudrianakis EN, and Wolffe AP (1996): An asymmetric model for the nucleosome: A binding site for linker histones inside the DNA gyres. *Science* 274: 614–617.

Reik A, Schutz G, and Stewart AF (1991): Glucocorticoids are required for establishment and maintenance of an alteration in chromatin structure: Induction leads to a reversible disruption of nucleosomes over an enhancer. *EMBO J.* 10: 2569–2576.

Richard-Foy H, and Hager GL (1987): Sequence-specific positioning of nucleosomes over the steroid-inducible MMTV promoter. *EMBO J.* 6: 2321–2328.

Richmond TJ, Finch JT, Rushton B, Rhodes D, and Klug A (1984): Structure of the nucleosome core particle at 7 A resolution. *Nature* 311: 532–537.

Rigaud G, Roux J, Pictet R, and Grange T (1991): In vivo footprinting of rat TAT gene: Dynamic interplay between the glucocorticoid receptor and a liver-specific factor. *Cell* 67: 977–986.

Rosenfeld, P.J, and Kelly TJ (1986): Purification of nuclear factor I by DNA recognition site affinity chromatography. *J. Biol. Chem.* 261: 1398–1408.

Roth, S.Y, and Allis CD (1992): Chromatin condensation: Does histone H1 dephosphorylation play a role? TIBS 17: 93–98.

Scarlett, C.O, and Robins DM (1995): In vivo footprinting of an androgen-dependent enhancer reveals an accessory element integral to hormonal response. *Mol. Endo.* 9: 413–423.

Schild C, Claret, F.-X, Wahli W, and Wolffe AP (1993): A nucleosome-dependent static loop potentiates estrogen-regulated transcription from the *Xenopus* vitellogenin B1 promoter *in vitro*. *EMBO J.* 12: 423–433.

Schmid E, Schmid W, Jantzen M, Mayer D, Jastorff B, and Schütz G (1987): Transcription activation of the tyrosine aminotransferase gene by glucocorticoids and cAMP in primary hepatocytes. *Eur. J. Biochem.* 165: 499–506.

Schütz G, Schmid W, Jantzen M, Danesch U, Gloss B, Strähle U, Becker P, and Boshart M (1986): Molecular basis for the hormonal regulation of the tyrosine aminotransferase and tryptophan oxygenase genes. *Ann. NY Acad. Sci.* 478: 93–100.

Seyfred, M.A, and Gorski J (1990): An interaction between the 5' flanking distal and proximal regulatory domains of the rat prolactin gene is required for transcriptional activation by estrogens. *Mol. Endo.* 4: 1226–1234.

Singh P, Coe J, and Hong W (1995): A role for retinoblastoma protein in potentiating transcriptional activation by the glucocorticoid receptor. *Nature* 374: 562–565.

Smith CL, Archer TK, Hamlin-Green G, and Hager GL (1993): Newly expressed progesterone receptor cannot activate stable, replicated mouse mammary tumor virus templates but acquires transactivation potential upon continuous expression. *Proc. Natl. Acad. Sci. USA* 90: 11202–11206.

Spindler SR, Crew MD, and Nyborg JK (1989): Thyroid hormone transcriptional regulatory region of the growth hormone gene. *Endocrine Research* 15: 475–493.

Stavenhagen, J.B, and Robins DM (1988): An ancient provirus has imposed androgen regulation on the adjacent mouse sex-limited protein gene. *Cell* 55: 247–254.

Strähle U, Schmid W, and Schutz G (1988): Synergistic action of the glucocorticoid receptor with transcription factors. *EMBO J.* 7: 3389–3395.

Travers AA (1992): The reprogramming of transcriptional competence. *Cell* 69: 573–575.

Truss M, Bartsch J, Schelbert A, Hache, R.JG, and Beato M (1995): Hormone induces binding of receptors and transcription factors to a rearranged nucleosome on the MMTV promoter *in vivo. EMBO* 14: 1737–1751.

Tsukiyama T, Daniel C, Tamkun J, and Wu C (1995): *ISWI*, a member of the *SWI2/SNF2* ATPase family, encodes the 140 kDa subunit of the nucleosome remodeling factor. *Cell* 83: 1021–1026.

Usala SJ, Young WS, III, Morioka H, and Nikodem VM (1988): The effect of thyroid hormone on the chromatin structure and expression of the malic enzyme gene in hepatocytes. *Mol. Endo.* 2: 619–626.

van Holde KE (1988): *Chromatin* Heidelberg: Springer-Verlag.

Wang W, Côté J, Xue Y, Zhou S, Khavari PA, Biggar SR, Muchardt C, Kalpana GV, Goff SP, Yaniv M, Workman JL, and Crabtree GR (1996a). Purification and biochemical heterogeneity of the mammalian SWI-SNF complex. *EMBO J.* 15: 5370–5382.

Wang W, Xue Y, Zhou S, Kuo A, Cairns BR, and Crabtree GR (1996b). Diversity and specialization of mammalian SWI/SNF complexes. *Genes Dev.* 10: 2117–2130.

Weintraub H (1983): Tissue-specific gene expression and chromatin structure. *Harvey Lect.* 79: 217–244.

Willis, S.D, and Seyfred MA (1996): Pituitary-specific chromatin structure of the rat prolactin distal enhancer element. *Nucl. Acids Res.* 24: 1065–1072.

Wolffe AP (1994): Nucleosome positioning and modification: Chromatin structures that potentiate transcription. *TIBS* 19: 240–244.

Wolffe AP (1995): *Chromatin Structure and Function* London: Academic Press.

Wong J, Shi YB, and Wolffe AP (1995): A role for nucleosome assembly in both silencing and activation of the Xenopus TR beta A gene by the thyroid hormone receptor. *Genes Dev.* 9: 2696–2711.

Wong NC, Raymond J, and Carr FE (1993): A liver-specific nuclear protein represses transcription of the S14 in vitro and in vivo. *J. Biol. Chem.* 268: 19431–19435.

Workman, J.L, and Buchman AR (1993): Multiple functions of nucleosomes and regulatory factors in transcription. *Trends Biochem. Sci.* 18: 90–95.

Wu C (1980): The 5′ ends of Drosophila heat shock genes in chromatin are hypersensitive to DNase I. *Nature* 286: 854–860.

Yamamoto KR (1985): Steroid receptor regulated transcription of specific genes and gene networks. *Annu. Rev. Genet.* 19: 209–252.

Yoshinaga SK, Peterson CL, Herskowitz I, and Yamamoto KR (1992): Roles of SWI1, SWI2, and SWI3 proteins for transcriptional enhancement by steroid receptors. *Science* 258: 1598–1604.

Young HA, Shih TY, Scolnick EM, and Parks WP (1977): Steroid induction of mouse mammary tumor virus: effect upon synthesis and degradation of viral RNA. *J. Virol.* 21: 139–146.

Zaret, K.S, and Yamamoto KR (1984): Reversible and persistent changes in chromatin structure accompany activation of a glucocorticoid-dependent enhancer element. *Cell* 38: 29–38.

Zlatanova J, and van Holde K (1992): Histone H1 and transcription: Still an enigma? *J. Cell Sci.* 103: 889–895.

9

Regulation of Glucocorticoid and Estrogen Receptor Activity by Phosphorylation

MICHAEL J. GARABEDIAN, INEZ ROGATSKY,
ADAM HITTELMAN, ROLAND KNOBLAUCH,
JANET M. TROWBRIDGE AND MARIJA D. KRSTIC

INTRODUCTION

Protein phosphorylation is a versatile posttranslational modification that can regulate enzymatic activities, cell proliferation, DNA replication, and gene expression (Boulikas 1995, Hofer 1996, Johnson and O'Reilly 1996, Karin 1994, Karin and Hunter 1995, Piwnica-Worms 1996). Steroid receptors, like many transcription factors, are phosphoproteins, and increasing evidence suggests that the activity of steroid receptors is regulated by phosphorylation (Bai and Weigel 1995, Blok et al 1996, Kuiper and Brinkmann 1994, Orti et al 1989, Weigel 1996). Treatment of cells with activators or inhibitors of protein kinases or phosphatases affects the transcriptional activity of steroid receptors (Edwards et al 1993, Edwards 1994, Katzenellenbogen 1996). In some cases, these treatments can activate the receptor in the absence of hormone (O'Malley et al 1995, Power et al 1992). Thus, there appears to be significant communication between signal transduction pathways that affect intracellular kinase and phosphatase activity and steroid receptor function.

No simple rules have emerged for predicting the mechanism by which phosphorylation alters protein function. For example, phosphorylation of the yeast transcription factor SWI 5 triggers its nuclear localization (Jans and Hubner 1996, Jans et al 1995, Moll et al 1993, Moll et al 1991), phosphorylation of mammalian serum response factor (SRF) alters its DNA binding kinetics (Gille et al 1996, Gille et al 1995, Papavassiliou 1994), while

Molecular Biology of Steroid and Nuclear Hormone Receptors
Leonard P. Freedman, Editor
© 1998 Birkhäuser Boston

phosphorylation of mammalian cAMP-response element binding protein (CREB) stimulates its association with a protein cofactor, CREB-binding protein (CBP), thus regulating its transcriptional activity (Chrivia et al 1993, Parker et al 1996). With respect to steroid receptors, it has been suggested that phosphorylation might modulate hormone binding, influence nucleocyto-plasmic shuttling of the protein, or affect transcription by altering DNA binding or interaction with the basal transcription machinery (Bodwell et al 1996, Bodwell et al 1993, DeFranco et al 1995, Orti et al 1992). The following is a review of the current literature regarding the role of phosphorylation in glucocorticoid and estrogen receptor signal transduction and transcriptional regulation.

MOLECULAR ORGANIZATION OF STEROID RECEPTORS

Steroid hormone receptors are ligand-dependent transcriptional regulatory proteins that affect many aspects of development and metabolism. The steroid receptor family includes the glucocorticoid receptor (GR), estrogen receptor (ER), androgen receptor (AR), progesterone receptor (PR), and mineralo-corticoid receptor (MR) (Beato et al 1995). Steroid hormones enter target cells via membrane transporters or pumps (Kralli et al 1995). Once inside the cell, the hormone binds to the unoccupied receptor and transforms it to an active DNA-binding form, which can interact directly with promoters to stimulate or repress transcription of target genes (Zilliacus et al 1995). Changes in gene expression lead to changes in mRNA and protein/enzyme levels which ultimately affect cellular metabolism.

Steroid receptors have a characteristic molecular organization, with indi-vidual domains performing distinct functions. The N-terminal segment contains a transcriptional activation domain. The central zinc-finger region includes the DNA binding and dimerization domains as well as a nuclear localization signal. The hormone binding region and an additional transcrip-tional regulatory domain are located at the C-terminus of the molecule. In the absence of ligand, the receptor is associated with the heat-shock protein, hsp90, which inactivates its DNA binding and transcriptional activation function.

LOCALIZATION AND IDENTIFICATION OF GR PHOSPHORYLATION SITES

The glucocorticoid receptor is phosphorylated in the absence of hormone; however, additional phosphorylation occurs in conjunction with agonist, but not antagonist, binding (Bodwell et al 1995, Orti et al 1993, Orti et al 1989, Tienrungroj et al 1987). Phosphoaminoacid analysis reveals that GR expressed in a variety of mammalian cells and in yeast (*S. cerevisiae*) is primarily phosphorylated on serine (S) residues, and that minor phosphorylation also occurs on threonine (T) residues (Krstic 1995). Deletion studies indicate that a majority of the phosphorylated residues lie within the N-terminus, the region of the receptor involved in transcriptional enhancement ($\tau 1$ or enh2) (Dalman et al 1988, Hoeck and Groner 1990, Hutchison et al 1993). Bodwell et al have identified seven phosphorylation sites in the mouse GR (positions S134, S162, T171, S224, S232, S246 and S327 in the rat GR numbering scheme) by direct sequencing of phosphorylated peptides that were isolated from the receptor overexpressed in WCL2 cells, a Chinese hamster ovary cell line (Bodwell et al 1991). Through peptide mapping and mutagenesis studies, Krstic (1995) has identified four major sites of phosphorylation on rat GR expressed in hepatoma cells and in yeast that coincide with a subset of sites identified in mouse GR and correspond to T171, S224, S232 and S246 (Figure 1). It has been shown that for the rat receptor, phosphorylation of S224 and S232 increases in the presence of hormone agonists, whereas residues T171 and S246 appear to be phosphorylated constitutively. In each case, residues T171, S224, S232 and S224 are followed by proline, thus corresponding to a motif typically modified by cyclin-dependent kinases (CDK consensus = S/T(P)–P–X–R/K) or mitogen-activated protein kinases (MAPK consensus = nonpolar–X– S/T(P) –P). Of the remaining phosphorylation sites, S134 is located in a consensus motif for casein kinase II (CK II consensus = S/T(P)–X–X–E/D), whereas the sequences surrounding S162 and S327 do not match any known kinase consensus motif. Recently, the rat GR has also been shown to be phosphorylated *in vitro* at S527 by DNA-dependent protein kinase (DNA-PK) (Giffen et al 1997, Giffen et al 1996).

Although specific residues become phosphorylated in the presence of agonists, no charge alterations are observed when receptor phosphorylation is examined by nonequilibrium isoelectric focusing, suggesting that GR must exist in a highly heterogeneous phosphorylation state (Bodwell et al 1996). It is tempting to speculate that specific subpopulations of receptors with different phosphorylation patterns may regulate distinct receptor activities or target specific genes for regulation. However, to date there is no experimental evidence to support this idea.

PHOSPHORYLATION OF GR DURING THE CELL Cycle

Two of the GR phosphorylation sites, S224 and S232, match the consensus sequence for the cyclin-dependent kinase, cyclin A/CDK2, which is involved

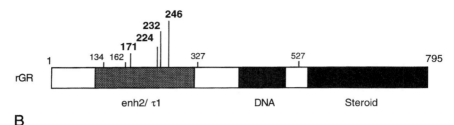

A

B

		Putative Kinase
S134	leu gly leu **ser(P)**[134] ser gly glu	CK II
S162	asn pro lys **ser(P)**[162] ser thr ser	?
T171	gly cys ala **thr(P)**[171] pro thr glu	MAPK/GSK3
S224	ser ala gly **ser(P)**[224] pro gly lys	CDK
S232	thr asn glu **ser(P)**[232] pro trp arg	CDK
S246	asn leu leu **ser(P)**[246] pro leu ala	MAPK
S327	ala ser phe **ser(P)**[327] gly thr asn	?
S527	ala gly val **ser(P)**[527] gln asp thr	DNA-PK

Figure 1.—Residues phosphorylated on the glucocorticoids receptor.
A, a schematic representation of the phosphorylation sites identified by Bodwell et al on the mouse GR using the rat GR numbering scheme. In bold are the residues phosphorylated in rat hepatoma cells or yeast (*S. cerevisiae*). B, shows the sequences surrounding the receptor phosphorylation sites and the putative protein kinases expected to target the motif: consensus site for casein kinase II [CK II = S/T(P)–X–X–E/D]; mitogen-activated protein kinase (MAPK consensus = non-polar–X–S/T(P)–P]; glycogen synthase kinase-3 [GSK = S/T(P)–P–X–S(P)] and cyclin-dependent kinases [CDK consensus = **S/T(P)**–P–K/R]. The consensus site for DNA-dependent protein kinase (DNA-PK) has not been determined. Question marks (?) represent phosphorylated sites that have no match to known consensus kinase target sites.

in cell cycle progression. Classic studies on the cell cycle dependence of glucocorticoid induction of tyrosine aminotransferase activity, as well as recent evidence on the induction of a variety of GR-responsive reporter constructs suggest that GR transcriptional activation is confined to the late G1- and S-phases of the cell cycle (Hsu et al 1992, Hu et al 1994). This cell cycle phase restriction on receptor transcriptional activity corresponds to a hormone-dependent increase in receptor phosphorylation. Conversely, a hormone-dependent increase in receptor phosphorylation is not observed in G2/M, and the receptor is unable to activate transcription in this stage of the cell cycle (Hsu et al 1992, Hu et al 1994). Receptor-mediated transcriptional repression of AP-1-dependent gene expression, on the other hand, operates throughout the cell cycle, suggesting that certain elements of receptor function, such as hormone binding and nuclear localization, operate in a cell-cycle-independent manner (Hsu and DeFranco 1995). Taken together, these results suggest a link to transcriptional activation, but not to repression, and receptor phosphorylation by cell cycle-dependent kinases (Figure 2).

Although hormone-dependent hyperphosphorylation of GR and transcriptional activation by the receptor fail to occur in the G2/M phase of the cell cycle, it appears that phosphorylation of the receptor in the absence of hormone is actually greater in cells synchronized in G2/M than in cells synchronized in G1/S (Hu et al 1994). This may result from increases in the activity of receptor kinases, and/or decreases in the receptor phosphatases, that occur during these stages of the cell cycle. This finding has prompted Munch and colleagues to speculate that the increased level of receptor phosphorylation observed in the absence of hormone in G2/M-arrested

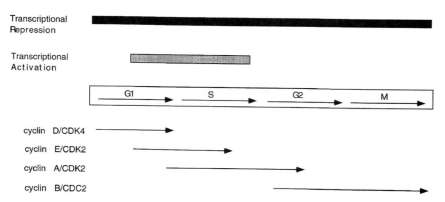

Figure 2.—Glucocorticoid-responsive periods during the cell cycle. Arranged in a linear fashion are the stages of the cell cycle. The periods of the cell cycle in which GR-mediated transcriptional activation (gray line) and repression (black line) operate are depicted above. Below the stages of the cell cycle are the periods during which particular cyclin-CDK complexes are active.

cells results in a form of GR that is insensitive to hormone and therefore fails to induce transcriptional activation, perhaps because of the receptor's inability to interact with factors involved in transcriptional enhancement (Bodwell et al 1996, Hu et al 1994). One might expect that this receptor hyperphosphorylation in G2/M would mimic the hormone-treated receptor in G1/S, thereby enabling GR to activate transcription. That this is not the case suggests that receptor phosphorylation in G1/S versus G2/M differs *quantitatively* and *qualitatively*, and implies that phosphorylation can negatively regulate receptor transcriptional activation. Thus, hormone-dependent transcriptional activity in G1/S, as well as the lack of receptor transcriptional enhancement in G2/M may be controlled by phosphorylation. This hypothesis remains to be tested.

Studies of GR phosphorylation by the DeFranco laboratory have shown that treatment of cells with okadaic acid (OA), an inhibitor of type I and II protein phosphatases, results in a receptor that accumulates in the cytoplasm and is unable to respond to glucocorticoid treatment (DeFranco et al 1991). Interestingly, the receptor isolated from OA-treated cells is hyperphosphorylated (which is reminiscent of the GR phosphorylation state in G2/M-synchronized cells) and refractory to glucocorticoid stimulation. Whether OA induces hyperphosphorylation of the receptor by inhibiting receptor phosphatases *in vivo* or by causing a G2/M arrest in the cells is not known. However, this intriguing correlation supports the notion that receptor hyperphosphorylation may lead to a glucocorticoid-insensitive state, perhaps through phosphorylation of sites that inhibit the ability of GR to associate with nuclear targets.

GR KINASES

Recent studies examining the kinases that phosphorylate GR suggest that two distinct protein kinase families, cyclin-dependent kinase (CDK) and mitogen-activated protein kinase (MAPK), are capable of phosphorylating GR *in vitro* and together produce a pattern of the phosphorylated receptor amino acids identical to that observed *in vivo* (Krstic et al 1997). MAPK phosphorylates receptor residues T171 and S246, *in vitro*, whereas cyclin E/CDK2 phosphorylates S224 and cyclin A/CDK2 modifies S224 and S232 *in vitro* (Figure 3A). An interdependency of the receptor sites phosphorylated by cyclin A/CDK2 has also been observed; S224 must first be phosphorylated, before phosphorylation of S232 can occur (Krstic et al 1997). This order of addition is consistent with phosphopeptide sequence data which have identified phosphopeptides containing the phosphorylated S232 residue only in conjunction with phosphoserine 224, and never by itself. It is interesting to note that this order, whereby S224 is first modified by cyclin E/CDK2 and subsequently

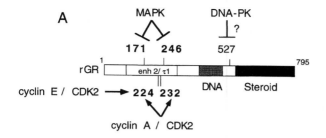

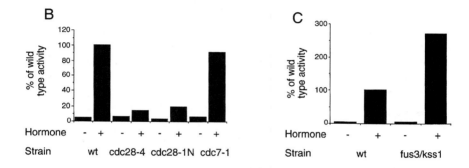

Figure 3.—Distinct protein kinase families target GR and differentially affect transcriptional enhancement *in vivo*.

A, a schematic representation of the receptor sites phosphorylated and the kinases that modify them *in vitro*. Threonine 171 and S246 are phosphorylated by MAPK, S224 is phosphorylated by cyclin E/CDK2 and cyclin A/CDK2, and S232 is phosphorylated by cyclin A/CDK2. Serine 527 is phosphorylated by DNA-PK *in vitro*. B, receptor-dependent transcriptional enhancement in yeast strains with defects in CDK activity. The glucocorticoid receptor and GRE-linked reporter gene were introduced into a wild-type strain and isogenic derivatives bearing two distinct alleles of the yeast CDK catalytic subunit homolog CDC28, cdc28-4 and 28-1N. In the presence of hormone, receptor-dependent transcriptional enhancement was decreased 4–5-fold in the cdc28 strains, relative to wild-type strain or relative to a strain with a defect in a non-cyclin-dependent kinase cdc7-1. C, effect of MAPK delections on receptor transcriptional activity. Receptor and reporter plasmids were introduced into a strain bearing deletions of the yeast MAPK homologs, FUS3 and KSS1, and hormone-dependent transcriptional activity was measured. In contrast to the decreased receptor transcriptional activity observed in the cdc28 strains, receptor transcriptional enhancement was increased in the absence of FUS3 and KSS1. Together, these genetic results suggest strongly that CDK and MAPK modulate receptor function, potentiating and attenuating transcriptional enhancement, respectively.

S232 is subsequently phosphorylated by cyclin A/CDK2, reflects the temporal order of expression of these kinases during the cell cycle: cyclin E/CDK2 appears late in G1, and cyclin A/CDK2 is active during S-phase. The combined action of cyclin E/CDK2 and cyclin A/CDK2 may contribute to hormone-induced phosphorylation and the increased receptor activity observed during the G1/S phase of the cell cycle. Thus, the convergence between the cell cycle–kinases that phosphorylate the receptor and cell cycle-dependent receptor phosphorylation may temporally link GR transcriptional regulation to the cell cycle.

The effects of mutations in MAPK and CDK on GR-dependent transcriptional activation *in vivo* have also been analyzed. Since GR phosphorylation is conserved between yeast and mammalian cells, receptor-dependent transcriptional enhancement was examined in yeast strains with defects in CDK and MAPK (Krstic et al 1997). It was found that mutations in the CDK catalytic subunit, p34^{CDC28}, or its regulatory cyclin subunits, reduced receptor-dependent transcriptional activation *in vivo*, indicating that CDK function is necessary for full receptor-mediated transcriptional enhancement (Figure 3B). In contrast, deletion of the MAPK homologs FUS3 and KSS1 in yeast increased transcriptional activation by the receptor, suggesting that phosphorylation of the receptor by MAPK decreases receptor transcriptional activity (Figure 3C). Together, these findings support the idea that the receptor is a common target for signaling by MAPK and CDKs with each exerting an opposite effect on receptor transcriptional enhancement. From these findings, a potential regulatory circuit combining the receptor kinases has been proposed (Figure 4). This hypothesis suggests that activators of the MAPK pathway, including growth factors and certain oncoproteins (such as H-Ras and v-Mos) or inhibitors of CDK function (such as TGF-β, p21, and p27) might attenuate receptor-induced transcriptional responses. In contrast, negative regulators of MAPK, such as protein kinase A (PKA), as well as activators of CDKs, such as the cyclins or CDK activating kinases (CAKs), should increase receptor transcriptional activity. Consistent with this regulatory scheme are the findings that H-Ras and v-Mos inhibit receptor-dependent transcriptional activation in mammalian cells (Borror et al 1995). In contrast, inhibition of MAPK activity, through activation of the PKA pathway, results in increased receptor transcriptional activation, although dramatic changes in GR phosphorylation are not observed (Moyer et al 1993). This apparent lack of receptor phosphorylation in response to PKA activators may reflect an inability of the two-dimensional phosphopeptide mapping method to detect quantitative differences in receptor phosphorylation, rather than an absence of attendant receptor phosphorylation *per se*. Alternately, PKA activators may not affect receptor action through direct receptor phosphorylation of inhibitory sites by MAPK, but instead may alter the phosphorylation state of cofactors that would mediate receptor-activated gene expression. Finally, it is conceivable that modifications of both the

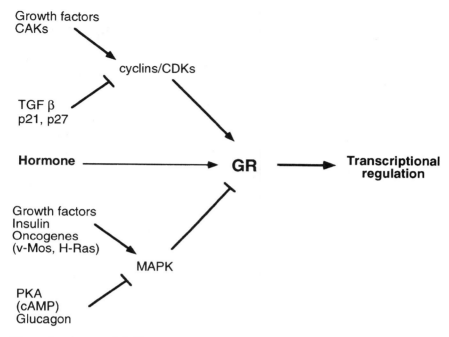

Figure 4.—A potential GR regulatory circuit.
Signaling pathways that may affect receptor kinase activity are shown. Activators of the MAPK pathway, including growth factors, insulin and certain oncoproteins, or inhibitors of CDK function, such as TGF-β, p21 and p27 might attenuate receptor-induced transcriptional responses. In contrast, negative regulators of MAPK, such as PKA, as well as activators of CDK, such as cyclins or CDK activating kinases (CAKs), should potentiate receptor action.

receptor and the cofactor are necessary for the observed increase in transcriptional response after treatment with PKA activators. It will be interesting to examine the mechanism of PKA activation of receptor transcription using receptor mutants with substitutions in the MAPK phosphorylation sites.

GR Phosphorylation: Site Conservation and Consequences of Phosphorylation Site Mutations for Receptor Transcriptional Activity

Many of the phosphorylated residues of GR are conserved across species. The equivalent positions of serines 134, 162, 224, 232, 246, and 527 from rat GR are conserved among human, primate, mouse, rat, and *Xenopus* receptors.

Threonine 171 in the rat receptor is conserved at the equivalent position in the mouse receptor, but corresponds to alanine and proline in human and primate GR, respectively. Although species-specific differences are evident, a vast majority of the receptor phosphorylation sites have been preserved throughout evolution. This suggests a common and conserved regulatory role for receptor phosphorylation.

Mutations of the GR phosphorylation sites that change the phosphorylated amino acids from serine or threonine to alanine, representing a nonphosphorylatable residue, or to aspartic acid, mimicking a constitutively phosphorylated site, have been generated and assayed for transcriptional activity in both mammalian cells and yeast. Mason and Housley tested whether the alanine or aspartic acid substitution in mouse GR affected transcriptional activation from a mouse mammary tumor virus promoter (MMTV-LTR-CAT) in COS-1 and L929 cells (Mason and Housley 1993). They found that individual serine to alanine or aspartic acid changes had very little effect on receptor transcriptional activation from this reporter construct in transient transfection experiments. Mutant receptors with five alanines at the most highly conserved phosphorylation sites reduced receptor transcriptional activity by only 20% relative to the wild-type receptor in their assay. In a separate study, Almlof et al investigated the role of phosphoaminoacids in transcriptional enhancement by mutating S224, S232, and S246 in human GR to alanines and assessing receptor transcriptional activity in yeast (Almlof et al 1995). Interestingly, S224A and S232A mutants exhibited a decreased (> 50%) level of transcriptional activity in yeast, suggesting that phosphorylation of S224 and S232 may increase receptor transcriptional enhancement. It should be noted, however, that the level of the S224A protein appears to be reduced, relative to wild-type, which may contribute to the observed decrease in transcriptional activity. The S246A mutation, on the other hand, resulted in a 2-fold increase in transcriptional activity relative to the wild-type receptor in yeast, suggesting that phosphorylation of S246 may inhibit receptor function. These findings are consistent with results from the receptor kinase deletion studies, which indicate that elimination of MAPK, which phosphorylates receptor residues T171 and S246, leads to increased receptor activity, whereas mutations in the CDKs that target S224 and S232 decrease receptor transcriptional enhancement. These findings support a role for phosphorylation in both positive and negative regulation of the receptor. Although Almlof et al conclude that phosphorylation is not important for receptor transcriptional activation in the contexts examined, it is important to consider that these experiments were performed under conditions in which the receptor is overexpressed, and that transcriptional activity has been assessed in only a few cell types and promoter contexts. Indeed, recent findings from the Weigel laboratory demonstrate that a mutation of serine 211 to alanine in the chicken PR reduces transcriptional activity, and that the magnitude of this decrease is dependent not only upon cell type and promoter context, but also on the level of receptor expression (Bai and Weigel 1996). Their work indicates that differences in the transcriptional activity of the mutant versus the wild-type receptor are more pronounced if the mutant protein is expressed at low, physiological levels, suggesting that overexpression of the receptor can override the effect of the mutation.

Recently, Webster et al examined the consequences of single or multiple phosphorylation site mutations of the mouse GR for receptor expression, ligand-dependent nuclear translocation, ligand-dependent down-regulation of the level of receptor protein, receptor half-life, and hormone-dependent transcriptional activation (Webster et al 1997). They found that single or multiple mutations had little effect on receptor expression, subcellular localization, ligand-dependent nuclear translocation, or transcriptional activation from the MMTV promoter in COS-1 cells. Interestingly, they demonstrated a marked decrease (> 50%) in the ability of mutant receptors S224A and S232A, as well as the double mutant S224A/S232A, to activate transcription from a minimal promoter containing a simple glucocorticoid response element. Virtually no loss of receptor activation was observed with the S246A mutant. Their results indicate that, depending on the type of promoter, GR phosphorylation can modulate the magnitude of the transcriptional response. In addition, they provide evidence that, compared to wild-type GR, a receptor mutated in seven of the eight receptor phosphorylation sites has a longer half-life and fails to undergo ligand-dependent destabilization, which suggests a role for phosphorylation in the regulation of receptor protein levels.

POTENTIAL MECHANISMS THROUGH WHICH PHOSPHORYLATION MAY REGULATE GR ACTIVITY

Recall that the majority of the phosphorylated residues in GR lie within the N-terminal transcriptional activation domain, which further suggests a role for receptor phosphorylation in transcriptional enhancement. Phenotypes of alanine substitutions at specific positions do modestly affect receptor transcriptional activity, but only in certain cell contexts. Deletion of potential receptor kinases CDK and MAPK regulates receptor transcriptional enhancement, positively and negatively, respectively.

How does phosphorylation regulate GR-mediated transcriptional enhancement? In principle, phosphorylation at multiple sites by different protein kinases could affect a range of distinct receptor activities, such as dimerization, DNA binding, protein stability, or interaction with other proteins, including those of the transcriptional machinery (Karin and Hunter 1995, Webster et al 1997). Each of these functions could be affected either positively or negatively, thereby adjusting the receptor response to

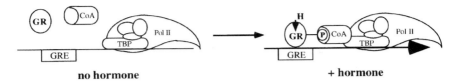

Figure 5.—A model for phosphate-mediated protein-protein interactions in GR transcriptional activation.

Diagramed are the receptor protein (GR), putative coactivator (CoA), and the basal transcription machinery including TATA binding protein (TBP) and RNA polymerase II (Pol II). Hormone-induced phosphorylation may promote protein-protein interactions between the receptor and the putative receptor cofactor and, in doing so, promote receptor-dependent transcriptional enhancement.

changing physiological conditions. For example, phosphorylation of GR might mediate an interaction with an accessory protein (Figure 5) (Janknecht and Hunter 1996a, Janknecht and Hunter 1996b). There are several examples of the modulatory effects of phosphates in protein-protein interactions, such as phosphorylation of CREB by PKA resulting in binding to CBP. Recently, using an elegant genetic screen, Iniguez-Lluhi et al have identified three amino acid residues (E219, F220, W234) in the N-terminal region of rat GR that are required for transcriptional activation and are in close proximity to the phosphorylation targets S224 and S232 (Iniguez-Lluhi et al 1997). These results support the idea that the regions surrounding highly conserved receptor phosphorylation sites are of critical importance to transcriptional activation and may therefore be accessible for interaction with other proteins. It is conceivable that GR phosphorylation at S224 and S232 enhances the putative protein-protein interactions involved in transcriptional activation by the receptor. Conversely, phosphorylation at T171 and S246, which appears to be inhibitory to receptor transcriptional activation, may disrupt protein-protein interaction or recruit inhibitory proteins that lower receptor-dependent transcriptional activity. Identifying proteins that interact with the receptor's N-terminal transcriptional activation domain, and examining whether receptor phosphorylation can affect these interactions, will be crucial to understanding the mechanism of receptor-mediated transcriptional activation.

ER Phosphorylation

Phosphorylation is also believed to play an important role in ER function. Mutagenesis and direct phosphopeptide sequencing studies have shown that ER contains several serines and a tyrosine residue that are phosphorylated *in vivo* (Figure 6) (Ali et al 1993, Arnold et al 1994, Arnold et al 1995, Lahooti et al 1995, Le Goff et al 1994). Tyrosine 537 (Y537) is located in the ligand binding domain and is phosphorylated in an estrogen-independent manner (Arnold et al 1995, Auricchio et al 1996). It has been proposed that modification of this tyrosine may play a role in regulating multiple ER activities, including ligand binding, dimerization, and DNA binding (Arnold and Notides 1995). Recent mutagenesis studies of Y537 demonstrate that the replacement of this residue with alanine or other nonhydrophobic residues generates an activated form of the receptor that stimulates transcription in the absence of ligand (White et al 1997). In contrast to the wild-type protein, the Y537A mutant interacts with the ER coactivator proteins SRC1 and RIP 140 in a ligand-independent manner. White et al propose that non-phosphorylated tyrosine 537 is required to keep the receptor in a transcriptionally inactive state. Phosphorylation of this residue may generate a conformation that allows for the recruitment of coactivators in the absence of ligand.

A majority of the ER phosphoserines are contained within the transcriptional activation domain of the receptor, the A/B region, and are modified upon estradiol binding, suggesting a role in transcriptional activation (Figure 6). Human ER transiently transfected into COS-1 cells is phosphorylated mainly on S118, but S104 and/or S106 also appear to be sites of receptor phosphorylation (Ali et al 1993, Joel et al 1995, Le Goff et al 1994). Identification of the phosphorylation sites in the mouse ER expressed in COS-1 cells also identifies S118 as well as S154 and S294, but not S104 and/or S106, as sites of phosphorylation (Lahooti et al 1995). Whether these discrepancies reflect methodological or species-specific differences remains unknown. Serine 167 appears to be the major site of phosphorylation in ER isolated from MCF-7 cells (Arnold et al 1994). Phosphorylation of other receptor sites in MCF-7 cells is also evident, but has not been fully characterized.

The sequence contexts of the putative ER phosphorylation sites reveal that serines 104, 106, 118 and 294 are potential targets for multiple ser/thr-pro-directed kinases, including MAPK, CDKs, and glycogen synthase kinase-3 (GSK3). Serine 154 lies within a sequence that may be modified by calmodulin-dependent kinase II (CaMK II). Serine 167 is flanked by consensus phosphorylation sites for both CaMK II and CK II, suggesting that this residue may be a common target for more than one kinase. MAPK has been shown to phosphorylate S118 *in vitro* and activators of MAPK are known to increase transcriptional activation by the ligand-bound ER *in vivo* (Arnold et al 1995, Kato et al 1995). CK II was shown to phosphorylate ER *in vitro* at S167 and to increase the receptor's ability to bind DNA (Arnold et al 1995, Tzeng and Klinge 1996). The DNA-PK has also been reported to phosphorylate ER *in vitro*, although neither the site(s) of phosphate addition nor the effect upon ER-dependent transcription has been determined (Arnold et al

1995). A CDK-independent effect of cyclin D1 as well as a CDK-dependent effect of cyclin A on ER transcriptional activation have also been reported (Trowbridge et al 1997, Zwijsen et al 1997).

Mutations of specific sites result in cell-type-specific phenotypic consequences for receptor function *in vivo*. For example, replacement of serine 118 with alanine reduces transcription by 15–75% (depending on promoter and cell type), with apparently no effect on DNA binding or nuclear localization of the receptor (Ali et al 1993, Lahooti et al 1995, Le Goff et al 1994). Replacement of serine 118 with an acidic residue increases transcriptional enhancement, implying that a negative charge at residue 118 may be important for receptor transcriptional activation (Ali et al 1993, Lahooti et al 1995, Le Goff et al 1994). A triple mutant, in which serines 104, 106, and 118 are replaced by alanines, displays a reduction in transcriptional enhancement in COS-1 cells by 50%, as compared to the wild-type receptor. The single serine 118 to alanine mutation in the same cells and promoter context reduces transcriptional enhancement by only 15% (Le Goff et al 1994).

LIGAND-INDEPENDENT ACTIVATION

Phosphorylation is a possible mechanism through which other signaling pathways may exert an effect on ER transcription in a ligand-dependent or

A

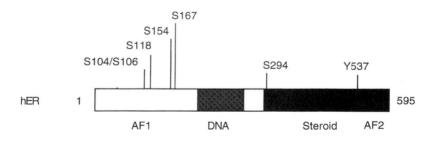

Figure 6A.—Residues phosphorylated on the estrogen receptor.
A, schematic representation of the phosphorylation sites identified on ER. Serines 104, 106, 118, 154, and 294 are phosphorylated in the presence of 17-β-estradiol in COS-1 cells. Serine 167 and tyrosine 537 are phosphorylated in MCF-7 cells in a ligand-independent manner.

B

Putative Kinase

S104	leu	asn	ser	val	ser(P)[104]	pro	ser	pro	leu	CDK?/GSK3?
S106	asn	val	ser	pro	ser(P)[106]	pro	leu	met	leu	CDK?/GSK3?
S118	pro	pro	gln	leu	ser(P)[118]	pro	phe	leu	gln	CDK?/MAPK
S154	try	arg	pro	asn	ser(P)[154]	asp	asn	arg	arg	CaMK II?
S167	glu	arg	leu	ala	ser(P)[167]	thr	asn	asp	lys	CaMK II?/CK II
S294	asn	leu	trp	pro	ser(P)[294]	pro	leu	met	ile	CDK?/GSK3?
Y537	val	val	pro	leu	tyr(P)[537]	asp	leu	leu	leu	Src family

Figure 6B.—Residues phosphorylated on the estrogen receptor.
Shows the sequences surrounding each phosphorylation site and the putative protein kinases expected to target the motif: consensus site for glycogen synthase kinase-3 [GSK3 = S/T(P)–P–X–S(P)]; mitogen activated protein kinase [MAPK consensus = non-polar-X-S/T(P)–P]; calmodulin-dependent protein kinase II [CaMK II = R–X–X–S/T(P)]; casien kinase II [CK II = S/T(P)–X–X–E/D]. Question marks alone (?) represent phosphorylated sites that have no match to consensus kinase phosphorylation sites. Unlike S118, which has been shown to be phosphorylated by MAPK *in vitro* and *in vivo*, kinases followed by question marks (?), such as GSK3 at S104 and CaMK II at S167, represent receptor phosphorylation sites that match consensus kinase phosphorylation sites, but have not been shown to be targets of these kinases *in vitro* or *in vivo*.

-independent manner. Synergistic activation of receptor-mediated transcription by estradiol and protein kinase A and C activators has been reported (Aronica and Katzenellenbogen 1993, Aronica et al 1994, Cho and Katzenellenbogen 1993, Ince et al 1994). The activation of ER in an estradiol-free context by dopamine and epidermal growth factor (EGF) also indicates a ligand-independent process (Ignar-Trowbridge et al 1996, Ignar-Trowbridge et al 1993, Smith et al 1993). Ligand-independent activation of PR by dopamine and EGF, and of AR by PKA activators has also been reported (Chauchereau et al 1994, Edwards et al 1993, Nazareth and Weigel 1996, Power et al 1991, Zhang et al 1994).

The best characterized ligand-independent interaction is that between ER and the EGF/RAS/MAPK signaling pathway (Bunone et al 1996, Kato et al 1995). Phosphorylation of the receptor by MAPK has been shown to lead to activation of unliganded ER by EGF, although the precise mechanism of this activation is not fully understood. The ligand-independent activation of ER is induced not only by EGF but also through the expression of constitutively active versions of the MAPK activators RAS and MEK1. Likewise, ER activation is blocked by expression of dominant negative forms of RAS and MEK1 *in vivo* (Bunone et al 1996).

EGF activation has also been shown to increase the amount of phosphate incorporated into ER *in vivo* (Bunone et al 1996). Both the ligand-independent transcriptional activation of ER by EGF, and EGF-stimulated phosphorylation are largely blocked in the ER mutant S118A. However, mimicking the effect of phosphorylation of ER by introducing an aspartic acid residue at serine 118 (S118D) does not lead to ligand-independent activity, which suggests that phosphorylation of S118 is necessary but not sufficient for ligand-independent receptor activity (Ali et al 1993, Bunone et al 1996). It is conceivable that cofactors involved in ER transcriptional activation are also modified by MAPK, which, together with ER phosphorylation, leads to ligand-independent receptor activity (Figure 7). Ligand-independent activity of ER provides a possible explanation of how EGF is able to mimic many of estrogen's physiological effects in uterine tissue, and it may also have important implications for estrogen-independent growth observed in some breast tumors (Korach 1994, Pietras et al 1995).

FUTURE STUDIES

Although steroid receptor phosphorylation has been the subject of much investigation, the mechanism by which phosphorylation affects receptor function remains unclear. Individual serine to alanine substitutions at most phosphorylated receptor residues only modestly affect transcriptional activation by GR and ER. However, this may be due in part to the insensitivity of the assays commonly used to examine receptor activity. That is, the use of transient transfections and synthetic reporter constructs may not accurately measure the contribution of single phosphorylation events to receptor function. Challenges for the future will include the examination of the transcriptional regulatory function of receptor phosphorylation site mutants in more physiological settings. Stable expression of the receptor phosphorylation site mutants at physiological levels and examination of their effects on the transcription of an endogenous receptor-responsive gene will undoubtedly reveal more about the role of phosphorylation in transcriptional regulation than current transient transfection studies.

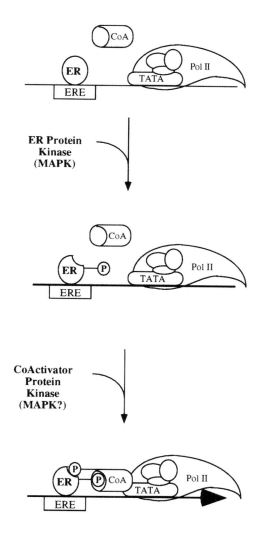

Figure 7.—A model for ligand-independent ER transcriptional activation. Diagramed are the receptor protein (ER), putative coactivator (CoA), and the basal transciption machinery including TATA binding protein (TBP) and RNA polymerase II (Pol II). In this model, phosphorylation of the receptor by MAPK at S118 as well as the modification of a putative cofactor, either by MAPK or another protein kinase activated by MAPK, would facilitate a protein-protein interaction between the receptor and cofactor, and thus promote receptor-dependent transcriptional enhancement in the absence of hormone.

An additional complexity confounding the interpretation of the mutagenesis studies for GR is the realization that phosphorylation may both positively and negatively affect receptor transcriptional enhancement. Mutations that simultaneously alter positive and negative regulatory phosphorylation sites may offset the effects of receptor phosphorylation on transcriptional enhancement, resulting in no apparent phenotype. Therefore, measuring transcriptional activation from receptors that harbor substitutions of activating (S224 and S232) or inhibitory (T171 and S246) sites exclusively may result in more pronounced phenotypes.

The study of receptor phosphorylation has also been hindered by the lack of a simple assay to detect changes in receptor phosphorylation. Certain proteins display a shift in their electrophoretic mobility on a one-dimensional SDS polyacrylamide gel in response to phosphorylation. However, this is not the case for most steroid receptors expressed in mammalian cells. Therefore, examination of receptor phosphorylation is a multistep process that requires (1) metabolic labeling of the receptor using large amounts of ^{32}P, (2) receptor isolation, (3) protease digestion, (4) separation, and (5) evaluation of the resulting phosphopeptides by HPLC or thin-layer chromatographic techniques. While informative, these studies have been limited to cultured cells and may not accurately reflect receptor phosphorylation in a given tissue or in response to a particular stimulus in the intact organism. Thus, the degree and sites of GR and ER phosphorylation in mammalian tissues have yet to be investigated. Novel reagents, such as phosphorylation site-specific antiserum, would provide the means to evaluate receptor phosphorylation during development, differentiation, or tumor progression using simple *in situ* and immunoblotting procedures.

The production of GR and ER "knock-out" mice has revealed important roles for GR in the development of the lung and brain, and for ER in reproductive tissues (Cole et al 1995, Korach et al 1996). To ultimately address the question of phosphorylation in receptor function, it may be necessary to perform "knock-in" experiments where the endogenous copy of the wild-type receptor is replaced with phosphorylation site mutants. The assessment of mice expressing a nonphosphorylated or constitutively phosphorylated version of these receptors might provide telling phenotype(s), that would be useful in determining the physiological role(s) of phosphorylation in receptor function.

SUMMARY

Although the precise role of phosphorylation is still a matter of speculation, it does appear that receptor phosphorylation functions to direct or refine receptor transcriptional activity in response to particular physiological requirements. The ability of steroid hormone receptors to respond to extra-cellular signals via phosphorylation permits a flexibility in receptor action that, independently or in conjunction with the steroid ligands, may be crucial in coordinating the various metabolic processes managed by these key regulatory molecules.

ACKNOWLEDGMENTS

We would like to thank Drs. Susan Logan and Samir Taneja for critically reading the manuscript. Michael J. Garabedian would like to acknowledge financial support from the Army Breast Cancer Research Fund (DAMD17-94-J-4454, DAMD17-96-1-6032), the Whitehead Fellowship for Junior Faculty in Biological Sciences, and the Kaplan Cancer Center.

REFERENCES

Ali S, Metzger D, Bornert JM, and Chambon, P (1993): Modulation of transcriptional activation by ligand-dependent phosphorylation of the human oestrogen receptor A/B region. *EMBO Journal*. 12: 1153–60.

Almlof T, Wright AP, and Gustafsson JA (1995): Role of acidic and phosphorylated residues in gene activation by the glucocorticoid receptor. *Journal of Biological Chemistry* 270: 17535–40.

Arnold SF, and Notides AC (1995): An antiestrogen: a phosphotyrosyl peptide that blocks dimerization of the human estrogen receptor. *Proceedings of the National Academy of Sciences of the United States of America* 92: 7475–9.

Arnold SF, Obourn JD, Jaffe H, and Notides AC (1994): Serine 167 is the major estradiol-induced phosphorylation site on the human estrogen receptor. *Molecular Endocrinology* 8: 1208–14.

Arnold SF, Obourn JD, Jaffe H, and Notides AC (1995): Phosphorylation of the human estrogen receptor by mitogen-activated protein kinase and casein kinase II: consequence on DNA binding. *Journal of Steroid Biochemistry & Molecular Biology* 55: 163–72.

Arnold SF, Obourn JD, Jaffe H, and Notides AC (1995): Phosphorylation of the human estrogen receptor on tyrosine 537 in vivo and by src family tyrosine kinases in vitro. *Molecular Endocrinology* 9: 24–33.

Arnold SF, Vorojeikina DP, and Notides AC (1995): Phosphorylation of tyrosine 537 on the human estrogen receptor is required for binding to an estrogen response element. *Journal of Biological Chemistry* 270: 30205–12.

Aronica SM, and Katzenellenbogen BS (1993): Stimulation of estrogen receptor-mediated transcription and alteration in the phosphorylation state of the rat uterine estrogen receptor by estrogen, cyclic adenosine monophosphate, and insulin-like growth factor-I. *Molecular Endocrinology* 7: 743–52.

Aronica SM, Kraus WL, and Katzenellenbogen BS (1994): Estrogen action via the cAMP signaling pathway: stimulation of adenylate cyclase and cAMP-regulated gene transcription. *Proceedings of the National Academy of Sciences of the United States of America* 91: 8517–21.

Auricchio F, Migliaccio A, Castoria G, Di Domenico M, Bilancio A, and Rotondi, A (1996): Protein tyrosine phosphorylation and estradiol action. [Review] [41 refs]. *Annals of the New York Academy of Sciences* 784: 149–72.

Bai W, and Weigel NL (1995): Phosphorylation and steroid hormone action. [Review] [115 refs]. *Vitamins & Hormones* 51: 289–313.

Bai W, and Weigel NL (1996): Phosphorylation of Ser211 in the chicken progesterone receptor modulates its transcriptional activity. *Journal of Biological Chemistry* 271: 12801–6.

Beato M, Herrlich P, and Schutz, G (1995): Steroid hormone receptors: many actors in search of a plot. [Review] [66 refs]. *Cell* 83: 851–7.

Blok LJ, de Ruiter PE, and Brinkmann AO (1996): Androgen receptor phosphorylation. [Review] [37 refs]. *Endocrine Research* 22: 197–219.

Bodwell JE, Hu JM, Hu LM, and Munck, A (1996): Glucocorticoid receptors: ATP and cell cycle dependence, phosphorylation, and hormone resistance. [Review] [35 refs]. *American Journal of Respiratory & Critical Care Medicine* 154: S2–6.

Bodwell JE, Hu JM, Orti E, and Munck, A (1995): Hormone-induced hyperphosphorylation of specific phosphorylated sites in the mouse glucocorticoid receptor. *Journal of Steroid Biochemistry & Molecular Biology* 52: 135–40.

Bodwell JE, Hu LM, Hu JM, Orti E, and Munck, A (1993): Glucocorticoid receptors: ATP-dependent cycling and hormone-dependent hyperphosphorylation. [Review] [52 refs]. *Journal of Steroid Biochemistry & Molecular Biology* 47: 31–8.

Bodwell JE, Orti E, Coull JM, Pappin DJ, Smith LI, and Swift, F (1991): Identification of phosphorylated sites in the mouse glucocorticoid receptor. *Journal of Biological Chemistry* 266: 7549–55.

Borror KC, Garabedian MJ, and DeFranco DB (1995): Glucocorticoid receptor phosphorylation in v-mos-transformed cells. *Steroids* 60: 375–82.

Boulikas, T (1995): Phosphorylation of transcription factors and control of the cell cycle. [Review] [524 refs]. *Critical Reviews in Eukaryotic Gene Expression* 5: 1–77.

Bunone G, Briand PA, Miksicek RJ, and Picard, D (1996): Activation of the unliganded estrogen receptor by EGF involves the MAP kinase pathway and direct phosphorylation. *EMBO Journal* 15: 2174–83.

Chauchereau A, Cohen-Solal K, Jolivet A, Bailly A, and Milgrom, E (1994): Phosphorylation sites in ligand-induced and ligand-independent activation of the progesterone receptor. *Biochemistry* 33: 13295–303.

Cho H, and Katzenellenbogen BS (1993): Synergistic activation of estrogen receptor-mediated transcription by estradiol and protein kinase activators. *Molecular Endocrinology* 7: 441–52.

Chrivia JC, Kwok RP, Lamb N, Hagiwara M, Montminy MR, and Goodman RH (1993): Phosphorylated CREB binds specifically to the nuclear protein CBP. *Nature* 365: 855–9.

Cole TJ, Blendy JA, Monaghan AP, Krieglstein K, Schmid W, Aguzzi A, Fantuzzi G, Hummler E, Unsicker K, and Schutz, G (1995): Targeted disruption of the glucocorticoid receptor gene blocks adrenergic chromaffin cell development and severely retards lung maturation. *Genes & Development* 9: 1608–21.

Dalman FC, Sanchez ER, Lin AL, Perini F, and Pratt WB (1988): Localization of phosphorylation sites with respect to the functional domains of the mouse L cell glucocorticoid receptor. *Journal of Biological Chemistry* 263: 12259–67.

Defranco DB, Madan AP, Tang Y, Chandran UR, Xiao N, and Yang, J (1995): Nucleocytoplasmic shuttling of steroid receptors. [Review] [86 refs]. *Vitamins & Hormones* 51: 315–38.

DeFranco DB, Qi M, Borror KC, Garabedian MJ, and Brautigan DL (1991): Protein phosphatase types 1 and/or 2A regulate nucleocytoplasmic shuttling of glucocorticoid receptors. *Molecular Endocrinology* 5: 1215–28.

Edwards DP, Weigel NL, Nordeen SK, and Beck CA (1993): Modulators of cellular protein phosphorylation alter the trans-activation function of human progesterone receptor and the biological activity of progesterone antagonists. *Breast Cancer Research & Treatment* 27: 41–56.

Edwards DR (1994): Cell signalling and the control of gene transcription. [Review] [62 refs]. *Trends in Pharmacological Sciences* 15: 239–44.

Giffin W, Kwast-Welfed J, Rodda DJ, Prefontaine GG, Traykova-Andonova M, Zhang Y, Weigel NL, Lefebvre YA, and Hache RJ, G (1997): Sequence-specific DNA Binding and Transcription Factor Phosphorylation by Ku Autoantigen/ DNA-dependent Protein Kinase. *Journal of Biological Chemistry* 272: 5647–58.

Giffin W, Torrance H, Rodda DJ, Prefontaine GG, Pope L, and Hache RJ (1996): Sequence-specific DNA binding by Ku autoantigen and its effects on transcription. *Nature* 380: 265–8.

Gille H, Kortenjann M, Strahl T, and Shaw PE (1996): Phosphorylation-dependent formation of a quaternary complex at the c-fos SRE. *Molecular & Cellular Biology* 16: 1094–102.

Gille H, Kortenjann M, Thomae O, Moomaw C, Slaughter C, Cobb MH, and Shaw PE (1995): ERK phosphorylation potentiates Elk-1-mediated ternary complex formation and transactivation. *EMBO Journal* 14: 951–62.

Hoeck W, and Groner, B (1990): Hormone-dependent phosphorylation of the glucocorticoid receptor occurs mainly in the amino-terminal transactivation domain. *Journal of Biological Chemistry* 265: 5403–8.

Hofer HW (1996): Conservation, evolution, and specificity in cellular control by protein phosphorylation. [Review] [27 refs]. *Experientia* 52: 449–54.

Hsu SC, and DeFranco DB (1995): Selectivity of cell cycle regulation of glucocorticoid receptor function. *Journal of Biological Chemistry* 270: 3359–64.

Hsu SC, Qi M, and DeFranco DB (1992): Cell cycle regulation of glucocorticoid receptor function. *EMBO Journal* 11: 3457–68.

Hu JM, Bodwell JE, and Munck, A (1994): Cell cycle-dependent glucocorticoid receptor phosphorylation and activity. *Molecular Endocrinology* 8: 1709–13.

Hutchison KA, Dalman FC, Hoeck W, Groner B, and Pratt WB (1993): Localization of the approximately 12 kDa M(r) discrepancy in gel migration of the mouse glucocorticoid receptor to the major phosphorylated cyanogen bromide fragment in the transactivating domain. *Journal of Steroid Biochemistry & Molecular Biology* 46: 681–6.

Ignar-Trowbridge DM, Pimentel M, Parker MG, McLachlan JA, and Korach KS (1996): Peptide growth factor cross-talk with the estrogen receptor requires the A/B domain and occurs independently of protein kinase C or estradiol. *Endocrinology* 137: 1735–44.

Ignar-Trowbridge DM, Teng CT, Ross KA, Parker MG, Korach KS, and McLachlan JA (1993): Peptide growth factors elicit estrogen receptor-dependent transcriptional activation of an estrogen-responsive element. *Molecular Endocrinology* 7: 992–8.

Ince BA, Montano MM, and Katzenellenbogen BS (1994): Activation of transcriptionally inactive human estrogen receptors by cyclic adenosine 3',5'-monophosphate and ligands including antiestrogens. *Molecular Endocrinology* 8: 1397–406.

Iniguez-Lluhi JA, Lou DY, and Yamamoto KR (1997): Three Amino Acid Substitutions Selectively Disrupt the Activation but Not the Repression Function of the Glucocorticoid Receptor N Terminus. *Journal of Biological Chemistry* 272: 4149–56.

Janknecht R, and Hunter, T (1996a): Transcription. A growing coactivator network [news; comment]. *Nature* 383: 22–3.

Janknecht R, and Hunter, T (1996b): Versatile molecular glue. Transcriptional control. [Review] [25 refs]. *Current Biology* 6: 951–4.

Jans DA, and Hubner, S (1996): Regulation of protein transport to the nucleus: central role of phosphorylation. [Review] [400 refs]. *Physiological Reviews* 76: 651–85.

Jans DA, Moll T, Nasmyth K, and Jans, P (1995): Cyclin-dependent kinase site-regulated signal-dependent nuclear localization of the SW15 yeast transcription factor in mammalian cells. *Journal of Biological Chemistry* 270: 17064–7.

Joel PB, Traish AM, and Lannigan DA (1995): Estradiol and phorbol ester cause phosphorylation of serine 118 in the human estrogen receptor. *Molecular Endocrinology* 9: 1041–52.

Johnson LN, and O'Reilly, M (1996): Control by phosphorylation. [Review] [36 refs]. *Current Opinion in Structural Biology* 6: 762–9.

Karin, M (1994): Signal transduction from the cell surface to the nucleus through the phosphorylation of transcription factors. [Review] [76 refs]. *Current Opinion in Cell Biology* 6: 415–24.

Karin M, and Hunter, T (1995): Transcriptional control by protein phosphorylation: signal transmission from the cell surface to the nucleus. [Review] [102 refs]. *Current Biology* 5: 747–57.

Kato S, Endoh H, Masuhiro Y, Kitamoto T, Uchiyama S, Sasaki H, Masushige S, Gotoh Y, Nishida E, Kawashima, H. et al (1995): Activation of the estrogen receptor through phosphorylation by mitogen-activated protein kinase. *Science* 270: 1491–4.

Katzenellenbogen BS (1996): Estrogen receptors: bioactivities and interactions with cell signaling pathways. [Review] [74 refs]. *Biology of Reproduction* 54: 287–93.

Korach KS (1994): Insights from the study of animals lacking functional estrogen receptor. [Review] [32 refs]. *Science* 266: 1524–7.

Korach KS, Couse JF, Curtis SW, Washburn TF, Lindzey J, Kimbro KS, Eddy EM, Migliaccio S, Snedeker SM, Lubahn DB, et al (1996): Estrogen receptor gene disruption: molecular characterization and experimental and clinical phenotypes. [Review] [50 refs]. *Recent Progress in Hormone Research* 51: 159–86.

Kralli A, Bohen SP, and Yamamoto KR (1995): LEM1, an ATP-binding-cassette transporter, selectively modulates the biological potency of steroid hormones. *Proceedings of the National Academy of Sciences of the United States of America* 92: 4701–5.

Krstic MK (1995): Functional Anaylsis of Glucocorticoid Receptor Phosphorylation. UCSF. Thesis.

Krstic MK, Rogatsky I, Yamamoto KR, and Garabedian MJ (1997): Mitogen-activated and cyclin-dependent protein kinases selectively and differentially modulate transcriptional enhancement by the glucocorticoid receptor. *Molecular and Cellular Biology* 17: 3947–54.

Kuiper GG, and Brinkmann AO (1994): Steroid hormone receptor phosphorylation: is there a physiological role?. [Review] [48 refs]. *Molecular & Cellular Endocrinology* 100: 103–7.

Lahooti H, White R, Hoare SA, Rahman D, Pappin DJ, and Parker MG (1995): Identification of phosphorylation sites in the mouse oestrogen receptor. *Journal of Steroid Biochemistry & Molecular Biology* 55: 305–13.

Le Goff P, Montano MM, Schodin DJ, and Katzenellenbogen BS (1994): Phosphorylation of the human estrogen receptor. Identification of hormone-regulated sites and examination of their influence on transcriptional activity. *Journal of Biological Chemistry* 269: 4458–66.

Mason SA, and Housley PR (1993): Site-directed mutagenesis of the phosphorylation sites in the mouse glucocorticoid receptor. *Journal of Biological Chemistry* 268: 21501–4.

Moll T, Schwob E, Koch C, Moore A, Auer H, and Nasmyth, K (1993): Transcription factors important for starting the cell cycle in yeast. *Philosophical Transactions of the Royal Society of London Series B: Biological Sciences*. 340: 351–60.

Moll T, Tebb G, Surana U, Robitsch H, and Nasmyth, K (1991): The role of phosphorylation and the CDC28 protein kinase in cell cycle-regulated nuclear import of the S. cerevisiae transcription factor SWI5. *Cell* 66: 743–58.

Moyer ML, Borror KC, Bona BJ, DeFranco DB, and Nordeen SK (1993): Modulation of cell signaling pathways can enhance or impair glucocorticoid-induced gene expression without altering the state of receptor phosphorylation. *Journal of Biological Chemistry* 268: 22933–40.

Nazareth LV, and Weigel NL (1996): Activation of the human androgen receptor through a protein kinase A signaling pathway. *Journal of Biological Chemistry* 271: 19900–7.

O'Malley BW, Schrader WT, Mani S, Smith C, Weigel NL, Conneely OM, and Clark JH (1995): An alternative ligand-independent pathway for activation of steroid receptors. [Review] [45 refs]. *Recent Progress in Hormone Research* 50: 333–47.

Orti E, Bodwell JE, and Munck, A (1992): Phosphorylation of steroid hormone receptors. [Review] [232 refs]. *Endocrine Reviews* 13: 105–28.

Orti E Hu, LM, and Munck, A (1993): Kinetics of glucocorticoid receptor phosphorylation in intact cells. Evidence for hormone-induced hyperphosphorylation after activation and recycling of hyperphosphorylated receptors. *Journal of Biological Chemistry* 268: 7779–84.

Orti E, Mendel DB, and Munck, A (1989): Phosphorylation of glucocorticoid receptor-associated and free forms of the approximately 90-kDa heat shock protein before and after receptor activation. *Journal of Biological Chemistry* 264: 231–7.

Papavassiliou AG (1994): The role of regulated phosphorylation in the biological activity of transcription factors SRF and Elk-1/SAP-1. [Review] [31 refs]. *Anticancer Research* 14: 1923–6.

Parker D, Ferreri K, Nakajima T, LaMorte VJ, Evans R, Koerber SC, Hoeger C, and Montminy MR (1996): Phosphorylation of CREB at Ser-133 induces complex formation with CREB-binding protein via a direct mechanism. *Molecular & Cellular Biology* 16: 694–703.

Pietras RJ, Arboleda J, Reese DM, Wongvipat N, Pegram MD, Ramos L, Gorman CM, Parker MG, Sliwkowski MX, and Slamon DJ (1995): HER-2 tyrosine kinase pathway targets estrogen receptor and promotes hormone-independent growth in human breast cancer cells. *Oncogene* 10: 2435–46.

Piwnica-Worms, H (1996): Reversible phosphorylation and mitotic control. [Review] [34 refs]. *Journal of Laboratory & Clinical Medicine* 128: 350–4.

Power RF, Conneely OM, and O'Malley BW (1992): New insights into activation of the steroid hormone receptor superfamily. [Review] [41 refs]. *Trends in Pharmacological Sciences* 13: 318–23.

Power RF, Mani SK, Codina J, Conneely OM, and O'Malley BW (1991): Dopaminergic and ligand-independent activation of steroid hormone receptors. *Science* 254: 1636–9.

Smith CL, Conneely OM, and O'Malley BW (1993): Modulation of the ligand-independent activation of the human estrogen receptor by hormone and antihormone. *Proceedings of the National Academy of Sciences of the United States of America* 90: 6120–4.

Tienrungroj W, Sanchez ER, Housley PR, Harrison RW, and Pratt WB (1987): Glucocorticoid receptor phosphorylation, transformation, and DNA binding. *Journal of Biological Chemistry* 262: 17342–9.

Trowbridge JM, Rogatsky I, and Garabedian MJ (1997): Regulation of estrogen receptor transcriptional enhancement by the cyclin A/cdk2 complex. *Proceedings of the National Academy of Sciences of the United States of America* (in press).

Tzeng DZ, and Klinge CM (1996): Phosphorylation of purified estradiol-liganded estrogen receptor by casein kinase II increases estrogen response element binding but does not alter ligand stability. *Biochemical & Biophysical Research Communications* 223: 554–60.

Webster JC, Jewell CM, Bodwell JE, Munck A, Sar M, and Cidlowski JA (1997): Mouse Glucocorticoid Receptor Phosphorylation Status Influences Multiple Functions of the Receptor Protein. *Journal of Biological Chemistry* 272: 9287–9293.

Weigel NL (1996): Steroid hormone receptors and their regulation by phosphorylation. [Review] [156 refs]. *Biochemical Journal* 319: 657–67.

White R, Sjoberg M, Kalkhoven E, and Parker MG (1997): Ligand-independent activation of the oestrogen receptor by mutation of a conserved tyrosine. *EMBO Journal* 16: 1427–35.

Zhang Y, Bai W, Allgood VE, and Weigel NL (1994): Multiple signaling pathways activate the chicken progesterone receptor. *Molecular Endocrinology* 8: 577–84.

Zilliacus J, Wright AP, Carlstedt-Duke J, and Gustafsson JA (1995): Structural determinants of DNA-binding specificity by steroid receptors. [Review] [114 refs]. *Molecular Endocrinology* 9: 389–400.

Zwijsen RM, Wientjens E, Klompmaker R, van der Sman J, Bernards R, and Michalides RJAM (1997): CDK-independent activation of estrogen receptor by cyclin D1. *Cell* 88: 405–15.

10

Monomeric Nuclear Receptors

Mitchell A. Lazar and Heather P. Harding

Nuclear Hormone Receptors: Fundamental Mediators of Cell Signaling

Cellular growth and differentiation are controlled by the concerted actions of many transcription factors. The biological activities of these transcription factors are in turn regulated by extracellular factors, through a variety of intricate signaling pathways. Many of these pathways require a cascade of events, including the synthesis of second messengers. In contrast, steroid and thyroid hormone receptors are particularly exciting because in many cases they are direct mediators of the ligand signal, with the ability to bind DNA and regulate transcription in addition to binding ligands with high affinity and specificity.

The Nuclear Hormone Receptor Superfamily

Advances in molecular biology have led to the cloning and characterization of receptors for all major steroid hormones, thyroid hormone (T_3), vitamin D, and retinoic acid (RA), and have revealed a tremendous amount about their structure (Mangelsdorf et al 1995). In particular, three remarkable and unexpected conclusions followed from the cloning studies. First and foremost, it is clear that receptors for molecules as diverse as glucocorticoids and thyroid hormones are evolutionarily related. This reflects the conservation of motifs that have proven efficacious for such functions as target gene

Molecular Biology of Steroid and Nuclear Hormone Receptors
Leonard P. Freedman, Editor
© 1998 Birkhäuser Boston

discrimination, dimerization, and transcriptional activation and repression. Second, for some hormones multiple forms of high-affinity nuclear receptor were discovered, resulting from multiple genes generating different subtypes, or from alternative splicing generating different isoforms (Lazar 1993, Mangelsdorf et al 1994). Third, perhaps the most unexpected and exciting consequence of the cloning of hormone receptors by different laboratories has been the discovery of "orphan receptors," which meet the structural criteria for inclusion in the thyroid/steroid hormone receptor superfamily despite the absence of a known hormone activator (Enmark and Gustafsson 1996, O'Malley and Conneely 1992). Finally, despite the clear similarities in structure and function among all nuclear receptors, there are also important differences in protein-protein interactions, transcriptional regulation, and DNA binding which contribute to the specificity of gene regulation by these receptors. This chapter discusses the subset of nuclear hormone receptors that are known to bind to DNA with high affinity as monomers.

DOMAIN STRUCTURE OF NUCLEAR HORMONE RECEPTORS

Nuclear hormone receptors are characterized by distinct domains, indicated in Figure 1. These domains mediate not only DNA and ligand binding, but also nuclear localization, dimerization, transactivation, and transrepression. These functional domains are remarkably modular, as illustrated by now-classical swapping experiments in which, for example, a chimeric receptor containing the glucocorticoid receptor (GR) A-B-C regions, including the DNA-binding domain (DBD), and the estrogen receptor (ER) D-E-F regions, including the ligand binding domain (LBD), causes normally gluco-corticoid-responsive genes to respond to estrogen (Giguere et al 1987, Green and Chambon 1987). These domains can also function when attached to heterologous functional domains from other proteins, including yeast and viral transcription factors (Ellers et al 1989).

DNA BINDING BY NUCLEAR HORMONE RECEPTORS

The DNA binding domain (DBD, also called Domain C) comprises 66–68 amino acids in the most highly conserved region of the nuclear hormone receptor superfamily. It contains two $(Cys)_4$ zinc-ordered modules that stabilize two α-helices in such a manner as to favor specific nucleotide contacts determined by a region of the DBD called the P-box (Danielsen et al 1989, Green et al 1988, Umesono and Evans 1989). The nuclear hormone receptor superfamily can be divided into two groups on the basis of primary

Domain:	A/B	C (DBD)	D		E		F

1	2	3	4	5	6	7	8	9	10

1. Activation function 1 (AF1) involved in transactivation in some receptors
2. Zn-module-containing DBD, including "P box" involved in half-site recognition
3. Dimerization domain
4. C-terminal extension ("T and A boxes," involved in recognition of bases 5' to half-site)
5. Nuclear localization signal
6. "CoR" box involved in binding corepressors for TR and RAR (not Rev-Erb)
7. Dimerization domain, including "ninth heptad" (actually "helix 10" in receptor crystal structure)
8. Ligand binding
9. Ligand-dependent activation function (AF2, not present in RevErb and RevErbβ/RVR)
10. "F" domain (unknown function) highly variable in size and sequence

Figure 1.—Domain structure of nuclear hormone receptors

DNA binding specificity as determined by the P-box: (1) the steroid hormone receptor subfamily, which recognizes the hexameric half-site AGAACA, and (2) the thyroid/retinoid receptor subfamily, which recognizes the hexamer AGGTCA (Stunnenberg 1993). Remarkably, almost all of the nearly 50 orphan receptors identified to date belong to the thyroid/retinoid receptor subfamily. It should be noted that the estrogen receptor cannot be classified by this simple rule because it has the thyroid/retinoid receptor P-box sequence, and therefore recognizes the AGGTCA half-site, but its other DNA-binding properties resemble those of other steroid hormone receptors. These properties include the propensity to homodimerize, and to bind with highest affinity to two half-sites arranged as inverted repeats (also called palindromes). In contrast, the thyroid/retinoid receptors interact with DNA as monomers, homodimers, and as heterodimers with other nuclear receptors, most commonly the retinoid X receptor (RXR) (Glass 1994). This chapter emphasizes the receptors that bind DNA as monomers.

THE EXTENDED DNA BINDING DOMAIN

Regions outside the highly conserved portion of the DBD are important for DNA binding by multiple nuclear receptors in the TR/RAR/orphan group. In particular a C-terminal extension of the DNA binding domain called the A and T boxes was first found by Wilson et al to be important for the binding of NGFI-B as a monomer and of RXR as a homodimer, respectively (Wilson et al 1992). NMR analysis of the RXR homodimer suggested that this C-terminal extension of the classical DBD forms a third alpha helix (the other two are in the zinc modules). In the case of RXR, this makes protein-protein contacts between the two DBDs of RXR homodimers rather than protein-DNA contacts (Lee et al 1993). In the crystal structure of the TR/RXR heterodimer, the C-terminal extension region of TR formed an extended alpha helix which reached into the minor groove between TR and RXR, but which did not make contacts with RXR (Rastinejad et al 1995). This region has been shown to play a role in stabilizing TR homodimers to everted repeats (Kurokawa et al 1993). For RAR/RXR heterodimers, the C-terminal extension has also been suggested to have differential roles in regulating the binding stability on DR2 versus DR5 elements (Predki et al 1994). In combination these results lead to the conclusion that the C-terminal extension has diverse receptor-specific roles in determining the receptors' preference for 5' extended core sites, as well as for stabilizing DNA binding either through protein-DNA or through protein-protein contacts.

MONOMERIC DNA BINDING BY NUCLEAR HORMONE RECEPTORS TO EXTENDED HALF-SITES

High-affinity monomeric binding requires not only the classical DBD, but also sequences just C-terminal to the DBD, called the T and A boxes, which form an extended α-helix and participate in minor groove interactions with specific base pairs just 5' to the AGGTCA half-site (Wilson et al 1992). This was initially shown for NGF-IB (Wilson et al 1992) and has also been shown for Steroidogenic Factor 1 (SF-1) (Wilson et al 1993a), RevErb (Harding and Lazar 1993), and ROR (Giguere et al 1995). NGFI-B requires AA immediately 5' to the AGGTCA for high-affinity binding, whereas the recognition site for SF-1 is TCAAGGTCA (Wilson et al 1993a). RevErb and ROR require a much longer (4–5 bp), A/T-rich 5' flank for higher-affinity binding (Giguere et al 1994, Harding and Lazar 1993). Thyroid hormone receptor (TR) also binds DNA as a monomer, with highest-affinity for the sequence TAAGGTCA (Katz and Koenig 1993). These monomer half-site preferences are summarized in Table 1. Thus, the extended DNA binding domain leads to preference for an extended half-site. The specific binding preferences are determined by the amino acids constituting the C-terminal extensions of monomeric nuclear hormone receptors, which are shown in Table 2.

THYROID HORMONE RECEPTORS

Although TR binds DNA with highest affinity as a heterodimer with RXR, TR was among the first nuclear receptors to be shown to bind DNA as a

Table I.—Extended half-sites recognized by monomeric nuclear receptors

TR	TA AGGTCA
NGFI-B/*nur77*	AA AGGTCA
SF1	TCA AGGTCA
RevErb/ROR	AANT AGGTCA

monomer (Forman et al 1992, Lazar et al 1991). TR also binds DNA as a homodimer, but TR homodimer binding is destabilized by ligand binding (Andersson et al 1992, Ribeiro et al 1992, Yen et al 1992). In contrast, both the TR monomer and TR/RXR heterodimer form stable complexes with DNA in the presence of T_3. The TR monomer will bind to the nuclear receptor hexameric half-site AGGTCA (Forman et al 1992, Lazar et al 1991), but unbiased binding site selection studies have shown that the octameric sequence TAAGGTCA is preferentially bound (Katz and Koenig 1993). Given the crystal structure of the TR DNA binding domain (Rastinejad et al 1995), as well as what is known about monomer binding by other receptors, it is likely that the polypeptide immediately C-terminal to the classical DNA binding domain of the TR makes minor groove contacts with the base pairs immediately 5′ to the AGGTCA. *In vivo*, the affinity of monomeric TR for the octameric site is sufficient to allow it to activate transcription from this site in the presence of T_3 (Katz and Koenig 1993).

In the absence of RXR, TRα preferentially binds to DNA as a monomer, whereas TRβ forms more stable, cooperative homodimers (Darling et al 1993, Lazar et al 1991). Since the major difference between TRα and TRβ is in their completely unique N-termini, it is likely that this region of the TR plays an important role in determining whether the TR will bind to DNA as a monomer or homodimer in the absence of RXR. In addition to this role of the N-terminus in regulating TR binding, the C-terminus of the TR may also contribute to the overall affinity of the protein for monomeric half-sites. This is clearly the case for TRα2, a C-terminal variant which lacks the T3-binding and AF2 transactivation functions (Izumo and Mahdavi 1988, Lazar et al 1988, Mitsuhashi et al 1988). The C-terminus of TRα2 is phosphorylated by casein kinase II (and/or a related kinase) on two adjacent serines (amino acids 474 and 475) (Katz et al 1995). Phosphorylation of these serines markedly reduces the affinity of the TRα2 monomer for DNA, suggesting an allosteric effect of phosphorylation on the structure of the DNA binding domain (Katz et al 1995).

Table 2.—T/A boxes of monomeric nuclear receptors

The "GM" preceding the C-terminal extension is conserved in all nuclear hormone receptors, and defines the end of the classical "C" domain. The T and A boxes, as described for NGFI-B and SF-1 (Wilson et al 1993a), are indicated

			T Box	A Box
TRα	GM	AMDLVLDDSKRV	AKRKLIE	QNRERRRKE...
TRβ	GM	ATDLVLDDSKRL	AKRKLIE	ENREKRRRE...
NGFI-B/*nur77*	GM	VKEVVRTDSLKG	RRGRLPS	KPKQPPDAS...
NURR-1	GM	VKEVVRTDSLKG	RRGRLPS	KPKSPQDPS...
SF-1	GM	RLEAVRADRMRG	GRNKFGP	MYKRDRALK...
FTZ-F1	GM	KLEAVRADRMRG	GRNKFGP	MYKRDRARK...
RevErb	GM	SRDAVRFGRIPK	REKQRML	AEMQSAMNL...
ROR	GM	SRDAVKFGRMSK	KQRDSLY	AEVQKHRMQ...
PPARγ	GM	SHNAIRFGRMPQ	AEKEKLL	AEISSDIDQ...

THE NGFI-B/nurr1 FAMILY OF MONOMER BINDING RECEPTORS

NGFI-B, also called *nurr77*, was originally isolated as a nerve-growth -factor-responsive gene and as a gene that was induced early in response to serum and growth factors (Hazel et al 1988, Milbrandt 1988). Other names include NAK1 and TR3 (Chang et al 1989, Nakai et al 1990). NGFI-B has also been implicated in T-cell apoptosis (Liu et al 1994, Woronicz et al 1994) and adrenal cortical gene regulation (Wilson et al 1993b), although mice engineered to be deficient in NGF-IB appear to have normal T-cell and adrenal function (Crawford et al 1995, Lee et al 1995a). In mammals there are two additional proteins encoded on different genes that are highly related to NGFI-B. These homologs, called Nurr1 (Law et al 1992) (also called RNR1 [Scearce et al 1993], NOT [Mages et al 1994], and HZF3 [LopesdaSilva et al 1995]) and Nor1 (Ohkura et al 1994), have biochemical properties similar to those of NGFI-B, including the ability to bind DNA as a monomer, but they are expressed with different developmental and tissue profiles. There are also homologs of these orphans in *Xenopus* (Smith et al 1993), *Drosophila* (Sutherland et al 1995), and *C. elegans* (Kostrouch et al 1995).

NGFI-B was the first nuclear receptor whose monomer binding was systematically studied, and delineation of its T and A boxes immediately C-terminal to the classic DBD marked the first realization that this region of nuclear receptors could increase the length of the specific half-site recognition sequence by making minor groove contacts. NGFI-B binds as a monomer to a single core site preceded by two adenines (AAAGGTCA), and the A box of NGFI-B was thought to contact residues within the minor groove of the DNA 5′ to the core element and stabilize the binding of a single DBD (Wilson et al 1992). Importantly, this C-terminal extension is modular in the same sense as other functional domains of the receptors. T/A box swaps between NGFI-B and SF-1 (Wilson et al 1993a) and between NGFI-B and RORα alter the DNA binding site specificity in a manner that correlates, for the most part, with the source of the T/A box. Furthermore, the A box amino acids are more important for discriminating between the 5′ AA and TCA bases of the receptor binding sites, whereas the T box was more involved with stabilization of binding. The A-box of NGFI-B can also be phosphorylated, which reduces the DNA binding affinity of the protein (Hirata et al 1993).

NGFI-B, Nurr1, and Nor1 activate transcription from their monomer binding site (Wilson et al 1993a, Zetterstrom et al 1996). This occurs in the absence of exogenous ligand and, as for other constitutively active orphan receptors (reviewed in Enmark and Gustafsson 1996), implies either that the proteins are transcriptionally active in the absence of any ligand or that there is an endogenous ligand present in the cell systems in which its transcriptional activity is assayed. Although it is active as a monomer, NGFI-B as well as Nurr1 can interact with RXR in solution, as well as on DNA where the RXR heterodimer is further activated by agonistic ligands for RXR, such as 9-cis retinoic acid (Forman et al 1995b, Perlmann and Jansson 1995). Thus it appears that under some circumstances RXR converts NGFI-B to a hetero-dimeric receptor that is activated by retinoid ligand. In contrast, Nor1 does not appear to interact with RXR, and thus cannot be activated by retinoid ligand (Zetterstrom et al 1996).

SF-1

SF-1 was originally isolated as a transcription factor capable of activating a variety of genes involved in steroid biosynthesis. These genes all had promoters that contained DNA sequences related to the consensus TCAAGGTCA, which were required for maximal gene expression (Lala et al 1995, Lala et al 1992). Remarkably, mice deficient in SF-1 not only have low levels of steroid hormones such as glucocorticoids and sex steroids, but lack the adrenal glands and gonads which normally produce these hormones (Luo et al 1994). This phenotype is similar to that of humans with congenital adrenal hypoplasia, which is due to mutation in a gene called DAX-1. DAX-1 encodes an unusual orphan receptor that lacks the classical DNA binding domain (Muscatelli et al 1994). At the time of this writing, it is uncertain whether SF-1 and DAX-1 interact in a genetic or physical sense. It is certain, however, that SF-1 binds to its consensus extended half-site with high affinity as a monomer and can function as a transcriptional activator in this context. As for other constitutively active orphan receptors, it is not known whether this represents true ligand-independence or a requirement for an as yet unknown ligand. SF-1 homologs exist in *Xenopus* (Ellinger et al 1994), *Drosophila* (where it is known as FTZ-F1; Lavorgna et al 1991), and the *C. elegans* genome as well.

THE RevErb/ROR FAMILY OF MONOMER BINDING RECEPTORS

RevErb is an orphan nuclear receptor that is unique because it is encoded on the noncoding strand of the TRα (c-erbAα) gene in multiple species (Lazar et al 1989, Miyajima et al 1989). The complementarity of the mRNAs for RevErb and TRα2 may influence their relative expression in a given cell (Lazar et al 1990). However, RevErb is highly conserved, and is also closely related to a subfamily of mammalian nuclear receptors, RevERbβ/RVR (also known as BD73) (Dumas et al 1994, Forman et al 1994, Retnakaran et al 1994), RORα (three isoforms as well as a fourth called RZRα) (Carlberg et al 1994, Giguere et al 1994), RORβ (Carlberg et al 1994) and RORγ (Hirose et al

1994). Interestingly, the C-terminal extensions of these receptors are also very similar to thoset of the peroxisome proliferator activated receptors (PPARs) (Dreyer et al 1992, Issemann and Green 1990, Kliewer et al 1994). The structural comparison of the RevErb/ROR family is shown in Figure 2.

RevErb and RevErbβ/RVR are most similar. RevErb and RevErbβ/RVR have the largest D domains in the nuclear receptor superfamily, and both lack the F domain, which in many other receptors contains the autonomous activation domain of AF2. Furthermore, the C-terminal extensions of RevErb and RevErbβ/RVR are nearly identical, as are the DNA binding properties of the proteins. These include high-affinity binding as a monomer to the site (A/T)(A/T)ANTAGGTCA. Interestingly, these receptors do not heterodimerize with RXR, but homodimerize specifically on tandem arrangements of monomer sites as direct repeats separated by two base pairs, referred to as the RevDR2 element (Harding and Lazar 1995, and unpublished data). The main differences between RevErb and RevErbβ/RVR is that their D domains, while both quite long, are not nearly as similar to each other as is the case for other nuclear receptor subtypes. Also, there is not evidence that the RevErbβ/RVR gene overlaps that of another nuclear receptor.

Both RevErb and RevErbβ/RVR repress transcription on monomer and homodimer binding sites. Repression as a monomer is primarily due to competition with other activating receptors that bind to the same site, including ROR (Dumas et al 1994, Forman et al 1994, Retnakaran et al

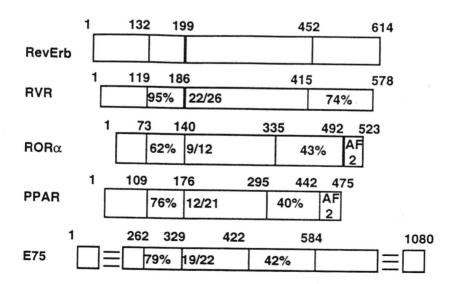

Figure 2.—The RevErb/ROR subfamily of nuclear receptors.
The percent identity of various functional domains of the proteins with RevErb is indicated, as is the number of identical amino acids in the C-terminal extensions.

1994). In contrast, repression as a homodimer is due to a more active regulation of basal transcription (Harding and Lazar 1995). Active repression requires recruitment of the nuclear receptor corepressor N-CoR to the dimeric binding site (Zamir et al 1996). Dimeric binding greatly facilitates the interaction between N-CoR and RevErb (Zamir et al 1997), and the interaction has different structural requirements than the interactions between N-CoR and TR or RAR (Horlein et al 1995, Zamir et al 1996). Naturally occurring RevDR2 sites have been described in the cellular retinol binding protein I gene (Harding and Lazar 1995) and in the RevErb gene itself (Adelmant et al 1996). E75 is a *Drosophila* homolog of RevErb with a nearly identical T/A region (Segraves and Hogness 1990) and, thus, similar DNA binding properties (H. P. Harding, W. Segraves, and M. A. Lazar, unpublished). A chicken homolog of RevErbβ/RVR has also been identified (Bonnelye et al 1994).

RORα is less similar to RevErb, but the initial 12 amino acids of its C-terminal extension (corresponding to the T box region of NGFI-B) are nearly identical to those of RevErb and RevErbβ/RVR (Table 2). As a result, RORα, like RevErb, displays high affinity monomeric binding to a similar extended half-site. In contrast to Rev-Erb, RORα is a transcriptional activator, and has an AF2-like sequence at its C-terminus. There are multiple RORα isoforms, called α1, α2, α3, and RZRα, with different DNA binding properties (Giguere et al 1994). There is also another ROR subtype, RZRb (Carlberg et al 1994), which is highly expressed in the nervous system. It should be noted that ROR/RZR subtypes have been reported to be nuclear receptors for melatonin (Becker-Andre et al 1994). However, this is very controversial, and probably not significant, because activation by melatonin has not been reproducible in other labs (Greiner et al 1996, Tini et al 1995), including our own (H. P. Harding and M. A. Lazar, unpublished). DHR3 is a *Drosophila* homolog of ROR with very similar DNA binding properties due to its nearly identical T/A region (Koelle et al 1992). The monomer binding site for ROR and related receptors has been shown to be (A/T)(A/T)-ANTAGGTCA, which is highly similar to that of RevErb/RVR (Giguere et al 1994). The C-terminal extension of ROR is a transferable domain imparting specificity to the recognition of the extended half-site (Giguere et al 1995), and the N-terminus of the RORα isoforms determines their DNA binding affinity (Giguere et al 1994), in part by influencing DNA bending (McBroom et al 1995).

In contrast to RevErb, RORα is a transcriptional activator. In mammalian cells transcriptional activation occurs in the absence of exogenous ligand. This could be due to the presence of an endogenous, intracellular ligand, but it is equally possible that no activating compound is required (autonomous activity) or that activation is caused by covalent modification, rather than by ligand binding. In any case, ROR activates transcription as a monomer (Giguere et al 1994). ROR does not heterodimerize with RXR. However,

RORα as well as RZRβ, have an increased ability to activate transcription from DNA sites to which two molecules of ROR can bind (Carlberg et al 1994, Greiner et al 1996, Harding et al 1997). These include widely spaced tandem repeats, as well as the RevDR2 site.

Another nuclear receptor whose C-terminal DBD extension is highly similar to that of RevErb and ROR is PPAR. PPAR is no longer technically an orphan receptor, since one subtype (γ) binds the antidiabetic thiazolidinediones (Lehmann et al 1995) as well as prostaglandin J2 (Forman et al 1995a, Kliewer et al 1995, Yu et al 1995) with high affinity, while another (α) is potentially a receptor for leukotriene (Devchand et al 1996) as well as 8S-HETE (Yu et al 1995). Thus PPARs are nuclear receptors for eicosanoids. Unlike RevErb and ROR, PPAR has little monomeric DNA binding affinity, and binds DNA much more avidly as a heterodimer with RXR (Kliewer et al 1992), which constrains its binding to DRs with one bp spacing. However, the similarity of the C-terminal extension of PPAR to those of RevErb and RORα suggests that PPAR would prefer to bind a similar extended half-site, and this is indeed the case (Palmer et al 1995; H. P. Harding and M. A. Lazar, unpublished data).

THE BIOLOGICAL SIGNIFICANCE OF MONOMER BINDING ORPHAN RECEPTORS

A number of the receptors discussed in this chapter have been shown to be developmentally regulated and/or to play a role in development. Most strikingly, the monomer binding orphan SF1 is critical for normal adrenal, gonadal, and pituitary development (Luo et al 1994). NGFI-B appears to play an important role in T-cell apoptosis (Liu et al 1994, Woronicz et al 1994), but the NGFI-B-deficient mouse is unaffected (Lee et al 1995a), suggesting redundancy with other family members. PPARα is important for normal liver development and function (Lee et al 1995b), and PPARγ expression is fat-specific (Chawla et al 1994, Tontonoz et al 1994) and required for adipocyte differentiation (Tontonoz et al 1995). Rev-Erb gene expression is also regulated during adipogenesis (Chawla and Lazar 1993), as well as myogenic differentiation (Downes et al 1995). RORα may regulate lens gene expression and possibly eye development (Tini et al 1995). However, RORα plays a major role in cerebellar development, illustrated graphically by the *staggerer* mouse, an experiment of nature that is due to a deletion in the RORα gene (Hamilton et al 1996). Finally, in lower species such as *Drosophila*, monomeric orphan receptors such as E75 and DHR3 are crucial for development, specifically during metamorphosis (Horner et al 1995, Thummel 1995).

Functional Significance of Monomeric DNA Binding by Nuclear Receptors

Monomeric binding by these nuclear receptors does appear to have functional significance. Thus, TR (plus thyroid hormone), ROR, NGFI-B, etcetera can activate transcription from a monomeric binding site, and RevErb/RVR can inhibit activation by competing for binding. However, most of the monomer binding nuclear receptors also function as higher-order units, either as heterodimers with RXR (TR, NGFI-B) or as homodimers (RevErb, ROR). For some functions, such as transcriptional repression, dimeric binding appears to be a requirement (Zamir et al 1997). In these cases, the high affinity of the nuclear receptor monomer for extended half-sites serves to increase the specificity of the receptor function. Thus, the RXR/Nurr1 heterodimer, as well as the RXR/TRα2 heterodimer, is specific for direct repeats in which the downstream half-site is the high-affinity monomer binding site for Nurr1 (Zetterstrom et al 1996) and TR (Reginato et al 1996, Yang et al 1996), respectively. Similarly, the high-affinity monomer binding sites in the RevDR2 create a subset of DR2 sites at which RevErb and ROR homodimers specifically regulate transcription. As mentioned earlier, a similar situation pertains to the PPAR/RXR heterodimer (Palmer et al 1995).

Conclusions

The extended DNA binding domain of monomeric nuclear receptors determines the specific 5' extension of the half-site. These nuclear receptors regulate transcription as monomers on the extended half-sites, as well as on a subset of dimeric sites dictated by their inherent DNA binding preferences. These features lend specificity to the function of a growing subset of the nuclear hormone receptor superfamily, including receptors that are required for normal growth and development in species as diverse as humans and flies.

ACKNOWLEDGMENTS

This work was supported by a grant to Mitchell A. Lazar from the National Institute for Diabetes, Digestive, and Kidney Diseases (DK 45586).

REFERENCES

Adelmant G, Begue A, Stehelin D, and Laudet V (1996): A functional Rev-erb alpha responsive element located in the human Rev-erb alpha promoter mediates a repressive function. *Proc. Natl. Acad. Sci. USA*. 93: 3553–3558.

Andersson ML, Nordstrom K, Demczuk S, Harbers M, and Vennstrom B (1992): Thyroid hormone alters the DNA binding properties of chicken thyroid hormone receptors alpha and beta. *Nuc. Acids Res.* 20: 4803–4810.

Becker-Andre M, Wiesenberg I, Schaeren-Wiemers N, Andre E, Missbach M, Saurat JH, and Carlberg C (1994): Pineal gland hormone melatonin binds and activates an orphan of the nuclear receptor superfamily. *Journal of Biological Chemistry* 269(46): 28531–4.

Bonnelye E, Vanacker J-M, Desbiens X, Begue A, Stehelin D, and Laudet V (1994): Rev-Erbβ, a new member of the nuclear receptor superfamily, is expressed in the nervous system during chicken development. *Cell Growth and Diff.* 5: 1357–1365.

Carlberg C, Huijsduijnen RHv, Staple JK, DeLamarter JF, and Becker-Andre M (1994): RZRs, a new family of retinoid-related orphan receptors that function as both monomers and homodimers. *Mol. Endocrinol.* 8: 757–770.

Chang C, Kokontis J, Liao SS, and Chang Y (1989): Isolation and characterization of human TR3 receptor: a member of the steroid receptor superfamily. *J. Steroid Biochem.* 34: 391–395.

Chawla A, and Lazar MA (1993): Induction of Rev-ErbAα, an orphan nuclear receptor encoded on the opposite strand of the α-thyroid hormone receptor gene, during adipocyte differentiation. *J. Biol. Chem.* 268: 16265–16269.

Chawla A, Schwarz EJ, Dimaculangan DD, and Lazar MA (1994): Peroxisome proliferator-activated receptor γ (PPARγ): Adipose predominant expression and induction early in adipocyte differentiation. *Endocrinology* 135: 798–800.

Crawford PA, Sadovsky Y, Woodson K, Lee SL, and Milbrandt J (1995): Adreno-cortical function and regulation of the steroid 21-hydroxylase gene in NGFI-B-deficient mice. *Mol. Cell. Biol.* 15: 4331–4336.

Danielsen M, Hinck L, and Ringold G (1989): Two amino acids within the knuckle of the first zinc finger specify DNA response element activation by the glucocorticoid receptor. *Cell* 57: 1131–1138.

Darling DS, Carter RL, Yen PM, Welborn JM, Chin WW, and Umeda PK (1993): Different dimerization activities of alpha and beta thyroid hormone receptor isoforms. *J. Biol. Chem.* 268: 10221–10227.

Devchand PR, Keller H, Peters JM, Vazquez M, Gonzalez FJ, and Wahli W (1996): The PPARα-leukotrience B4 pathway to inflammation control. *Nature* 384: 39–43.

Downes M, Carozzi AJ, and Muscat GEO (1995): Constitutive expression of the orphan receptor, Rev-ErbAα, inhibits muscle differentiation and abrogates the expression of the *myoD* gene family. *Mol. Endocrinol.* 9: 1666–1678.

Dreyer C, Krey G, Keller H, Givel F, Helftenbein G, and Wahli W (1992): Control of the peroxisomal β-oxidation pathway by a novel family of nuclear hormone receptors. *Cell.* 68: 879–887.

Dumas B, Harding H, Choi H-S, Lehmann K, Chung M, Lazar MA, and Moore DD (1994): A new orphan member of the nuclear hormone receptor superfamily closely related to Rev-Erb. *Mol. Endocrinol.* 8: 996–1005.

Ellers M, Picard D, Yamamoto KR, and Bishop JM (1989): Chimaeras of Myc oncoprotein and steroid receptors cause hormone-dependent transformation of cells. *Nature* 340: 66–68.

Ellinger ZH, Hihi AK, Laudet V, Keller H, Wahli W, and Dreyer C (1994): FTZ-F1-

related orphan receptors in Xenopus laevis: transcriptional regulators differentially expressed during early embryogenesis. *Mol. Cell. Biol.* 14: 2786–2797.

Enmark E, and Gustafsson J-A (1996): Orphan nuclear receptors—the first eight years. *Mol. Endocrinol.* 10: 1293–1307.

Forman BM, Casanova J, Raaka BM, Ghysdael J, and Samuels HH (1992): Half-site spacing and orientation determines whether thyroid hormone and retinoic acid receptors and related factors bind to DNA response elements as monomers, homodimers, or heterodimers. *Mol. Endocrinol.* 6: 429–442.

Forman BM, Chen J, Blumberg B, Kliewer SA, Henshaw R, Ong ES, and Evans RM (1994): Cross-talk among RORα1 and the Rev-erb family of orphan nuclear receptors. *Mol. Endocrinol.* 8: 1253–1261.

Forman BM, Tontonoz P, Chen J, Brun RP, Spiegelman BM, and Evans RM (1995a): 15-deoxy, delta 12, 14-prostaglandin J2 is a ligand for the adipocyte determination factor PPARγ. *Cell.* 83: 803–812.

Forman BM, Umesono K, Chen J, and Evans RM (1995b): Unique response pathways are established by allosteric interactions among nuclear hormone receptors. *Cell.* 81: 541–550.

Giguere V, McBroom LDB, and Flock G (1995): Determinants of target gene specificity for RORα1: Monomeric DNA binding by an orphan nuclear receptor. *Mol. Cell. Biol.* 15: 2517–2526.

Giguere V, Ong ES, Segui P, and Evans RM (1987): Identification of a receptor for the morphogen retinoic acid. *Nature* 330: 624–629.

Giguere V, Tini M, Flock G, Ong E, Evans RM, and Otulakowski G (1994): Isoform-specific amino-terminal domains dictate DNA-binding properties of RORα, a novel family of orphan hormone nuclear receptors. *Genes Dev.* 8: 538–553.

Glass CK (1994): Differential recognition of target genes by nuclear receptor monomers, dimers, and heterodimers. *Endocrine Rev.* 15: 391–407.

Green S, and Chambon P (1987): Oestradiol induction of a glucocorticoid-responsive gene by a chimeric receptor. *Nature* 325: 75–78.

Green S, Kumar V, Theulaz I, Wahli W, and Chambon P (1988): The N-terminal DNA-binding 'zinc finger' of the oestrogen and glucocorticoid receptors determines target gene specificity. *EMBO J.* 7: 3037–3044.

Greiner EF, Kirfel J, Greschik H, Dorflinger U, Becker P, Mercep A, and Schule R (1996): Functional analysis of retinoid Z receptor beta, a brain-specific nuclear orphan receptor. *Proc. Natl. Acad. Sci. USA.* 93: 10105–10110.

Hamilton BA, Frankel WY, Kerrebrock AW, Hawkins TL, FitzHugh W, Kusumi K, Russell LB, Mueller KL, vanBerkel V, Birren BW, Kruglyak L and Lander ES (1996): Disruption of the nuclear hormone receptor RORα in staggerer mice. *Nature* 379: 736–739.

Harding HP, Atkins GB, Jaffe A, Seo W, and Lazar MA (1997): Transcriptional activation and repression by RORα, an orphan nuclear receptor required for cerebellar development. *Mol. Endocrinol.* 11: 1737–1746.

Harding HP, and Lazar MA (1993): The orphan receptor Rev-ErbAα activates transcription via a novel response element. *Mol. Cell. Biol.* 13: 3113–3121.

Harding HP, and Lazar MA (1995): The monomer-binding orphan receptor Rev-Erb represses transcription as a dimer on a novel direct repeat. *Mol. Cell. Biol.* 15: 4791–4802.

Hazel TG, Nathans DF, and Lau LF (1988): A gene inducible by serum growth factors encodes a member of the steroid and thyroid hormone receptor superfamily. *Proc. Natl. Acad. Sci. USA.* 85: 8444–8448.

Hirata Y, Kiuchi K, Chen HC, Milbrandt J, and Guroff G (1993): The phosphorylation and DNA binding of the DNA-binding domain of the orphan nuclear receptor NGFI-B. *J. Biol. Chem.* 268: 24808–24812.

Hirose T, Smith RJ, and Jetten AM (1994): RORgamma: the third member of the ROR/RZR orphan receptor subfamily that is highly expressed in skeletal muscle. *Biochem. Biophys. Res. Commun.* 205: 1975–1983.

Horlein AJ, Naar AM, Heinzel T, Torchia J, Gloss B, Kurokawa R, Ryan A, Kamei Y, Soderstrom M, Glass CK, and Rosenfeld MG (1995): Ligand-independent repression by the thyroid hormone receptor mediated by a nuclear receptor co-repressor. *Nature* 377: 397–404.

Horner MA, Chen T, and Thummel CS (1995): Ecdysteroid regulation and DNA binding properties of Drosophila nuclear hormone receptor superfamily members. *Dev. Biol.* 168: 490–502.

Issemann I, and Green S (1990): Activation of a member of the steroid hormone receptor superfamily by peroxisome proliferators. *Nature* 347: 645–650.

Izumo S, and Mahdavi V (1988): Thyroid hormone receptor α isoforms generated by alternative splicing differentially activate myosin HC gene transcription. *Nature* 334: 539–542.

Katz D, Reginato MJ, and Lazar MA (1995): Functional regulation of thyroid hormone receptor variant TRα2 by phosphorylation. *Mol. Cell. Biol.* 15(5): 2341–2348.

Katz RW, and Koenig RJ (1993): Nonbiased identification of DNA sequences that bind thyroid hormone receptor alpha 1 with high affinity. *J. Biol. Chem.* 268: 19392–19397.

Kliewer SA, Forman BM, Blumberg B, Ong ES, Borgmeyer U, Mangelsdorf DJ, Umesono K, and Evans RM (1994): Differential expression and activation of a family of murine peroxisome proliferator-activated receptors. *Proc. Natl. Acad. Sci. USA.* 91: 7355–7359.

Kliewer SA, Lenhard JM, Willson TM, Patel I, Morris DC, and Lehmann JM (1995): A prostaglandin J2 metabolite binds peroxisome proliferator-activated receptor γ and promotes adipocyte differentiation. *Cell.* 83: 813–819.

Kliewer SA, Umesono K, Noonan DJ, Heyman RA, and Evans RM (1992): Convergence of 9-cis retinoic acid and peroxisome proliferator signalling pathways through heterodimer formation of their receptors. *Nature* 358: 771–774.

Koelle MR, Segraves WA, and Hogness DS (1992): DHR3: a Drosophila steroid receptor homolog. *Proc. Natl. Acad. Sci. USA.* 89: 6167–6171.

Kostrouch Z, Kostrouchova M, and Rall JE (1995): Steroid/thyroid hormone receptor genes in *Caenorhabditis elegans*. *Proc. Natl. Acad. Sci. USA.* 92: 156–159.

Kurokawa R, Yu VC, Naar A, Kyakumoto S, Han Z, Silverman S, Rosenfeld MG, and Glass CK (1993): Differential orientations of the DNA-binding domain and carboxy-terminal dimerization interface regulate binding site selection by nuclear receptor heterodimers. *Genes Dev.* 7: 1423–1435.

Lala DS, Ikeda Y, Luo X, Baity LA, Meade JC, and Parker KL (1995): A cell-specific nuclear receptor regulates the steroid hydroxylases. *Steroids* 60: 10–14.

Lala DS, Rice DA, and Parker KL (1992): Steroidogenic factor I, a key regulator of steroidogenic enzyme expression, is the mouse homolog of fushi tarazu-factor I. *Mol. Endo.* 6(8): 1249–1258.

Lavorgna G, Ueda H, Clos J, and Wu C (1991): FTZ-F1, a steroid hormone receptor-like protein implicated in the activation of fushi tarazu. *Science* 252(5007): 848–851.

Law SW, Conneely OM, DeMayo FJ, and O'Malley BW (1992): Identification of a new brain-specific transcription factor, NURR1. *Mol. Endocrinol.* 6: 2129–2135.

Lazar MA (1993): Thyroid hormone receptors: Multiple forms, multiple possibilities. *Endocrine Rev.* 14: 184–193.

Lazar MA, Berrodin TJ, and Harding HP (1991): Differential DNA binding by monomeric, homodimeric, and potentially heteromeric forms of the thyroid hormone receptor. *Mol. Cell. Biol.* 11: 5005–5015.

Lazar MA, Hodin RA, Cardona GR, and Chin WW (1990): Gene expression from the c-erbAα/Rev-ErbAα genomic locus: Potential regulation of alternative splicing by complementary transcripts from opposite DNA strands. *J. Biol. Chem.* 265: 12859–12863.

Lazar MA, Hodin RA, Darling DS, and Chin WW (1988): Identification of a rat c-erbAα-related protein which binds deoxyribonucleic acid but does not bind thyroid hormone. *Mol. Endocrinol.* 2: 893–901.

Lazar MA, Hodin RA, Darling DS, and Chin WW (1989): A novel member of the thyroid/steroid hormone receptor family is encoded by the opposite strand of the rat c-erbAα transcriptional unit. Mol. Cell. Biol. 9: 1128–1136.

Lee MS, Kliewer SA, Provencal J, Wright PE, and Evans RM (1993): Structure of the retinoid X receptor alpha DNA binding domain: a helix required for homodimeric DNA binding. *Science* 260: 1117–1121.

Lee SL, Wesselschmidt RL, Linette GP, Kanagawa O, and Milbrandt JH (1995a): Unimpaired thymic and peripheral T cell death in mice lacking the nuclear receptor NGFI-B (Nur77). *Science* 269: 532–535.

Lee SST, Pineau T, Drago J, Lee EJ, Owens JW, Kroetz DL, Fernandez-Salguero PM, Westphal H, and Gonzalez FJ (1995b): Targeted disruption of the a-isoform of the peroxisome proliferator-activated receptor gene in mice results in abolishment of the pleiotropic effects of peroxisome proliferators. *Mol. Cell. Biol.* 15: 3012–3022.

Lehmann JM, Moore LB, Smith-Oliver TA, Wilkison WO, Willson TM, and Kliewer SA (1995): An antidiabetic thiazolidinedione is a high affinity ligand for the nuclear peroxisome proliferator-activated receptor γ (PPARγ). *J. Biol. Chem.* 270: 12953–12956.

Liu Z-G, Smith SW, McLaughlin KA, Schwartz LM, and Osborne BA (1994): Apoptotic signals delivered through the T-cell receptor of a T-cell hybrid require the immediate-early gene nur77. *Nature* 367: 281–284.

LopesdaSilva S, vanHorssen A, Chang C, and Burbach JP (1995): Expression of nuclear hormone receptors in the rat supraoptic nucleus. *Endocrinology* 136: 2276–2283.

Luo X, Ikeda Y, and Parker KL (1994): A cell-specific nuclear receptor is essential for adrenal and gonadal development and sexual differentiation. *Cell.* 77(4): 481–90.

Mages HW, Rilke O, Bravo R, Senger G, and Kroczek RA (1994): NOT, a human immediate-early response gene closely related to the steroid/thyroid hormone receptor NAK1/TR3. *Mol. Endocrinol.* 8: 1583–1591.

Mangelsdorf DJ, Thummel C, Beato M, Herrlich P, Schutz G, Umesono K, Blumberg B, Kastner P, Mark M, Chambon P, and Evans RM (1995): The nuclear receptor superfamily: the second decade. *Cell.* 83: 835–839.

Mangelsdorf DJ, Umesono K, and Evans RM (1994): The Retinoid Receptors. *The Retinoids: Biology, Chemistry, and Medicine, 2nd edition.* New York, Raven Press, Ltd.

McBroom LDB, Flock G, and Giguere V (1995): The nonconserved hinge region and distinct amino-terminal domains of the RORα orphan nuclear receptor isoforms are required for proper DNA bending and RORα-DNA interactions. *Mol. Cell. Biol.* 15: 796–808.

Milbrandt J (1988): Nerve growth factor induces a gene homologous to the glucocorticoid receptor gene. *Neuron.* 1: 183–188.

Mitsuhashi TG, Tennyson GE, and Nikodem VM (1988): Alternative splicing generates messages encoding rat c-erbA proteins that do not bind thyroid hormone. *Proc. Natl. Acad. Sci. USA.* 85: 5804–5808.

Miyajima N, Horiuchi R, Shibuya Y, Fukushige S, Matsubara K, Toyoshima K, and Yamamoto T (1989): Two erbA homologs encoding proteins with different T3 binding capacities are transcribed from opposite DNA strands of the same genetic locus. *Cell.* 57: 31–39.

Muscatelli F, Strom TM, Walker AP, Zanaria E, Recan D, Meindl A, Bardoni B, Guioli S, Zehetner G, Rabl W, et al (1994): Mutations in the DAX-1 gene give rise to both X-linked adrenal hypoplasia congenita and hypogonadotropic hypogonadism. *Nature* 372: 672–676.

Nakai A, Kartha S, Sakurai A, Toback FG, and DeGroot LJ (1990): A human early response gene homologous to murine nur77 and rat NGFI-B, and related to the nuclear receptor superfamily. *Mol. Endocrinol.* 4: 1438–1443.

O'Malley BW, and Conneely OM (1992): Orphan receptors: in search of a unifying hypothesis for activation. *Mol. Endocrinol.* 6: 1359–1361.

Ohkura N, Hijikuro M, Yamamoto A, and Miki K (1994): Molecular cloning of a novel thyroid/steroid receptor superfamily gene from cultured rat neuronal cells. *Biochem. Biophys. Res. Commun.* 205: 1959–1965.

Palmer CNA, Hsu M-H, Griffin KJ, and Johnson EF (1995): Novel sequence determinants in peroxisome proliferator signalling. *J. Biol. Chem.* 270: 16114–16121.

Perlmann T, and Jansson L (1995): A novel pathway for vitamin A signaling mediated by RXR heterodimerization with NGFI-B and NURR1. *Genes Dev.* 9: 769–782.

Predki PF, Zamble D, Sarkar B, and Giguere V (1994): Ordered binding of retinoic acid and retinoid X receptors to asymmetric response elements involves determinants adjacent to the DNA-binding domain. *Mol. Endocrinol.* 8: 31–39.

Rastinejad F, Perlmann T, Evans RM, and Sigler PB (1995): Structural determinants of nuclear receptor assembly on DNA direct repeats. *Nature* 375: 203–211.

Reginato MJ, Zhang J, and Lazar MA (1996): DNA-independent and DNA-dependent mechanisms regulate the differential heterodimerization of the isoforms of the thyroid hormone receptor with retinoid X receptor. *J. Biol. Chem.* 271: 28109–18205.

Retnakaran R, Flock G, and Giguere V (1994): Identification of RVR, a novel orphan nuclear receptor that acts as a negative transcriptional regulator. *Mol. Endocrinol.* 8: 1234–1244.

Ribeiro RCJ, Kushner PJ, Apriletti JW, West BL, and Baxter JD (1992): Thyroid hormone alters in vitro binding of monomers and dimers of thyroid hormone receptors. *Mol. Endocrinol.* 6: 1142–1152.

Scearce LM, Laz TM, Hazel TG, Lau LF, and Taub R (1993): RNR-1, a nuclear receptor in the NGFI-B/nur77 family that is rapidly induced in regenerating liver. *J. Biol. Chem.* 268: 8855–8861.

Segraves WA, and Hogness DS (1990): The E75 ecdysone-inducible gene responsible for the 75B early puff in *Drosophila* encodes two new members of the steroid receptor superfamily. *Genes Dev.* 4: 204–219.

Smith TS, Matharu PJ, and Sweeney GE (1993): Cloning and sequencing of a Xenopus homologue of the inducible orphan receptor NGFI-B. *Biochim. Biophis. Acta.* 1173: 239–242.

Stunnenberg HG (1993): Mechanisms of transactivation by retinoic acid receptors. *BioEssays.* 15: 309–315.

Sutherland JD, Kozlova T, Tzertinis G and Kafatos FC (1995): Drosophila hormone receptor 38: a second partner for Drosophila USP suggests an unexpected role for nuclear receptors of the nerve growth factor-induced protein B type. *Proc. Natl. Acad. Sci. USA.* 92: 7966–7970.

Thummel CS (1995): From embryogenesis to metamorphosis: the regulation and function of Drosophila nuclear receptor superfamily members. *Cell.* 83: 871–877.

Tini M, Fraser RA, and Giguere V (1995): Functional interactions between retinoic acid receptor-related orphan nuclear receptor (RORα) and the retinoic acid receptors in the regulation of the γF-crystallin promoter. *J. Biol. Chem.* 270: 20156–20161.

Tontonoz P, Hu E, Devine J, Beale EG, and Spiegelman BM (1995): PPAR gamma 2 regulates adipose expression of the phosphoenolpyruvate carboxykinase gene. *Mol. Cell. Biol.* 15: 351–357.

Tontonoz P, Hu E, and Spiegelman BM (1994): Stimulation of adipogenesis in fibroblasts by PPARγ2, a lipid-activated transcription factor. Cell. 79: 1147–1156.

Umesono K, and Evans RM (1989): Determinants of target gene specificity for steroid/thyroid hormone receptors. *Cell.* 57: 1139–1146.

Wilson TE, Fahrner TJ, and Milbrandt J (1993a): The orphan receptors NGFI-B and steroidogenic factor 1 establish monomer binding as a third paradigm of nuclear receptor-DNA interaction. *Mol. Cell. Biol.* 13: 5794–5804.

Wilson TE, Mouw AR, Weaver CA, Milbrandt J, and Parker KL (1993b): The orphan nuclear receptor NGFI-B regulates expression of the gene encoding steroid 21-hydroxylase. *Mol. Cell. Biol.* 13: 861–868.

Wilson TE, Paulsen RE, Padgett KA, and Milbrandt J (1992): Participation of non-zinc finger residues in DNA binding by two nuclear orphan receptors. *Science* 256: 107–110.

Woronicz JD, Calnan B, Ngo V, and Winoto A (1994): Requirement for the orphan steroid receptor Nur77 in apoptosis of T-cell hybridomas. *Nature* 367: 277–281.

Yang Y-Z, Burgos-Trinadad M, Wu Y, and Koenig RJ (1996): Thyroid hormone receptor variant a2. Role of the ninth heptad in DNA binding, heterodimerization with retinoid X receptor, and dominant negative activity. *J. Biol. Chem.* 271: 28235–28242.

Yen PM, Darling DS, Carter RL, Forgione M, Umeda PK, and Chin WW (1992):

Triiodothyronine (T3) decreases binding to DNA by T3-receptor homodimers but not receptor-auxiliary protein heterodimers. *J. Biol. Chem.* 267: 3565–3568.

Yu K, Bayona W, Kallen CB, Harding HP, Ravera C, McMahon J, Brown M, and Lazar MA (1995): Differential activation of peroxisome proliferator-activated receptors by eicosanoids. *J. Biol. Chem.* 270: 23975–23983.

Zamir I, Harding HP, Atkins GB, Horlein A, Glass CK, Rosenfeld MG, and Lazar MA (1996): A nuclear hormone receptor corepressor mediates transcriptional silencing by receptors with different repression domains. *Mol. Cell. Biol.* 16: 5458–5465.

Zamir I, Zhang J, and Lazar MA (1997): Stoichiometric and steric principles governing repression by nuclear hormone receptors. *Genes & Dev.* 11: 835–846.

Zetterstrom RH, Solomin L, Mitsiadis T, Olson L, and Perlmann T (1996): Retinoid X receptor heterodimerization and developmental expression distinguish the orphan nuclear receptors NGFI-B, Nurr1, and Nor1. *Mol. Endocrinol.* 10: 1656–1666.

11

Orphan Nuclear Receptors and Their Ligands

BARRY MARC FORMAN

ORPHAN NUCLEAR RECEPTORS

Nuclear hormone receptors constitute a superfamily of related transcription factors that mediate the actions of nuclear-acting hormones including gluco-corticoids, mineralocorticoids, estrogens, progestins, androgens, thyroid hormones, vitamin D, and retinoic acid (Mangelsdorf and Evans 1995). As described in earlier chapters of this book, members of this superfamily are characterized by their conserved DNA-binding, ligand-binding, and C-terminal transactivation domain (AF2) (Figure 1). By 1988, nuclear receptors had been identified for all known nuclear acting hormones. It appeared that all members of this superfamily had been identified until Giguere et al (Giguere et al 1988) reported a new and unexpected addition to the nuclear receptor superfamily. Over the next few years this group of receptor-like molecules rapidly expanded and collectively became known as orphan nuclear receptors. Now, almost a decade later, the orphan receptors have expanded to include over 30 distinct mammalian genes (Mangelsdorf et al 1995). As genome sequencing projects continue, it is likely that additional orphan receptors will be identified.

Nuclear receptors have been identified in vertebrate organisms, in insects, and in worms. A multiple sequence alignment was performed to compare the sequence relationship among all known vertebrate nuclear receptor proteins. This relationship is illustrated by the dendogram in Figure 2. The conserva-tion of orphan receptors through evolutionary time implies that they play important roles in a wide variety of biological systems. Some of these activities are listed in Figure 2, as are the known ligands and activators that mediate these activities. The remainder of this review will focus on the orphan nuclear

Molecular Biology of Steroid and Nuclear Hormone Receptors
Leonard P. Freedman, Editor
© 1998 Birkhäuser Boston

Figure 1.—Domain structure of nuclear hormone receptors.
The conserved DNA and ligand binding domains are illustrated. The DNA binding domain facilitates binding of the receptor to specific response elements in target genes. This region contains three helical motifs that are stabilized by the coordination of two zinc atoms. The first helix serves as the major DNA-recognition helix, while the third helix makes contacts in the minor groove upstream of the response element core. The ligand binding domain is a complex region containing multiple embedded subdomains. These include a dimerization motif, a ligand-dependent transcriptional activation domain (AF2) and a transcriptional corepressor interaction domain. Ligand binding results in a conformational change that both dissociates the transcriptional corepressor and recruits several coactivator proteins and histone acetylases. By integrating these activities, nuclear receptors and their ligands can alter both chromatin structure and the rate of transcription initiation by RNA polymerase II.

Figure 2.—Orphan nuclear receptors.
This dendogram presents a graphical view of the sequence relationship among vertebrate nuclear hormone receptors. The insect ecdysone receptor was included for comparative purposes. The dendogram was prepared by multiple sequence alignment (University of Wisconsin Genetics Computer Group, PILEUP program) of the full-length receptor protein sequence. It should be noted that the orphan receptors DAX-1 and SHP lack the conserved DBD (DNA binding domain). When known, the ligands, activators, and biological functions have been listed. The human ortholog of each receptor was used wherever possible in this dendogram; other species were utilized in the following cases: bovine: -SF-1α; mouse: RXRγ, TLX, GCNF, SF-1β; *xenopus*: ONR, HNF-4β; *drosophila*: EcR. Accession numbers are as follows: GR, X03225; MR, M16801; AR, M21748; PR, M15716; ERα, M11457; ERβ, X99101; RARα, X06538; RARβ, X07282; RARγ, M57707; T3Rα, M24748; T3Rβ, M26747; VDR, J03258; PPARα, S74349; PPARγ, L40904, PPARδ, L07592; FXR, U68233; LXRα, U22662; LXRβ, U07132; CARα, Z30425; ONR, X75163; NGFI-Bα, D49728; NGFI-Bβ, S77154; NFGI-Bγ, U12767; RORα, U04897; RORβ, Y08639; RORγ, U16997; RevEerbα, X72631; RevEerbβ, L31785; RXRα, X52773; RXRβ, X63522; RXRγ, M84819; COUPα, X16155; COUPβ, X12794; COUPγ, M62760; TR2α, M29960; TR2β, U10990; TLX, S77482; HNF-4α, X76930; HNF-4β, Z49827; ERRα, X51416; ERRβ, X51417; GCNF, U09563; SF-1α, D13569; SF-1β, M81385; EcR, M74078; SHP, L76571; DAX-1, S74720. Note that instead of PPARδ, *xenopus* may possess a unique PPAR subtype, PPARβ (M84162), that is not included in this comparison.

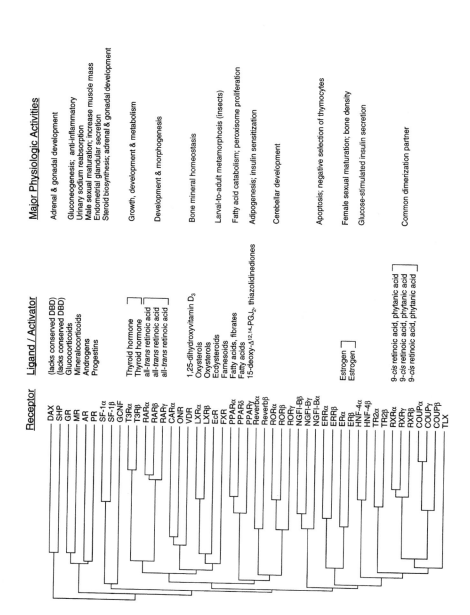

receptors. I will outline the methods available to identify ligands for orphan receptors and will end with a discussion of orphan receptors whose biological properties are now beginning to emerge.

A New Endocrinology

An obvious question that arises is whether or not ligands exist for the orphan nuclear receptors. A comparison of the 3-dimensional structures of the retinoid x (RXR), thyroid hormone (T_3R), and all-trans retinoic acid receptors (RAR) indicates that the ligand binding domains of these receptors fold into a characteristic three-dimensional structure with a well-defined ligand binding pocket (Bourguet et al 1995, Renaud et al 1995, Wagner et al 1995, Wurtz et al 1996). See Chapter 3 for a more detailed discussion. In effect, the primary amino acid sequence that defines the ligand binding domain is predictive of a common tertiary structure. This would imply that at least some, if not most, orphan receptors may possess yet to be identified ligands.

If orphan nuclear receptors do possess ligands, one may wonder why they have not been identified. When considering this question, we should recall how the classically defined endocrine hormones were discovered. Most of these hormones were described during the early part of this century by observing physiological deficits that resulted from removal of specific endocrine glands or tissues (Goodman et al 1996). Active hormones were then identified by isolating factors that restored normal physiologic functioning to the affected animal. This type of approach was highly successful and led to the identification of most nuclear-acting hormones. Despite its success, it has several limitations. For example, a specific physiological deficit may not be observed upon removal of any single tissue, if the relevant signaling molecule is produced in other tissues or throughout the body. In general, products of intermediary metabolism fall into this category. This raises the possibility that fatty acids, eicosanoids, mevalonate derivatives, steroid precursors, amino acids, sugars, and other metabolites could potentially serve as ligands for orphan receptors. In yeast and bacteria, transcriptional regulation by intermediary metabolites is a common paradigm. For example, lactose and tryptophan modulate transcriptional rates by binding directly to DNA binding proteins. It seems reasonable to imagine that transcriptional regulation by intermediary metabolites may also occur in higher animals (Tomkins 1975).

There are other reasons why classical approaches may have failed to identify additional signaling molecules. If the ligand functions in an autocrine fashion, then the same tissue will produce the ligand and transduce its signal.

In this case, test animals lacking a specific tissue would not be able to respond to the signal even if it were provided exogenously. Other limitations will arise if the active compound is unstable or if its concentration is low. Signaling molecules like these could easily be lost during purification. It is reasonable to hypothesize that ligands for orphan nuclear receptors may possess some of these properties. Identification of signaling molecules of this nature will require novel approaches. One possibility is to establish sensitive screening assays that are based on the potent signaling properties of orphan receptors. In this way, orphan nuclear receptors promise to pave the way for a new endocrinology.

SCREENING FOR NOVEL LIGANDS

In order to utilize orphan receptors as a screening tool, sensitive assays must be established. The cell-based transcription assay is the most common approach. In this assay, cells are cotransfected with a receptor expression vector and a reporter construct, then treated with potential ligands. When the proper controls are included, the presence of a ligand can be inferred from activation of the reporter construct. This assay is extremely sensitive when luciferase is used as a reporter construct; activity can be detected from as little as 100 expressing cells. This sensitivity allows the assays to be performed in a 96-well plate, allowing a large number of compounds or biological fractions to be rapidly screened for ligand-inducible transcriptional activity.

Despite its sensitivity, this assay has several drawbacks. Chief among these is that the assay is indirect; i.e., it does not distinguish between an indirect activator and a *bona fide* ligand. It is thus possible to identify a compound as transcriptionally active when in reality it is an inactive precursor that must be metabolized to an active ligand. Alternatively, the active fraction could contain an enzymatic inhibitor that causes accumulation of the actual ligand. This cell-based assay should be viewed as a useful screen for the identification of potential ligands. In all cases, a ligand binding experiment must be performed to determine if the compound acts as a true ligand. Nonetheless, even if a true ligand is not identified initially, the identification of an activator can provide important clues to the nature of the ultimate ligand. Thus, despite its limitations, the transcription-based assay is likely to remain an essential tool in the characterization of orphan receptor ligands.

A complementary approach that has been successfully used is the "ligand-trap" assay (Levin et al 1992). This approach can be applied when an activator for an orphan receptor has already been identified in the cell-based assay. In this approach, cells expressing the orphan receptor are treated with the metabolic precursor, and a ligand is identified or "trapped" as a result of its ability to interact directly with the receptor. This approach has not been widely used, because a metabolic precursor must first be available, and even then, it can be difficult to distinguish between specifically and non-specifically bound compounds. A challenge for the future will be the development of sensitive assays that directly measure the binding of ligands to orphan nuclear receptors.

In the following sections I will focus on the progress that has been made toward the identification of ligands for specific orphan nuclear receptors. Readers who are interested in other aspects of orphan nuclear receptor biology are referred to recent reviews (Mangelsdorf and Evans 1995, Enmark and Gustafsson 1996) and to other chapters in this book.

RETINOID X RECEPTOR (RXR)

In 1990, RXR was identified in the Evans laboratory as an orphan nuclear receptor that was activated in cells exposed to high concentrations of all-trans retinoic acid (at-RA) (Mangelsdorf et al 1990). However, since at-RA failed to bind to RXR, it was hypothesized that the actual ligand may be a metabolite of at-RA. Indeed, in 1992 Heyman et al and Levin et al (Heyman et al 1992, Levin et al 1992) reported that cells treated with at-RA produced 9-cis retinoic acid (9c-RA), which bound to and activated RXR at concentrations of approximately 10 nM. This raised the possibility that 9c-RA (Figure 3) may be a naturally occurring ligand for RXR. However, since cells are not normally exposed to such high levels of at-RA, it is unclear if sufficient amounts of 9c-RA are normally present in cells. Indeed, recent evidence suggests that if 9c-RA is naturally occurring, its endogenous concentrations in several tissues may be too low to act as a biologically relevant ligand for RXR (Pappas et al 1993, Horton and Maden 1995, Costaridis et al 1996, Kitareewan et al 1996).

What then is the endogenous ligand for RXR? Recent data suggest that phytenic acid and/or phytanic acid (Figure 3) may be an appropriate candidate (Kitareewan et al 1996, Lemotte et al 1996). These terpenoids can bind to and activate RXR with an affinity of approximately 5 μM. Although phytanic and phytenic acids are about 2–3 orders of magnitude less potent than 9c-RA, they could represent a physiologically relevant ligand because their circulating concentrations—approximately 5 μM—are similar to its dissociation constant for RXR. Given that these compounds originate from dietary sources, phytanic and phytenic acids may represent nutrients that activate RXR-dependent signaling pathways.

Although it remains unclear which endogenous signaling molecules are actually bound to RXR *in vivo*, the identification of 9c-RA as an RXR ligand was an important discovery because it provided a tool to explore the

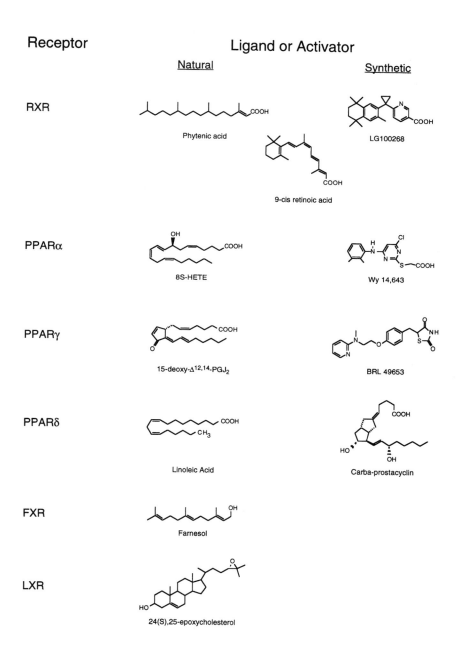

Figure 3.—Structures of known ligands and activators for orphan nuclear receptors.

Table 1.—Orphan nuclear receptors. This table provides a list of the known vertebrate orphan nuclear receptors. In cases where they are known, the ligands, activators, and/ or biological functions of these receptors are listed. Other receptors have been identified in vertebrates (MB67α, β; ONR; SHP; COUPα, β, γ; Rev-erbα, β; GCNF; ERRα, β; TR2-11α,β; TR2-11α, β; TLX), but less is known about their function. References and accession numbers for these receptors have been reported elsewhere (Mangelsdorf et al 1995). Note that PPARβ may be unique to *xenopus* (Kliewer et al 1994).

Receptor		Ligand/Activator	Function
RXR	α, β, γ	9-cis retinoic acid, phytanic acid	Common dimerization partner
PPAR	α	8S-HETE, fibrates, fatty acids	Fatty acid catabolism; peroxisome proliferation
	β		
	γ	15d-$\Delta^{12,14}$-PGJ$_2$ thiazolidine-diones fatty acids	Adipogenesis; insulin sensitization
	δ		
FXR		farnesoids	
LXR	α, β	oxysterols	
SF-1	α		Adrenal/gonadal development; steroid biosynthesis
	β		
DAX-1			Adrenal hypoplasia congenita
ROR	α		Cerebellar development
	β		
	γ		
HNF-4	α		Maturity-onset diabetes of the young
	β		
NGFI-b	α		Apoptosis; negative selection of thymocytes
	β		
	γ		

molecular and physiologic consequences of RXR activation. In practice, the use of 9c-RA to probe RXR action has been limited by the observation that 9c-RA is also an agonist for the at-RA receptor (RAR) (Boehm et al 1994a). However, this problem was overcome by the development of synthetic retinoids that are RXR-selective (Figure 3) (Lehmann et al 1992, Boehm et al 1994b, Boehm et al 1995).

An understanding of the biological activities of RXR agonists is crucial because of the central role that RXR plays in nuclear receptor signaling. Biochemical studies indicate that RXR can bind to DNA as a homodimer and activate transcription in response to RXR agonists. Perhaps more importantly, RXR serves as a common heterodimerization partner that promotes high-affinity DNA binding by RAR, T_3R, vitamin D receptor (VDR), the peroxisome proliferator activated receptor (PPAR), and several orphan receptors including FXR, LXR/RLD, CAR, ONR, and NGFI-B (Mangelsdorf and Evans 1995). The physiological role of RXR homodimers is unclear, since homodimers are not as stable as the above RXR • receptor heterodimers. Thus, endogenous RXR ligands could potentially have their greatest effects through a variety of heterodimeric signaling pathways.

With this in mind, Kurokawa et al and Forman et al (Kurokawa et al 1994, Forman et al 1995c) have investigated the potential of RXR to activate through various RXR-receptor heterodimers. Surprisingly, while RXR itself is a *bona fide* receptor, these studies have indicated that RXR cannot be transcriptionally activated when complexed with T_3R or unliganded RARα. The mechanism underlying this restriction is twofold. First, heterodimerization with T_3R and unliganded RARα introduces an allosteric change in RXR that prevents its binding to synthetic retinoids. Second, unliganded T_3R and RARα recruit transcriptional corepressors (SMRT/N-Cor), which further prevent RXR-dependent transcriptional activation (Chen and Evans 1995, Horlein et al 1995, Kurokawa et al 1995, Chen et al 1996, Sande and Privalsky 1996). These studies suggest that RXR acts as a "silent partner" in certain heterodimeric contexts. It should be noted that this block to RXR function is dependent on the particular heterodimerization partner. For example, in contrast to unliganded-RARα • RXR complexes, RXR ligands potentiate signaling through liganded-RARα-RXR (Forman et al 1995c, Roy et al 1995, Minucci et al 1997) or through unliganded-RARβ • RXR heterodimers (La Vista-Picard et al 1996). Similarly, RXR ligands appear to bind to VDR • RXR heterodimers and promote dissociation of this complex (MacDonald et al 1993, Cheskis and Freedman 1994).

The mechanistic studies described above would predict that RXR-selective ligands may not necessarily have the same physiologic effects as their partner ligands. Indeed, *in vivo* studies have failed to identify a broad range of activities for RXR agonists. The most notable property of RXR-selective retinoids is their ability to lower triglycerides and to promote insulin sensitization in diabetic mice (Mukherjee et al 1997). These activities are reminiscent of PPARα and PPARγ ligands (see below) and suggests that *in vivo*, RXR ligands may function in part through RXR • PPAR heterodimers.

One important caveat to the interpretation of these results is that they were performed with synthetic RXR agonists. In principle, one cannot be sure that endogenous RXR ligands will behave in the same way as the synthetic agonists. This is not simply a theoretical concern, as a synthetic RXR-selective ligand has recently been identified that acts as both an RXR homodimer antagonist and an agonist of the RAR • RXR and PPAR • RXR heterodimers (Lala et al 1996). Clearly, it will be important to identify the endogenous RXR ligand(s) and to examine their activities.

An alternate approach to studying the physiologic functions of RXR is to utilize gene-targeting strategies to develop mice that lack functional RXR (Sucov et al 1994, Kastner et al 1995, Kastner et al 1996, Krezel et al 1996). Such studies have shown that RXRα is required for cardiac and liver organogenesis, whereas RXRβ is required for spermatogenesis. It remains unclear whether these phenotypes are due to a deficient response to RXR ligands or due to a disruption of a specific heterodimeric complex. No apparent deficit has yet been observed in animals lacking a functional RXRγ.

PEROXISOME PROLIFERATOR ACTIVATED RECEPTOR (PPAR)

PPARα

This receptor was identified in 1990 through a series of elegant experiments from the laboratory of Stephen Green (Issemann and Green 1990). It had been known for some time that a class of synthetic compounds, the peroxisome proliferators, could promote an increase in the size and number of hepatic peroxisomes in rodents (Schoonjans et al 1996). This peroxisome proliferation was accompanied by an increase in the transcription rate of several genes required for the catabolism of fatty acids (FAs) via β- and ω-oxidation pathways. This led the Green laboratory to hypothesize that peroxisome proliferators may act through an orphan nuclear receptor. Indeed, a cDNA clone was identified in liver that encoded a nuclear receptor, PPARα, which was responsive to peroxisome proliferators. One subset of peroxisome proliferators includes the hypolipidemic fibrate drugs. Several fibrates, including gemfibrozil, are currently being used as serum triglyceride lowering agents. Thus, the identification of PPARα as a fibrate-responsive transcription factor was an important milestone, as it suggested that this orphan receptor may play an important role in lipid homeostasis. A requirement for PPARα in mediating the effect of peroxisome proliferators *in vivo* was confirmed in mice with a homozygous disruption of the PPARα gene (Lee et al 1995b).

The experiments described above suggested important molecular links among peroxisome proliferators, the PPARα orphan receptors and the induction of FA oxidation pathways. However, the endogenous signals that regulate PPARα remained unknown. In 1992, Gottlicher et al demonstrated that several naturally occurring long-chain FAs activate PPARα (Gottlicher et al 1992). Soon a large number of apparently unrelated compounds (FAs, eicosanoids, fibrates, peroxisome proliferators, inhibitors of mitochondrial β-oxidation) were found to activate PPARα, typically at micromolar doses (Keller et al 1993, Gulick et al 1994, Kliewer et al 1994, Schoonjans et al 1996). This apparent lack of specificity led to the suggestion that all of these compounds act indirectly. For example, they might inhibit endogenous β-oxidation pathways, thereby leading to the accumulation of the true PPARα ligand (Gottlicher et al 1993). This hypothesis was fueled by the inability to demonstrate direct binding of any of these compounds to PPARα.

We felt that negative data obtained by classical ligand-binding studies were potentially misleading. One reason for this is that the FAs that activate PPARα do so at concentrations of approximately 10–100 μM. It is very difficult to perform ligand-binding studies in this concentration range because FAs tend to exhibit an enormous amount of nonspecific binding. To overcome these limitations, we developed an assay that does not utilize a radiolabeled ligand. Our approach relied on the ability of nuclear receptor ligands to induce conformational changes that promote dimerization and subsequent DNA binding. In previous studies, ecdysone (Yao et al 1993), vitamin D (Cheskis and Freedman 1994), and 9c-RA (Zhang et al 1992) were shown to enhance the dimerization and DNA-binding activities of their respective receptors. Accordingly, we examined whether PPARα activators could promote dimerization of PPAR$\alpha \bullet$ RXR heterodimers (Forman et al 1997). Indeed, we found that fibrates and specific long-chain polyunsaturated FAs (Figure 3) bound to PPARα and promoted dimerization at micromolar concentrations. The highest-affinity natural ligand was 8S-hydroxyeicosate-traenoic acid (8S-HETE, $K_d = 100$ nM) (Figure 3) (Forman et al 1997), a metabolite of arachidonic acid known to selectively activate PPARα (Yu et al 1995). Taken together, these findings indicated that PPARα transduces the transcriptional effects of dietary FAs and their metabolites. Thus, PPARα provides an excellent example of a mammalian transcription factor that is directly regulated by intermediary metabolites. The demonstration of a direct interaction between fibrates and PPARα has important implications for drug development, as this receptor may now be utilized as a target for the rapid identification of more potent and selective hypolipidemic agents.

A previous report suggested that leukotriene B$_4$ binds *xenopus* PPARα with an affinity of approximately 100 nM (Devchand et al 1996). However, nonspecific binding to PPARα was not accounted for, and half-maximal displacement required 10–50 μM of unlabeled leukotriene B$_4$. In addition, we were unable to detect activation of or binding to PPARα with 10 μM leukotriene B$_4$ (Forman et al 1997). Thus, we believe that leukotriene B$_4$ is not a physiologically relevant ligand for mammalian PPARα.

PPARγ

The γ1 isoform of PPAR was cloned in several labs by low stringency hybridization with PPARα probes. Around the same time, the Spiegelman lab was seeking to isolate transcription factors that directed adipocyte-specific expression of aP2, the adipocyte fatty acid binding protein (Spiegelman and

Flier 1996). Analysis of an adipocyte-specific enhancer in the aP2 gene led to the identification of PPARγ2, an N-terminal splice variant of PPARγ1 that is highly expressed in adipose tissue. PPARγ appears to be a key adipocyte protein, as it is induced early in the course of adipogenesis (Chawla et al 1994, Tontonoz et al 1994b). Indeed, forced expression of PPARγ in fibroblasts initiates a transcriptional cascade that leads to the development of mature adipocytes (Tontonoz et al 1994c). These studies suggest that endogenous signals that modulate PPARγ activity may serve a primary role in regulating adipogenesis and lipid homeostasis.

Since naturally occurring FAs could activate PPARα (Keller et al 1993, Gulick et al 1994, Kliewer et al 1994, Schoonjans et al 1996), we asked whether these activators could also activate PPARγ. Surprisingly, we found that most PPARα activators were selective and did not effectively activate PPARγ (Kliewer et al 1994, Forman et al 1997). However, an analysis of arachidonate metabolites identified 15-deoxy-$\Delta^{12,14}$-prostaglandin J_2 (15d-J_2) (Figure 3) as a natural PPARγ ligand and an activator of adipogenesis. These findings suggest that 15d-J_2, or a related eicosanoid, may represent an endogenous regulator of adipogenesis and lipid homeostasis.

Our demonstration that 15d-J_2 is a PPARγ ligand provided the first example of a nuclear signaling pathway for a prostaglandin. All previously identified prostaglandin receptors had been shown to be cell-surface receptors (Narumiya 1995). The demonstration of a nuclear pathway is of interest when one considers the biosynthetic pathway leading to prostaglandin production. The first step in the conversion of arachidonic acid to prostaglandins is mediated by cyclooxygenase (COX). Two cyclooxygenase subtypes have been identified. Interestingly, the COX-2 subtype is localized to both the nucleus and the cytoplasm (Morita et al 1994). It thus appears that prostaglandin production may occur in the nucleus, suggesting that COX-2 could be utilized to produce nuclear signaling molecules such as 15d-J_2. Future studies will be required to determine whether COX-2 contributes to the production of ligands for PPARγ or other nuclear receptors.

Antidiabetic Drugs and PPARγ

The ability of 15d-J_2 to activate PPARγ response elements was reminiscent of the properties of pioglitazone, an antidiabetic drug of the thiazolidinedione class (Sohda et al 1990) that activates the aP2 (adipocyte fatty acid binding protein) gene PPRE (Harris and Kletzien 1994, Tontonoz et al 1994a, Tontonoz et al 1994b) and induces adipogenesis. This led us to examine the possibility that pioglitazone and other antidiabetic thiazolidinediones (Figure 3, BRL 49653) might be PPARγ agonists. Indeed, we (Forman et al 1995b) and others (Lehmann et al 1995, Willson et al 1996) have found a good correlation between the antidiabetic potency of thiazolidinediones and their affinities for PPARγ. These results suggest a specific molecular linkage between 15d-J_2, adipogenesis, and insulin signaling and thus identify new and unexpected avenues of prostanoid research.

The mechanism by which PPARγ activation leads to insulin sensitization remains unclear. PPARγ levels are highest in adipose tissue, but the major insulin-sensitizing effect of the thiazolidinediones is in skeletal muscle. Thus, it is unclear which tissue is the immediate target of thiazolidinedione action. If adipose tissue is the target, then one must explain how this tissue can promote insulin sensitization in muscle. One possibility is that PPARγ activation in fat inhibits secretion of an inhibitor of insulin action. Candidate inhibitory factors include FAs/triglycerides (Rebrin et al 1995) and tumor necrosis factor α (Hotamisligil and Spiegelman 1994), as both have been implicated in mediating insulin resistance. Consistent with this hypothesis is the observation that thiazolidinediones lower FAs/triglyceride levels *in vivo* (Zhang et al 1996) and block tumor necrosis factor α action in cells (Ohsumi et al 1994, Szalkowski et al 1995).

It is interesting to note that PPARγ is highly expressed in both white and brown fat, which have opposing metabolic activities; brown fat mediates energy dissipation, whereas white fat is used as an energy storage depot. Thus, if adipose tissue is the target for thiazolidinedione action, it will be necessary to distinguish between the effects of PPARγ activation in brown and white adipose tissue (Tai et al 1996).

Although 15d-J_2 is a natural PPARγ ligand, its levels have yet to be directly measured in any tissue. Thus, as with RXR and 9c-RA, it remains possible that other molecules serve as endogenous ligands. It will be important to determine whether 15d-J_2 is present at physiologically relevant concentrations, and if its levels are altered in individuals with non-insulin-dependent (type II) diabetes and/or hyperplastic obesity.

PPARδ

No function has yet been ascribed to PPARδ, a ubiquitously expressed member of the PPAR subfamily. We have found that PPARα and PPARδ bind to a broad and overlapping array of polyunsaturated FAs, as well as to carbaprostacyclin, a synthetic prostanoid (Forman et al 1997). The ability to bind to such a broad array of ligands is unique among the nuclear receptors, and suggests that PPARα/δ have evolved to sense a variety of changes in FA status and dietary inputs.

Our data indicate that PPARα/δ ligands are distinct from those of PPARγ. The ability of these receptors to respond to distinct metabolic cues provides a potential mechanism for the animal to establish a balance between FA breakdown and storage. Thus, overall FA homeostasis may be determined by the relative levels of PPARα/δ and PPARγ ligands.

PPAR, Dietary FAs, and Human Disease

Inappropriately high levels of triglycerides and nonesterified FAs are a common component of obesity, insulin resistance, hypertension, and hyperlipidemia (Durrington 1993, Reaven 1994). This constellation of abnormalities often develops in the same individual, and they are ominous signs of impending coronary heart disease, a major cause of death in industrialized societies. The PPARs and/or their ligands may be central players in these metabolic disorders, since PPARα and PPARγ agonists represent useful pharmacologic interventions.

Since PPARγ ligands induce adipogenesis, they have the potential to promote terminal differentiation of tumors derived from immature adipocyte precursors. Indeed, it has recently been shown that PPARγ ligands can induce liposarcoma-cell differentiation in cell culture experiments. This raises hope that PPARγ ligands might serve as useful differentiation agents (Tontonoz et al 1997) for this otherwise untreatable malignancy.

FARNESOID X RECEPTOR (FXR)

FXR (Forman et al 1995a) was identified as an orphan nuclear receptor expressed in the liver, kidney, intestine, and adrenal cortex. One common property of all these tissues is that they utilize the mevalonate biosynthetic pathway at a very high rate. This raised the possibility that FXR could be responsive to products of mevalonate metabolism. We demonstrated that FXR was activated by several farnesoid metabolites of mevalonic acid, including farnesol (Figure 3) and juvenile hormone III (Forman et al 1995a). These findings suggest that FXR is a key mediator of a novel transcriptional control pathway that is responsive to farnesoid-like metabolites. It remains unclear if farnesol and juvenile hormone III act as FXR ligands, or if they activate indirectly. A major challenge for the future will be the identification of an FXR ligand and its responsive genes, and ultimately an analysis of the physiological functions of this receptor.

LIVER X RECEPTOR (LXR/RLD)

LXR/RLD (Apfel et al 1994, Willy et al 1995) is a constitutively active member of the orphan nuclear receptor family; i.e., when expressed in cells, this receptor displays transcriptional activity in the absence of an exogenous ligand. One possibility is that constitutive activity may reflect the production of an endogenous LXR/RLD ligand. We have found that inhibition of mevalonic acid biosynthesis suppressed the constitutive activity of LXR RLD (Barry M. Forman and Ronald M. Evans, unpublished data), suggesting that mevalonic acid or its metabolites may activate LXR/RLD. We (Barry M. Forman and Ronald M. Evans, unpublished data) and others (Janowski et al 1996, Lehmann et al 1997) have shown that several oxysterol metabolites of mevalonic acid including 20(R)-, 22(R)-, 24(S)-hydroxycholesterol, and 24(S),25-epoxycholesterol (Figure 3) activate LXR/RLD. Surprisingly, the constitutive activity of LXR/RLD was inhibited by geranylgeraniol (Barry M. Forman and Ronald M. Evans, unpublished data), an endogenous metabolite of mevalonic acid. These findings suggest that RLD/LXR represents a central component of a novel signaling pathway that is differentially regulated by multiple products of mevalonate metabolism.

A physiologic role for LXR/RLD is unknown. However, an LXR/RLD binding site was observed in the 7α-hydroxylase gene promoter (Lehmann et al 1997). As this enzyme represents the rate-limiting step in bile acid biosynthesis, it is possible that LXR/RLD may regulate the production of bile acids, thus facilitating cholesterol excretion and intestinal fat absorption.

STEROIDOGENIC FACTOR 1α (SF-1α) AND DAX-1

SF-1α plays a central role in modulating reproductive function. This orphan receptor binds as a monomer to response elements upstream of the cytochrome P450 steroid hydroxylase, the steroidogenic acute regulatory protein, and the mullerian inhibiting substance genes (Shen et al 1994, Luo et al 1995a, Luo et al 1995b, Parker et al 1996). As a result, SF-1 can regulate steroid hormone biosynthesis and mullerian duct regression. SF-1 is expressed in mouse embryos at the time of adrenal and gonadal development, suggesting that it plays important roles in the differentiation of these steroidogenic organs. Indeed, gene-targeting studies indicate that SF-1α is required for adrenal and gonadal development, for pituitary gonadotrope function, and for formation of the ventromedial hypothalamic nucleus as well (Luo et al 1994). Thus, SF-1 plays multiple roles in endocrine responses that are essential for reproduction.

Interestingly, patients with X-linked adrenal hypoplasia congenita display a similar constellation of abnormalities as mice lacking SF-1. The gene responsible for adrenal hypoplasia congenita has been shown to be DAX-1 (Muscatelli et al 1994, Zanaria et al 1994), an unusual orphan nuclear receptor that lacks the conserved DNA binding domain. Like SF-1, DAX-1 is expressed in the adrenal and gonads. These similarities suggest that the SF-1 and DAX-1 signaling pathways may converge at some level. A ligand has not been identified for SF-1 or DAX-1; however, it is tempting to speculate that SF-1 and/or DAX-1 may recognize an intermediate in steroid hormone biosynthesis. In this way, either SF-1 or DAX-1 could sense changes in sterol levels and respond by altering the rate of transcription of genes required for steroid biosynthesis.

RORα

Genetic evidence suggests that RORα (Giguere et al 1994) may be required for cerebellar development and/or function. Specifically, *staggerer* mice were found to harbor a deletion within the RORα gene that prevents translation of the putative ligand binding domain (Hamilton et al 1996). Homozygous *staggerer* mice are characterized by severe ataxia secondary to a cell-autonomous defect in the development of purkinje cells. These cells display immature morphology and are reduced in number and size. In addition, RORα appears to function in mature purkinje cells, since *staggerer* heterozygotes show accelerated dendritic atrophy.

RORα binds to DNA as a monomer and can constitutively activate transcription (see Chapter 10). It is unclear whether this activation represents true ligand-independent activity or activation by an endogenously produced ligand. It has been suggested that melatonin may be a ligand for RORα. However, activation of RORα by melatonin has not been reproducible (Barry M. Forman and Ronald M. Evans, unpublished data; Vincent Giguere, personal communication), and melatonin binding sites have not been demonstrated in certain RORα-expressing tissues (Hazlerigg et al 1996).

HEPATOCYTE NUCLEAR FACTOR 4α (HNF-4α)

HNF-4α was first defined as a liver-enriched transcription factor that binds to key regulatory sites in the transthyretin and apolipoprotein CIII genes (Sladek et al 1990). Purification and cloning of HNF-4α revealed that it is a constitutively active orphan nuclear receptor that binds DNA as a homodimer. Mice with a disrupted HNF-4α gene have impaired gastrulation that results in embryonic lethality (Chen et al 1994). In the adult, HNF-4α is expressed in the liver, kidney and intestine. HNF-4α appears to be required for glucose-stimulated insulin secretion, since mutations in HNF-4α are associated with the development of maturity-onset diabetes of the young (Yamagata et al 1996). It is unclear if there are sufficient levels of HNF-4α in the pancreatic β-cell to account for a direct effect of this receptor on pancreatic insulin secretion. Alternately, control of insulin secretion by

HNF-4α could occur via an indirect effect of HNF-4α in the liver and/or other tissues.

The observation that HNF-4α regulates insulin action is of interest because HNF-4α homodimers and PPAR $\bullet$ RXR heterodimers bind to an overlapping array of response elements. Since PPAR $\bullet$ RXR heterodimers have already been implicated in regulating insulin response and lipid homeostasis, it appears that, together, HNF-4α and PPAR $\bullet$ RXR may coordinately regulate these processes. Clearly, an HNF-4α ligand, should it exist, will contribute greatly to our understanding of the interrelationships between lipid metabolism and insulin action.

NERVE GROWTH FACTOR INDUCIBLE-B (NGFI-B)

As its name implies, NGFI-B (nur77) was first identified as a transcript that is rapidly induced upon stimulation of cells with NGF (Hazel et al 1988, Milbrandt 1988) or other growth factors and signals (Davis and Lau 1994). Subsequently, the laboratories of Osborne and Winoto demonstrated that NGFI-B was rapidly induced by activation of the CD3/T cell antigen receptor in T cells (Liu et al 1994, Woronicz et al 1994). This induction was rapid, and, more importantly, NGFI-B activity was required for subsequent apoptosis. These results provide strong evidence that NGFI-B is a key intermediate in this T cell receptor signaling pathway. Paradoxically, animals lacking NGFI-B display unimpaired thymocyte death (Lee et al 1995a). This may be due to an unknown compensatory mechanism, since mice expressing dominant negative inhibitors of NGFI-B are deficient in negative selection of self-reactive thymocytes (Calnan et al 1995, Zhou et al 1996). NGFI-B appears to induce apoptosis, at least in part, by upregulating Fas ligand expression (Weih et al 1996).

A detailed understanding of the mechanisms by which NGFI-B induces apoptosis will require an identification of its immediate target genes and its ligand, if any. Like SF-1 and ROR, NGFI-B can bind to DNA as a monomer (Wilson et al 1991, Wilson et al 1993), where it appears to activate transcription constitutively. Interestingly, NGFI-B can also interact with RXR (Forman et al 1995c, Perlmann and Jansson 1995), to form a complex that is responsive to RXR ligands. Thus, NGFI-B can function through RXR-dependent and RXR-independent pathways. The relative contribution of each pathway to the apoptotic process is currently unknown.

FUTURE DIRECTIONS

It is clear that since their discovery in 1988, the orphan nuclear receptors have provided unexpected insights into a variety of biological events. A major challenge for the future will be the continued identification of orphan receptor ligands, their target genes, and their physiologic functions and mechanisms of action. These goals will be made easier by several technological advances. First, an understanding of physiological function will be facilitated by conditional gene-targeting strategies that allow a specific gene to be excised at specific times in specific tissues (Feil et al 1996, No et al 1996, Tsien et al 1996). Second, target gene identification will be facilitated by the use of microarray-based expression monitoring (Schena et al 1996) and the eventual completion of various genomic and cDNA sequencing projects. When these two technologies are combined, it will eventually be possible to rapidly monitor the effect of a ligand on the expression of all known genes. Finally, small-molecule combinatorial chemistry (Baldwin and Henderson 1996) should make it possible to develop synthetic agonists and antagonists for orphan receptors. These synthetic drugs should provide useful reagents to probe the effects of receptor activation or inhibition *in vivo*.

The identification of endogenous ligands remains a difficult, yet essential task. Facile screening for these compounds will require novel technologies that are based on ligand binding rather than on transcriptional activation strategies. Even when a ligand is purified, an unequivocal structural assignment remains challenging when the ligand is present in minute quantities. Thus, sensitive mass spectrometry and nuclear magnetic resonance technologies will be useful. Despite these challenges, it is clear that the orphan nuclear receptors will continue to pave the way toward a new endocrinology.

REFERENCES

Apfel PW, Benbrook D, Lernhardt O, Gilles S, and Pfahl M (1994): A novel orphan receptor specific for a subset of thyroid hormone-responsive elements and its interaction with the retinoid/thyroid hormone receptor subfamily. *Mol. Cell Biol.* 14(10): 7025–35.

Baldwin JJ, and Henderson I (1996): Recent advances in the generation of small-molecule combinatorial libraries: encoded split synthesis and solid-phase synthetic methodology. *Med. Res. Rev.* 16(5): 391–405.

Boehm MF, McClurg MR, Pathirana C, Mangelsdorf D, White SK, Hebert J, Winn D, Goldman ME, and Heyman RA (1994a): Synthesis of high specific activity [3H]-9-cis-retinoic acid and its application for identifying retinoids with unusual binding properties. *J. Med. Chem.* 37(3): 408–14.

Boehm MF, Zhang L, Badea BA, White SK, Mais DE, Berger E, Suto CM, Goldman ME, and Heyman RA (1994b): Synthesis and structure-activity relationships of novel retinoid X receptor-selective retinoids. *J. Med. Chem.* 37(18): 2930–41.

Boehm MF, Zhang L, Zhi L, McClurg MR, Berger E, Wagoner M, Mais DE, Suto CM, Davies JA, Heyman RA, et al (1995): Design and synthesis of potent retinoid X receptor selective ligands that induce apoptosis in leukemia cells. *J. Med. Chem.* 38(16): 3146–55.

Bourguet W, Ruff M, Chambon P, Gronemeyer H, and Moras D (1995): Crystal structure of the ligand-binding domain of the human nuclear receptor RXR-alpha [see comments]. *Nature* 375(6530): 377–82.

Calnan BJ, Szychowski S, Chan FK, Cado D, and Winoto A (1995): A role for the orphan steroid receptor Nur77 in apoptosis accompanying antigen-induced negative selection. *Immunity* 3(3): 273–82.

Chawla A, Schwarz EJ, Dimaculangan DD, and Lazar MA (1994): Peroxisome proliferator-activated receptor (PPAR) gamma: adipose-predominant expression and induction early in adipocyte differentiation. *Endocrinology* 135(2): 798–800.

Chen JD, and Evans RM (1995): A transcriptional co-repressor that interacts with nuclear hormone receptors [see comments]. *Nature* 377(6548): 454–7.

Chen JD, Umesono K, and Evans RM (1996): SMRT isoforms mediate repression and anti-repression of nuclear receptor heterodimers. *Proc. Natl. Acad. Sci. USA* 93(15): 7567–71.

Chen WS, Manova K, Weinstein DC, Duncan SA, Plump AS, Prezioso VR, Bachvarova RF, and Darnell JE, Jr. (1994): Disruption of the HNF-4 gene, expressed in visceral endoderm, leads to cell death in embryonic ectoderm and impaired gastrulation of mouse embryos. *Genes. Dev.* 8(20): 2466–77.

Cheskis B, and Freedman LP (1994): Ligand modulates the conversion of DNA-bound vitamin D3 receptor (VDR) homodimers into VDR-retinoid X receptor heterodimers. *Mol. Cell Biol.* 14(5): 3329–38.

Costaridis P, Horton C, Zeitlinger J, Holder N, and Maden M (1996): Endogenous retinoids in the zebrafish embryo and adult. *Dev. Dyn.* 205(1): 41–51.

Davis IJ, and Lau LF (1994): Endocrine and neurogenic regulation of the orphan nuclear receptors Nur77 and Nurr-1 in the adrenal glands. *Mol. Cell Biol.* 14(5): 3469–83.

Devchand PR, Keller H, Peters JM, Vazquez M, Gonzalez FJ, and Wahli W (1996): The PPAR-Alpha-Leukotriene B4 Pathway to Inflammation Control. *Nature*. 384(6604): 39–43.

Durrington PN (1993): Diabetes, hypertension and hyperlipidaemia. *Postgrad. Med. J.* 69 Suppl 1: S18–25; discussion S25–9.

Enmark E, and Gustafsson JA (1996): Orphan nuclear receptors. The first eight years. *Molecular Endocrinology* 10(11): 1293–1307.

Feil R, Brocard J, Mascrez B, LeMeur M, Metzger D, and Chambon P (1996): Ligand-activated site-specific recombination in mice. *Proc. Natl. Acad. Sci. USA* 93(20): 10887–90.

Forman BM, Chen J, and Evans RM (1997): Hypolipidemic drugs, polyunsaturated fatty acids and eicosanoids are ligands for PPARα and PPARδ. *Proc. Natl. Acad. Sci. USA* In Press:.

Forman BM, Goode E, Chen J, Oro AE, Bradley DJ, Perlmann T, Noonan DJ, Burka LT, McMorris T, Lamph WW, Evans RM, and Weinberger C (1995a): Identification of a nuclear receptor that is activated by farnesol metabolites. *Cell* 81(5): 687–93.

Forman BM, Tontonoz P, Chen J, Brun RP, Spiegelman BM, and Evans RM (1995b): 15-Deoxy-delta 12, 14-prostaglandin J2 is a ligand for the adipocyte determination factor PPAR gamma. *Cell* 83(5): 803–12.

Forman BM, Umesono K, Chen J, and Evans RM (1995c): Unique response pathways are established by allosteric interactions among nuclear hormone receptors. *Cell* 81(4): 541–50.

Giguere V, Tini M, Flock G, Ong E, Evans RM, and Otulakowski G (1994): Isoform-specific amino-terminal domains dictate DNA-binding properties of RORα, a novel family of orphan nuclear hormone receptors. *Genes. Dev.* 8:538–553.

Giguere V, Yang N, Segui P, and Evans RM (1988): Identification of a new class of steroid hormone receptors. *Nature.* 331(6151): 91–4.

Goodman LS, Gilman A, Hardman JG, Gilman AG, and Limbird LE (1996): *Goodman & Gilman's the pharmacological basis of therapeutics.*

Gottlicher M, Demoz A, Svensson D, Tollet P, Berge RK, and Gustafsson JA (1993): Structural and metabolic requirements for activators of the peroxisome proliferator-activated receptor. *Biochem. Pharmacol.* 46(12): 2177–84.

Gottlicher M, Widmark E, Li Q, and Gustafsson JA (1992): Fatty acids activate a chimera of the clofibric acid-activated receptor and the glucocorticoid receptor. *Proc. Natl. Acad. Sci. USA* 89(10): 4653–7.

Gulick T, Cresci S, Caira T, Moore DD, and Kelly DP (1994): The peroxisome proliferator-activated receptor regulates mitochondrial fatty acid oxidative enzyme gene expression. *Proc. Natl. Acad. Sci. USA* 91(23): 11012–6.

Hamilton BA, Frankel WN, Kerrebrock AW, Hawkins TL, FitzHugh W, Kusumi K, Russell LB, Mueller KL, van Berkel V, Birren BW, Kruglyak L, and Lander ES (1996): Disruption of the nuclear hormone receptor RORalpha in staggerer mice. *Nature* 379(6567): 736–9.

Harris PK, and Kletzien RF (1994): Localization of a pioglitazone response element in the adipocyte fatty acid-binding protein gene. *Mol. Pharmacol.* 45(3): 439–45.

Hazel TG, Nathans D, and Lau LF (1988): A gene inducible by serum growth factors encodes a member of the steroid and thyroid hormone receptor superfamily. *Proc Natl. Acad. Sci. USA* 85(22): 8444–8.

Hazlerigg DG, Barrett P, Hastings MH, and Morgan PJ. (1996): Are nuclear receptors involved in pituitary responsiveness to melatonin? *Mol. Cell. Endocrinol.* 123(1): 53–9.

Heyman RA, Mangelsdorf DJ, Dyck JA, Stein RB, Eichele G, Evans RM, and Thaller C (1992): 9-cis retinoic acid is a high affinity ligand for the retinoid X receptor. *Cell.* 68(2): 397–406.

Horlein AJ, Naar AM, Heinzel T, Torchia J, Gloss B, Kurokawa R, Ryan A, Kamei Y, Soderstrom M, Glass CK, et al (1995): Ligand-independent repression by the thyroid hormone receptor mediated by a nuclear receptor co-repressor [see comments]. *Nature* 377(6548): 397–404.

Horton C, and Maden M (1995): Endogenous distribution of retinoids during normal development and teratogenesis in the mouse embryo. *Dev. Dyn.* 202(3): 312–23.

Hotamisligil GS, and Spiegelman BM (1994): Tumor necrosis factor alpha: a key component of the obesity-diabetes link. *Diabetes* 43(11): 1271–8.

Issemann I, and Green S (1990): Activation of a member of the steroid hormone receptor superfamily by peroxisome proliferators. *Nature* 347(6294): 645–50.

Janowski BA, Willy PJ, Devi TR, Falck JR, and Mangelsdorf DJ. (1996): An oxysterol signalling pathway mediated by the nuclear receptor LXR alpha. *Nature* 383(6602): 728–31.

Kastner P, Mark M, and Chambon P (1995): Nonsteroid nuclear receptors: Ahat are genetic studies telling us about their role in real life? *Cell.* 83(6): 859–69.

Kastner P, Mark M, Leid M, Gansmuller A, Chin W, Grondona JM, Decimo D, Krezel W, Dierich A, and Chambon P (1996): Abnormal spermatogenesis in RXR beta mutant mice. *Genes. Dev.* 10(1): 80–92.

Keller H, Dreyer C, Medin J, Mahfoudi A, Ozato K, and Wahli W (1993): Fatty acids and retinoids control lipid metabolism through activation of peroxisome proliferator-activated receptor-retinoid X receptor heterodimers. *Proc. Natl. Acad. Sci. USA* 90(6): 2160–4.

Kitareewan S, Burka LT, Tomer KB, Parker CE, Deterding LJ, Stevens RD, Forman BM, Mais DE, Heyman RA, McMorris T, and Weinberger C (1996): Phytol metabolites are circulating dietary factors that activate the nuclear receptor RXR. *Molecular Biology of the Cell.* 7(8): 1153–1166.

Kliewer SA, Forman BM, Blumberg B, Ong ES, Borgmeyer U, Mangelsdorf DJ, Umesono K and Evans RM (1994): Differential expression and activation of a family of murine peroxisome proliferator-activated receptors. *Proc. Natl. Acad. Sci. USA* 91(15): 7355–9.

Krezel W, Dupe V, Mark M, Dierich A, Kastner P, and Chambon P (1996): RXR gamma null mice are apparently normal and compound RXR alpha $+/-$/RXR beta $-/-$/RXR gamma $-/-$ mutant mice are viable. *Proc. Natl. Acad. Sci. USA* 93(17): 9010–4.

Kurokawa R, DiRenzo J, Boehm M, Sugarman J, Gloss B, Rosenfeld MG, Heyman RA, and Glass CK (1994): Regulation of retinoid signalling by receptor polarity and allosteric control of ligand binding. *Nature* 371(6497): 528–31.

Kurokawa R, Soderstrom M, Horlein A, Halachmi S, Brown M, Rosenfeld MG, and Glass CK (1995): Polarity-specific activities of retinoic acid receptors determined by a co-repressor [see comments]. *Nature* 377(6548): 451–4.

La Vista-Picard N, Hobbs PD, Pfahl M, Dawson MI, and Pfahl M (1996): The receptor-DNA complex determines the retinoid response: a mechanism for the diversification of the ligand signal. *Mol. Cell Biol.* 16(8): 4137–46.

Lala DS, Mukherjee R, Schulman IG, Koch SS, Dardashti LJ, Nadzan AM, Croston GE, Evans RM, and Heyman RA (1996): Activation of specific RXR heterodimers by an antagonist of RXR homodimers. *Nature* 383(6599): 450–3.

Lee SL, Wesselschmidt RL, Linette GP, Kanagawa O, Russell JH, and Milbrandt J. (1995a): Unimpaired thymic and peripheral T cell death in mice lacking the nuclear receptor NGFI-B (Nur77). *Science* 269(5223): 532–5.

Lee SST, Pineau T, Drago J, Lee EJ, Owens JW, Kroetz DL, Fernandez-Salguero PM, Westphal H, and Gonzalez FJ. (1995b): Targeted disruption of the alpha Isoform of the peroxisome proliferator-activated receptor gene in mice results in abolishment of the pleiotropic effects of peroxisome proliferators. *Mol. Cell. Biol.* 15(6): 3012–3022.

Lehmann JM, Jong L, Fanjul A, Cameron JF, Lu XP, Haefner P, Dawson MI, and Pfahl M (1992): Retinoids selective for retinoid X receptor response pathways. *Science.* 258(5090): 1944–6.

Lehmann JM, Kliewer SA, Moore LB, Smith-Oliver TA, Oliver BB, Su J-L, Sundseth SS, Winegar DA, Blanchard DE, Spencer TA, and Willson TM (1997): Activation of nuclear receptor LXR by oxysterols defines a new hormone response pathway. *J. Biol. Chem.* 272 (In Press).

Lehmann JM, Moore LB, Smith-Oliver TA, Wilkison WO, Willson TM, and Kliewer SA (1995): An antidiabetic thiazolidinedione is a high affinity ligand for peroxisome proliferator-activated receptor gamma (PPAR gamma). *J. Biol. Chem.* 270(22): 12953–6.

Lemotte PK, Keidel S, and Apfel CM (1996): Phytanic acid is a retinoid X receptor ligand. *Eur. J. Biochem.* 236(1): 328–33.

Levin AA, Sturzenbecker LJ, Kazmer S, Bosakowski T, Huselton C, Allenby G, Speck J, Kratzeisen C, Rosenberger M, Lovey A, et al (1992): 9-cis retinoic acid stereoisomer binds and activates the nuclear receptor RXR alpha. *Nature* 355(6358): 359–61.

Liu ZG, Smith SW, McLaughlin KA, Schwartz LM, and Osborne BA (1994): Apoptotic signals delivered through the T-cell receptor of a T-cell hybrid require the immediate-early gene nur77. *Nature* 367(6460): 281–4.

Luo X, Ikeda Y, Lala DS, Baity LA, Meade JC, and Parker KL (1995a): A cell-specific nuclear receptor plays essential roles in adrenal and gonadal development. *Endocr. Res.* 21(1–2): 517–24.

Luo X, Ikeda Y, and Parker KL (1994): A cell-specific nuclear receptor is essential for adrenal and gonadal development and sexual differentiation. *Cell.* 77(4): 481–90.

Luo X, Ikeda Y, and Parker KL (1995b): The cell-specific nuclear receptor steroidogenic factor 1 plays multiple roles in reproductive function. *Philos. Trans. R. Soc. Lond. B. Biol. Sci.* 350(1333): 279–83.

MacDonald PN, Dowd DR, Nakajima S, Galligan MA, Reeder MC, Haussler CA, Ozato K, and Haussler MR (1993): Retinoid X receptors stimulate and 9-cis retinoic acid inhibits 1,25-dihydroxyvitamin D3-activated expression of the rat osteocalcin gene. *Mol. Cell. Biol.* 13(9): 5907–17.

Mangelsdorf DJ, and Evans RM (1995): The RXR heterodimers and orphan receptors. *Cell.* 83(6): 841–50.

Mangelsdorf DJ, Ong ES, Dyck JA, and Evans RM (1990): Nuclear receptor that identifies a novel retinoic acid response pathway. *Nature* 345(6272): 224–9.

Mangelsdorf DJ, Thummel C, Beato M, Herrlich P, Schutz G, Umesono K, Blumberg B, Kastner P, Mark M, Chambon P, and M. ER (1995): The nuclear receptor superfamily: the second decade. *Cell.* 83: 835–839.

Milbrandt J. (1988): Nerve growth factor induces a gene homologous to the glucocorticoid receptor gene. *Neuron.* 1(3): 183–8.

Minucci S, Leid M, Toyama R, Saint-Jeannet JP, Peterson VJ, Horn V, Ishmael JE, Bhattacharyya N, Dey A, Dawid IB, and Ozato K (1997): Retinoid X receptor (RXR) within the RXR-retinoic acid receptor heterodimer binds its ligand and enhances retinoid-dependent gene expression. *Molecular and Cellular Biology.* 17(2): 644–655.

Morita I, Schindler M, Regier MK, Otto JC, Hori T, DeWitt DL, and Smith WL (1994): Different intracellular locations for prostaglandin endoperoxide H synthase-1 and -2. *J. Biol. Chem.* 270(18): 10902–8.

Mukherjee R, Davies PJA, D. C, E. B, Cesario R, Jow L, Hammann LG, Boehm MF, Mondon CE, Nadzan AM, Paterniti JR, and Heyman RA (1997): Sensitization of diabetic and obese mic to insulin by retinoid X receptor agonists. *Nature* (In Press).

Muscatelli F, Strom TM, Walker AP, Zanaria E, Recan D, Meindl A, Bardoni B, Guioli S, Zehetner G, Rabl W, et al (1994): Mutations in the DAX-1 gene give rise to both X-linked adrenal hypoplasia congenita and hypogonadotropic hypogonadism. *Nature* 372(6507): 672–6.

Narumiya S (1995): Structures, properties and distributions of prostanoid receptors. *Adv. Prostaglandin Thromboxane Leukot. Res.* 23:17–22.

No D, Yao TP, and Evans RM (1996): Ecdysone-inducible gene expression in mammalian cells and transgenic mice. *Proc. Natl. Acad. Sci. USA* 93(8): 3346–51.

Ohsumi J, Sakakibara S, Yamaguchi J, Miyadai K, Yoshioka S, Fujiwara T, Horikoshi H, and Serizawa N (1994): Troglitazone prevents the inhibitory effects of inflammatory cytokines on insulin-induced adipocyte differentiation in 3T3-L1 cells. *Endocrinology* 135(5): 2279–82.

Pappas RS, Newcomer ME, and Ong DE (1993): Endogenous retinoids in rat epididymal tissue and rat and human spermatozoa. *Biol. Reprod.* 48(2): 235–47.

Parker KL, Ikeda Y, and Luo X (1996): The roles of steroidogenic factor-1 in reproductive function. *Steroids* 61(4): 161–5.

Perlmann T, and Jansson L (1995): A novel pathway for vitamin A signaling mediated by RXR heterodimerization with NGFI-B and NURR1. *Genes. Dev.* 9(7): 769–82.

Reaven GM (1994): Syndrome X: 6 years later. *J. Intern Med. Suppl.* 736:13–22.

Rebrin K, Steil GM, Getty L, and Bergman RN (1995): Free fatty acid as a link in the regulation of hepatic glucose output by peripheral insulin. *Diabetes* 44(9): 1038–45.

Renaud JP, Rochel N, Ruff M, Vivat V, Chambon P, Gronemeyer H, and Moras D (1995): Crystal structure of the RAR-gamma ligand-binding domain bound to all-trans retinoic acid. *Nature* 378(6558): 681–9.

Roy B, Taneja R, and Chambon P (1995): Synergistic activation of retinoic acid (RA)-responsive genes and induction of embryonal carcinoma cell differentiation by an RA receptor alpha (RAR alpha)-, RAR beta-, or RAR gamma-selective ligand in combination with a retinoid X receptor-specific ligand. *Mol. Cell. Biol.* 15(12): 6481–7.

Sande S, and Privalsky ML (1996): Identification of TRACs (T3 receptor-associating cofactors), a family of cofactors that associate with, and modulate the activity of, nuclear hormone receptors. *Mol. Endocrinol.* 10(7): 813–25.

Schena M, Shalon D, Heller R, Chai A, Brown PO, and Davis RW (1996): Parallel human genome analysis: microarray-based expression monitoring of 1000 genes. *Proc. Natl. Acad. Sci. USA* 93(20): 10614–9.

Schoonjans K, Staels B, and Auwerx J. (1996): The peroxisome proliferator activated receptors (PPARS) and their effects on lipid metabolism and adipocyte differentiation. *Biochim. Biophys. Acta.* 1302(2): 93–109.

Shen WH, Moore CC, Ikeda Y, Parker KL, and Ingraham HA (1994): Nuclear receptor steroidogenic factor 1 regulates the mullerian inhibiting substance gene: a link to the sex determination cascade. *Cell.* 77(5): 651–61.

Sladek FM, Zhong WM, Lai E, and Darnell JE, Jr. (1990): Liver-enriched transcription factor HNF-4 is a novel member of the steroid hormone receptor superfamily. *Genes. Dev.* 4(12B): 2353–65.

Sohda T, Momose Y, Meguro K, Kawamatsu Y, Sugiyama Y and Ikeda H (1990): Studies on antidiabetic agents. Synthesis and hypoglycemic activity of 5-[4-(pyridylalkoxy)benzyl]-2,4-thiazolidinediones. *Arzneimittelforschung.* 40(1): 37–42.

Spiegelman BM and Flier JS (1996): Adipogenesis and obesity: rounding out the big picture. *Cell.* 87(3): 377–89.

Sucov HM, Dyson E, Gumeringer CL, Price J, Chien KR, and Evans RM (1994):

RXR alpha mutant mice establish a genetic basis for vitamin A signaling in heart morphogenesis. *Genes. Dev.* 8(9): 1007–18.

Szalkowski D, White-Carrington S, Berger J, and Zhang B (1995): Antidiabetic thiazolidinediones block the inhibitory effect of tumor necrosis factor-alpha on differentiation, insulin-stimulated glucose uptake, and gene expression in 3T3-L1 cells. *Endocrinology.* 136(4): 1474–81.

Tai TAC, Jennermann C, Brown KK, Oliver BB, MacGinnitie MA, Wilkison WO, Brown HR, Lehmann JM, Kliewer SA, Morris DC, and Graves RA (1996): Activation of the nuclear receptor peroxisome proliferator-activated receptor gamma promotes brown adipocyte differentiation. *J. Biol. Chem.* 271(47): 29909–14.

Tomkins GM (1975): The metabolic code. *Science.* 189(4205): 760–3.

Tontonoz P, Graves RA, Budavari AI, Erdjument-Bromage H, Lui M, Hu E, Tempst P, and Spiegelman BM (1994a): Adipocyte-specific transcription factor ARF6 is a heterodimeric complex of two nuclear hormone receptors, PPAR gamma and RXR alpha. *Nucleic Acids Res.* 22(25): 5628–34.

Tontonoz P, Hu E, Graves RA, Budavari AI, and Spiegelman BM (1994b): mPPAR gamma 2: tissue-specific regulator of an adipocyte enhancer. *Genes Dev.* 8(10): 1224–34.

Tontonoz P, Hu E, and Spiegelman BM (1994c): Stimulation of adipogenesis in fibroblasts by PPAR gamma 2, a lipid-activated transcription factor. *Cell.* 79(7): 1147–56.

Tontonoz P, Singer S, Forman BM, Sarraf P, Fletcher JA, Fletcher CDM, Brun RP, Mueller E, Altiok S, Oppenheim H, Evans RM, and Spiegelman BM (1997): Terminal differentiation of human liposarcoma cells induced by ligands for the peroxisome proliferator-activated receptor γ and the retinoid X receptor. *Proc. Natl. Acad. Sci. USA* 94:237–241.

Tsien JZ, Chen DF, Gerber D, Tom C, Mercer EH, Anderson DJ, Mayford M, Kandel ER, and Tonegawa S (1996): Subregion- and cell type-restricted gene knockout in mouse brain. *Cell.* 87(7): 1317–1326.

Wagner RL, Apriletti JW, McGrath ME, West BL, Baxter JD, and Fletterick RJ. (1995): A structural role for hormone in the thyroid hormone receptor. *Nature* 378(6558): 690–7.

Weih F, Ryseck RP, Chen L, and Bravo R (1996): Apoptosis of nur77/N10-transgenic thymocytes involves the Fas/Fas ligand pathway. *Proc. Natl. Acad. Sci. USA* 93(11): 5533–8.

Willson TM, Cobb JE, Cowan DJ, Wiethe RW, Correa ID, Prakash SR, Beck KD, Moore LB, Kliewer SA, and Lehmann JM (1996): The structure-activity relationship between peroxisome proliferator-activated receptor gamma agonism and the antihyperglycemic activity of thiazolidinediones. *J. Med. Chem.* 39(3): 665–8.

Willy PJ, Umesono K, Ong ES, Evans RM, Heyman RA and Mangelsdorf DJ. (1995): LXR, a nuclear receptor that defines a distinct retinoid response pathway. *Genes Dev.* 9(9): 1033–45.

Wilson TE, Fahrner TJ, Johnston M, and Milbrandt J. (1991): Identification of the DNA binding site for NGFI-B by genetic selection in yeast. *Science* 252(5010): 1296–300.

Wilson TE, Fahrner TJ, and Milbrandt J. (1993): The orphan receptors NGFI-B and steroidogenic factor 1 establish monomer binding as a third paradigm of nuclear receptor-DNA interaction. *Mol. Cell. Biol.* 13(9): 5794–804.

Woronicz JD, Calnan B, Ngo V, and Winoto A (1994): Requirement for the orphan steroid receptor Nur77 in apoptosis of T-cell hybridomas. *Nature* 367(6460): 277–81.

Wurtz JM, Bourguet W, Renaud JP, Vivat V, Chambon P, Moras D, and Gronemeyer H (1996): A canonical structure for the ligand-binding domain of nuclear receptors. *Nat Struct Biol.* 3(2): 206.

Yamagata K, Furuta H, Oda N, Kaisaki PJ, Menzel S, Cox NJ, Fajans SS, Signorini S, Stoffel M, and Bell GI (1996): Mutations in the hepatocyte nuclear factor-4alpha gene in maturity-onset diabetes of the young (MODY1) [see comments]. *Nature.* 384(6608): 458–60.

Yao TP, Forman BM, Jiang Z, Cherbas L, Chen JD, McKeown M, Cherbas P, and Evans RM (1993): Functional ecdysone receptor is the product of EcR and Ultraspiracle genes. *Nature.* 366(6454): 476–9.

Yu K, Bayona W, Kallen CB, Harding HP, Ravera CP, McMahon G, Brown M, and Lazar MA (1995): Differential activation of peroxisome proliferator-activated receptors by eicosanoids. *J. Biol. Chem.* 270(41): 23975–83.

Zanaria E, Muscatelli F, Bardoni B, Strom TM, Guioli S, Guo W, Lalli E, Moser C, Walker AP, McCabe ER, et al (1994): An unusual member of the nuclear hormone receptor superfamily responsible for X-linked adrenal hypoplasia congenita. *Nature* 372(6507): 635–41.

Zhang B, Graziano MP, Doebber TW, Leibowitz MD, White-Carrington S, Szalkowski DM, Hey PJ, Wu M, Cullinan CA, Bailey P, Lollmann B, Frederich R, Flier JS, Strader CD, and Smith RG (1996): Down-regulation of the expression of the obese gene by an antidiabetic thiazolidinedione in Zucker diabetic fatty rats and db/db mice. *J. Biol. Chem.* 271(16): 9455–9.

Zhang XK, Lehmann J, Hoffmann B, Dawson MI, Cameron J, Graupner G, Hermann T, Tran P, and Pfahl M (1992): Homodimer formation of retinoid X receptor induced by 9-cis retinoic acid. *Nature* 358(6387): 587–91.

Zhou T, Cheng J, Yang P, Wang Z, Liu C, Su X, Bluethmann H, and Mountz JD (1996): Inhibition of Nur77/Nurr1 leads to inefficient clonal deletion of self-reactive T cells. *J. Exp. Med.* 183(4): 1879–92.

Index

DATE DUE

MAR 1 1 1999	
JUN 3 0 2000	
SEP 0 5 2000	
JAN 1 1 2001	
RET'D JUN 1 8 2001	
MAY 2 8 2002	
OCT 2 2 2007	
FEB 2 9 2008	